RADIO RECOMBINATION LINES: 25 YEARS OF INVESTIGATION

ASTROPHYSICS AND SPACE SCIENCE LIBRARY

A SERIES OF BOOKS ON THE RECENT DEVELOPMENTS
OF SPACE SCIENCE AND OF GENERAL GEOPHYSICS AND ASTROPHYSICS
PUBLISHED IN CONNECTION WITH THE JOURNAL
SPACE SCIENCE REVIEWS

Editorial Board

R. L. F. BOYD, *University College, London, England*

W. B. BURTON, *Sterrewacht, Leiden, The Netherlands*

C. DE JAGER, *University of Utrecht, The Netherlands*

J. KLECZEK, *Czechoslovak Academy of Sciences, Ondřejov, Czechoslavakia*

Z. KOPAL, *University of Manchester, England*

R. LÜST, *European Space Agency, Paris, France*

L. I. SEDOV, *Academy of Sciences of the U.S.S.R., Moscow, U.S.S.R.*

Z. ŠVESTKA, *Laboratory for Space Research, Utrecht, The Netherlands*

PROCEEDINGS
VOLUME 163

RADIO RECOMBINATION LINES: 25 YEARS OF INVESTIGATION

PROCEEDINGS OF THE 125TH COLLOQUIUM OF THE
INTERNATIONAL ASTRONOMICAL UNION,
HELD IN PUSCHINO, U.S.S.R., SEPTEMBER 11–16, 1989

Edited by

M. A. GORDON

National Radio Astronomy Observatory, Tucson, Arizona, U.S.A.

and

R. L. SOROCHENKO

*P. N. Lebedev Physical Institute, U.S.S.R. Academy of Sciences,
Moscow, U.S.S.R.*

KLUWER ACADEMIC PUBLISHERS

DORDRECHT / BOSTON / LONDON

Library of Congress Cataloging in Publication Data

```
International Astronomical Union. Colloquium (125th : 1989 :
  Pushchino, Moscow, R.S.F.S.R.)
    Radio recombination lines : 25 years of investigation :
  proceedings of the 125th Colloquium of the International
  Astronomical Union, held in Puschino, U.S.S.R., September 11-16,
  1989 / edited by M.A. Gordon, R.L. Sorochenko.
        p.    cm. -- (Astrophysics and space science library ; v. 163)
    Includes bibliographical references and indexes.
    ISBN 0-7923-0804-2 (U.S. : alk. paper)
    1. Radio recombination lines--Congresses.   I. Gordon, M. A. (Mark
  A.), 1937-      .  II. Sorochenko, R. L.  III. Title.  IV. Series.
  QB462.5.I56  1989
  522'.682--dc20                                                  90-4713
```

ISBN 0–7923–0804–2

Published by Kluwer Academic Publishers,
P.O. Box 17, 3300 AA Dordrecht, The Netherlands.

Kluwer Academic Publishers incorporates
the publishing programmes of
D. Reidel, Martinus Nijhoff, Dr W. Junk and MTP Press.

Sold and distributed in the U.S.A. and Canada
by Kluwer Academic Publishers,
101 Philip Drive, Norwell, MA 02061, U.S.A.

In all other countries, sold and distributed
by Kluwer Academic Publishers Group,
P.O. Box 322, 3300 AH Dordrecht, The Netherlands.

Printed on acid-free paper

All Rights Reserved
© 1990 Kluwer Academic Publishers
No part of the material protected by this copyright notice may be reproduced or
utilized in any form or by any means, electronic or mechanical,
including photocopying, recording or by any information storage and
retrieval system, without written permission from the copyright owner.

Printed in the Netherlands

TABLE OF CONTENTS

Foreword ix

List of Participants xi

GENERAL REVIEWS

Postulation, Detection, and Observations of Radio Recombination Lines
    R. L. Sorochenko 1

Cavity Quantum Electrodynamics: Rydberg Atoms in Atomic Physics and Quantum Optics
    G. Rempe & H. Walther 19

THEORY OF RADIO RECOMBINATION LINES

Review of the Populations of Highly-Excited States of Atoms in Low Density Plasmas
    S. A. Gulyaev 37

The Broadening of Radio Recombination Lines by Ion Collisions: New Theoretical Results
    D. Hoang-Binh 51

RADIO RECOMBINATION LINES FROM HII REGIONS

High Resolution Radio Recombination Line Observations
    P. R. Roelfsema 59

Radio Recombination Lines from Compact HII Regions
    G. Garay 73

Radio Recombination Line Emission from Ultra-Compact HII Regions
    E. Churchwell, C. M. Walmsley, D. O. S. Wood, & H. Steppe 83

Radio Recombination Lines at Millimeter Wavelengths in HII Regions
    M. A. Gordon 93

Maps of the 64α Radio Recombination Lines in Orion A
    T. L. Wilson & L. Filges 105

Estimate of Electron Densities in HII Regions from Observations of Pairs of α-Type Recombination Radio Lines
    A. F. Dravskikh & Z. V. Dravskikh 115

Interferometric Observations of HII, CII, and H^0 Regions in Orion B
    K. R. Anantharamaiah, W. M. Goss, & P. E. Dewdney 123

The Helium Abundance in the HII Region DR21
    A. P. Tsivilev 131

RRLs FROM PLANETARY NEBULAE AND STELLAR ENVELOPES

Radio Recombination Lines from Stellar Envelopes: Planetary Nebulae
    Y. Terzian 141

Radio Recombination Lines from Compact Planetary Nebulae
    G. Garay 155

Radio Recombination Line Maser Emission in MWC349
    J. Martín-Pintado, R. Bachiller, & C. Thum 161

Recombination Emission from CII-Regions Around Be-Stars
    S. P. Tarafdar & K. M. V. Apparao 169

LOW FREQUENCY RECOMBINATION LINES

Review of Decameter Wave Recombination Lines: Problems and Methods
    A. A. Konovalenko 175

Low Frequency Radio Recombination Lines Towards Cas A
    R. L. Sorochenko & G. T. Smirnov 189

Interferometric Observations of Carbon Recombination Lines Towards Cassiopeia A at 332 MHz
    H. E. Payne, K. R. Anantharamaiah, & W. C. Erickson 203

Detection of Carbon Recombination Lines at Decameter Wavelengths in Some Galactic Objects
    A. A. Golynkin & A. A. Konovalenko 209

RRLs from the Local Interstellar Medium
    N. G. Bochkarev 219

LARGE-SCALE PROPERTIES OF THE GALAXY

Recombination Lines and Galactic Structure
    F. J. Lockman 225

Study of the H166α Recombination Line in the Southern Milky Way
    J. C. Cersosimo 237

Radio Recombination Line Imaging of Sgr A
    W. M. Goss, J. H. van Gorkom, D. A. Roberts, & J. P. Leahy 249

Limits of the Temperature and Filling Factor of the Warm Ionized Medium Towards the Galactic Centre
    K. R. Anantharamaiah, H. E. Payne, & D. Bhattacharya 259

RADIO RECOMBINATION LINES FROM EXTRAGALACTIC OBJECTS

VLA Observations of Recombination Lines from the Star Burst Galaxy NGC253
    K. R. Anantharamaiah & W. M. Goss 267

CONCLUSION OF THE COLLOQUIUM

Radio Recombination Lines at 25
   P. A. Shaver 277

INDEXES

Author 285

Citation 287

Subject 295

FOREWORD

The year 1989 was the 25th anniversary of the astronomical detection of radio lines from highly excited atoms. In the years following, these lines—known as Radio Recombination Lines (RRLs)—have become a powerful tool to astronomers for distinguishing thermal sources from nonthermal ones, for studying star formation regions, for investigating the diffuse interstellar medium, and for exploring galactic structure.

This book contains the proceedings of the IAU Colloquium 125, focussed on the 25 years of astronomical research with RRLs since their discovery in 1965. The colloquium was held on September 12–14, 1990 in Puschino, USSR—a research station of the USSR Academy of Sciences located about 100 km south of Moscow on the banks of the Oka river.

For many of us this was an historic occasion. First, it was an opportunity to meet people whom we had come to know only by name in the scientific literature. Second, it was an opportunity to appreciate the enormous scope of RRL astronomical research over the last 25 years. And, third, it was an opportunity to remind ourselves that RRL research also involves research into the fundamental physics of atoms and not just astronomy.

The Colloquium covered a wide range of topics. This book contains papers dealing with research into Rydberg atoms both in the laboratory and in the interstellar medium of our galaxy and others. It contains papers dealing with the interaction of radiation and atomic systems, as well as with the effects of inadiabatic collisons between these atoms and both ions and electrons. It deals with astronomical observations of atoms with "diameters" ranging from $0.08\mu m$ to $50\mu m$—a factor of 625 in size. It deals with RRLs in absorption, in emission, and as true masers. And, it deals with plasmas with temperatures ranging from 10 to greater than 10^4 Kelvins, and with an even larger range of volume densities.

Although in some respects a sequel to the Ottawa Workshop on RRLs held on August 24–25, 1979, this meeting involved a much larger number of people representing many countries. Additionally, it dealt with the exciting new topics of low frequency RRLs discovered in 1980 and of the maser RRLs from the star MWC349 discovered in 1989. The advent of aperture synthesis telescopes and large single-element telescopes are now making possible RRL studies with high angular resolution, studies which will enhance the wonderful results already in hand. Our hope is that the papers in this volume will serve as a platform from which to search new horizons in RRL research.

For this meeting the Scientific Organizing Committee consisted of K. R. Anantharamaiah (India), J. Caswell (Australia), M. A. Gordon (USA), W. M. Goss (USA), D. Hoang-Binh (France), P. G. Mezger (FRG), P. A. Shaver (FRG), and R. L. Sorochenko (USSR). Shaver and Sorochenko served as Co-Chairmen.

We thank the Scientific Organizing Committee and the meeting participants for an excellent program, the Local Organizing Committee for a well-organized meeting, Jolanda Karada of Kluwer Academic Publishers for help with publication, and the USSR Academy of Sciences and the International Astronomical Union for their financial support.

M. A. Gordon	R. L. Sorochenko
National Radio Astronomy Observatory	P. N. Lebedev Physical Institute
Tucson, Arizona, USA	USSR Academy of Sciences
	Moscow, USSR

LIST OF PARTICIPANTS

Argentina
Cersosimo, J. T.

Chile
Garay, G.

Federal Republic of Germany
Rempe, G.
Shaver, P. A.
Walmsley, C. M.
Wilson, T. L.

France
Hoang-Binh, D.

India
Anantharamaiah, K. R.
Qaiyum, A.
Shankar, R.
Tarafdar, S. P.

Netherlands
Roelfsema, P. R.

Spain
Martín-Pintado, J.

USA
Chaisson, E. J.
Churchwell, E. B.
Gordon, M. A.
Goss, W. M.
Lockman, F. J.
Payne, H. E.
Terzian, Y.
Wood, D. O. S.

USSR
Abramenkov, E. A.
Avedisova, V.
Beigman, I. L.
Bochkarev, N. G.
Berulis, I. I.
Golynkin, A. A.
Gulyaev, S. A.
Dagkesamanskij, R. D.
Draviskikh, A. F.
Ershov, A. A.
Kardeshev, N. S.
Konovalenko, A. A.
Lekht, E. E.
Losinkaja, T. A.
Malafeev, V. A.
Ponomarenko, N. I.
Smirnov, G. T.
Sorochenko, R. L.
Stepkin, S. V.
Shmeld, I.
Strelnizkij, V. S.
Tsivilev, A. P.
Waltz, I. A.

POSTULATION, DETECTION AND OBSERVATIONS OF RADIO RECOMBINATION LINES
(REVIEW)

R.L. SOROCHENKO
Lebedev Physical Institute of the USSR Academy of Sciences
Moscow, Leninsky Prospect 53
USSR

ABSTRACT. Radio recombination lines (RRL) detected 25 years ago are ⋯
investigated in a wide, from mm to dam, range of wavelengths It has ⋯
cleared out that in Galaxy conditions an atom as a quantum system may
exist with excitation level n of up to 1000 reaching the giant size of
0.1mm. RRL proved to be a new powerful tool for astrophysical research.

1. PREDICTION AND DETECTION OF RADIO RECOMBINATION LINES

Atomic spectral radiation caused by transitions between levels with
different principal quantum numbers (n) was detected about 100 years
ago. The lines were the well known Lyman, Balmer and Paschen line series
emitted by hydrogen in UV, visible and IR ranges. From this observations
Bohr (1913) developed his quantum theory of atom, in which the spectral
lines' frequencies are defined by

$$\nu = R\,(1/n_1^2 - 1/n_2^2) \cong 2R\,\Delta n/n^3 \quad (\text{if } n \gg 1) \qquad (1)$$

where n_2 and n_1 - are the principal quantum numbers of upper and lower
levels respectively, $\Delta n = n_2 - n_1$, and R is Rydberg's constant (for hydrogen
$R = 3.288057 \times 10^{15}$ Hz).

Bohr's theory explained the observed series as transitions to the
first three atomic levels ($n_1 = 1,2,3$), and it predicted new lines.
Although it did not restrict[1] the number of atomic levels nor ⋯
number of line series, the theory gave no indication of how many series
could be detected in practice. Progress of experimental studies towards
the longer wavelength was very slow. The fourth series, the Bracket
series ($\lambda_{5-4} = 4.05\mu m$) was detected nine years after Bohr's theory had appeared; the fifth series, the Pfundt ($\lambda_{6-5} = 7.46\mu m$) after 11 years; and
the sixth with $\lambda_{7-6} = 12.3\,\mu m$, only after 40 years, as a result of
minute spectrum measurements of gas discharge (Humphrey, 1953).

With these measurements the classical laboratory spectroscopy ran
out of abilities. Only new research technique could find new series.
Here is a situation where astronomy can not only solve a problem in
physics but gain a new field of astronomical research at the same time.

Van de Hulst (1945) was the first to note the possibility of radioline radiation from transitions between highly excited levels of atoms in the ISM. In his classical paper, where the 21cm line was predicted, Van de Hulst also considered radiation from ionized hydrogen for both free-free and bound-bound transitions. The expressions he obtained for the integrated intensity and the Doppler broadening of the radio lines well agree with modern concepts. But, Van de Hulst estimated the Stark broadening incorrectly and, as a result, he concluded that excited hydrogen radio lines (EHRR) were too dispersed and were therefore unobservable. For the same reasons, Wild (1952) later came to the same pessimistic conclusion.

In 1959 Kardashev came to the opposite conclusion. Independantly from Van de Hulst* he had made calculations of EHRR intensity and linewidth, and deduced that in HII regions such lines maybe detectable. He showed that n,n-1 transitions would be the most probable and the resulting radio lines are observable by normal radioastronomical techniques (Kardashev, 1959). Kardashev's prediction drew attention of radioastronomers in many countries and stimulated subsequent experimental and theoretical investigations into EHRR.

The first attempt to detect EHRR was undertaken in Pulkovo by Egorova and Ryzkov (1960).They searched for the hydrogen radio line $n_{272}-n_{271}$ (λ=91.2cm) in the Galactic plane, l_1=60°-115°, but did not detect it. Additional theoretical calculations made it possible to define the intensity of expected radio lines more accurately and thereby optimise a search in wavelength and sources (Sorochenko, 1965). Fig.1 shows the result of these calculations. In the mm range where the linewidth is defined only by Doppler broadening $T_L \sim \lambda$. In the cm range T_L reaches a maximum and then begins to drop, with increasing n because of Stark broadening. This calculations include worked out by Griem (1960) theory of spectral line broadening in plasmas. Detection of the expected radio lines was determined to be most probable in cm range at λ=2-5 cm in the bright and spatially extendeded HII regions, the Omega and Orion nebulae.

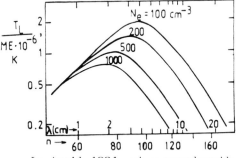

Fig.1 Brightness temperature in line's centre normalised by emission measure as a function of wavelength, level number n and HII electron density.

In April 1964, at a wavelength of 3.38cm and with the 22-m radiotelescope of Lebedev Physical institute in Pushchino, Sorochenko and Borodzich (1965) detected the excited hydrogen radio line n=91-90 (ν=8872.5

---

* Van de Hulst's paper was published in 1945 in very rare edition which was absent from soviet libraries. It became known considerably later in USSR in english edition (Sullivan, 1982).

MHz) in the spectrum of the Omega nebula. The line was distinctly seen even in single spectrograms. Observations carried out in the next three months showed the Doppler shift of line frequency to agree with the orbital Earth rotation and proved conclusively the cosmic origin of the line.

In the same time the Pulkovo Observatory resumed it's attempts to detect the EHRR. Their search was conducted in the range of 5cm. The initial attempts undertaken in December 1963 appeared to have detected

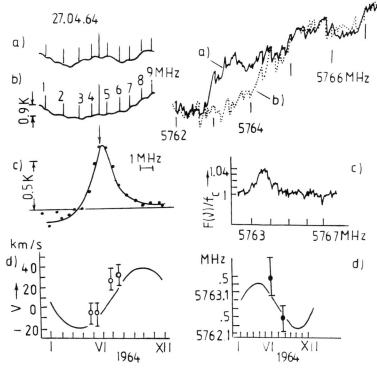

Fig.2 First spectrograms of excited hydrogen radiolines and the observed radial velocity (frequency) shift due to Earth rotation.

On the left, Pushchino H90α line observations. a)Omega nebula's spectrogram. b) test spectrogram for antenna off the source; c)averaged result of seven measurements towards Omega and five test measurements made on April 1964. The abscissa is frequency, the ordinate is the antenna temperature, big mark is calculated line frequency and the vertical dashes are of 1MHz; d) measured Doppler shift during the year 1964. The curve indicates the nebulae's calculated radial veloc. relative to Earth.

On the right, Pulkovo observations of H104α a) spectrogram towards Omega; b)test spectrogram; c)averaged 12 spectrograms obtained in May 1964. The abscissa is the frequency, the ordinate is the ratio of nebula's spectral to continuum flux densities; d)measured Doppler shift during the year 1964. The curve shows the calculated frequency shift.

the line n=105-104 at the frequency of 5763MHz (λ=5.21cm) (Dravskikh and Dravskikh, 1964). In May-June of 1964 the hydrogen radio line n=105 104 was definitely detected in Omega nebula following improvements to the equipment which increasing it's sensitivity (Dravskikh et al,, 1965).

The first spectrograms of the excited hydrogen lines H90α* and H104α which were detected with a high signal/noise ratio, obtained in the Lebedev Physical Institute and in Pulkovo, are given in Fig.2.

Both soviet groups reported the discovery of exited hydrogen radio lines at the 12th IAU Gen. Assembly on 31 of August 1964 (Dravskikh, Dravskikh and Kolbasov; Sorochenko and Borodzich,1964) 25 years ago!

In July of 1965 Höglund and Mezger (1965) detected H109α radio line in Orion, Omega and 9 other HII regions with the 42-m radiotelescope of National Radioastronomical Observatory. By the end of 1965 Lilley et al. (1966a) of Harvard detected two more lines H156α and H158α in the range of 18cm. Soon, EHRR observations were expanded to the wavelength of 75cm, when the H253α line was detected by Penfield, Palmer and Zuckerman (1967); the H104α line was detected in 8 sources (Gudnov and Sorochenko, 1967) and the H109α line, in 16 (Mezger and Hoglund, 1967).

In 1966 Lilley et al. (1966b) reported detection of three exited helium radio lines He156α, He158α and He159α in Omega. The possibility of helium radio lines detection had also been predicted by Kardashev(1959). In 1967 Palmer et al. (1967) detected a radio line with frequency a bit higher than the He109α frequency in NGC 2024 and IC 1795. Goldberg and Dupree (1967) identified this line as radiation from excited carbon C109α line. Fig.3 shows the IC1795 spectrum with hydrogen, helium and the new lines.

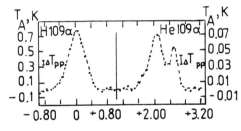

Fig.3 Spectrum of the source 1795 with H109α, He109α lines and new one later identified as C109α. The abscissa is the frequency relative to the rest frequency H109α (MHz) (Palmer et al., 1967).

These obsevations gave impetus to broad research into excited atomic radio lines or radio recombination lines (RRLs) as they were called because the emission is preceded by the recombination process.

2. RRLs: NEW RESULTS IN ATOMIC PHYSICS.

2.1. Stark Broadening.

The first observations of excited hydrogen radio lines showed that the line has no broadening and disperse as a function of n as was expected from the theory (Sorochenko and Borodzich, 1965; Hoglund and Mezger,

---

*Here and farther on we use the adopted convention for highly exited radio lines: the element's name by Mendeleev's table, the principal quantum number of the lower level, line's order (by Greek letters)

1965; Lilley et al.,1966a). All the observed RRLs up to n=158 showed pure Doppler broadening. The failure to detect Stark effects contradicted both Kardashev's early estimations of the broadening and the later more rigorous theory by Griem (1960) which formed the basis for the width and intensity calculations of EHRR (Sorochenko, 1965).

New theoretical consideration (Griem, 1967; Minaeva, Sobelman and Sorochenko, 1967) showed that, for an astrophysical plasma with low densities and high velocities of the exciting particles, an impact approximation is the appropriate method to calculate Stark broadening. Here during the elastic interactions between highly excited atoms and charged particles the compensation of Stark effect occures: cl neighbouring levels distort similarly, and their energy difference is changed much less than the energies of the levels themselves. The inelastic interaction dominate the classical elastic ones, thereby reducing the Stark broadening of RRLs.

The revised Stark theory decreased the expected line broadening, and explained the first observational results in Omega nebula But subsequent observations led to new difficulties. Obsevations of the H220α lines in Orion (Pedlar and Davies, 1971) required the nebula's density $N_e < 200 cm^{-3}$. Contrarily the optical observations of forbidden lines gave a density $N_e = 10^4 - 10^5 cm^{-3}$ (Osterbrock and Flatter, 1959). Furthermore, the same high densities were necessary to explain the intensities, of RRLs observed at many different wavelenghts (Sorochenko and Berulis, 1969). To demonstrate this Fig.4 shows all the observation data obtained for Orion up to 1971 and theoretical results expected for different densities.

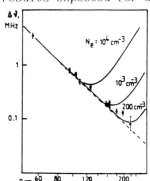

Fig.4 The RRL line width as a function of n in Orion. The dotted line is the Doppler width determined from H56α line. The curves are the calculated values of the line widths due to Stark broadening for different densities N_e.

A number of other experimentors also came to conclusion that the new theory overestimates the Stark broadening. Churchwell (1971) observed linewidths to increase with n, which he explained by Stark broadening, but the broadening's magnitude was substantially less than expected. Davies (1971) concluded that theory evidently overestimated Stark broadening by an order of the value.

The possibility of agreement between theory and experiment arose when the effect of Stark broadening was shown to be considerably decreased by inhomogeneous densities within nebulae (Brocklehurst and Seaton, 1972; Gulyaev and Sorochenko, 1974; Lockman and Brown. 1975).

This hypothesis was confirmed by special Stark broadening observations carried out at two radiotelescopes: 100-m in Effelsberg and 22-m in Pushchino (Smirnov, Sorochenko and Pankonin, 1984). At RT-22 at a wavelength of 8.2mm ($\Delta\varphi_A$=1.9 ang.min.), they precisely measured the H56α line shape towards the central part of Orion nebula from which the Doppler core of the line was determined. In the same direction with the Effelsberg telescope at λ=3.3cm, the lines' profiles of H90α, H114β, H128γ, H141δ, H152ε and H161ζ were measured with the same angular resolutiion as RT-22. Unlike the previous experiments the RRLs were observed from the same volume of gas and comparison of the H56α line unbroadiened because of atom and electron interactions with the lines of higher levels, should permit the separation of Stark and Doppler broadeining.

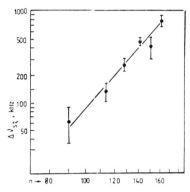

Fig.5 The measured dependence of Stark broadening on level number. The curve is the theoretical dependence $\Delta\nu \sim n^{4.4}$.

Fig.5 shows the results of the measurements. The observed experimental dependence of RRL Stark broadening on level number n well agrees with the revised broadening theory and is quantified (Smirnov, Sorochenko and Pankonin, 1984; Smirnov, 1985) by the relationship:

$$\Delta\nu_{st} = 8.3(n/100)^{4.4} N_e, \text{ Hz} \qquad (2)$$

2.2 Intensity of Radio Recombination Lines. Maser Effect.

In addition to the absense of Stark broadening the RRL data at first posed a second problem: the radio lines intensity. Kardashev's pioneering work (1959) showed that the ratio of radio line and continuum intensities could be used to determine the electron temperature of HII regions as $(T_L/T_c)\Delta\nu_D \sim T_e^{-1}$ (T_L - line's brightness temperature, T_c - the continuum temperature and $\Delta\nu_D$ is the Doppler linewidth).

Even in the first measurements there was a contradiction between the T_e values gained by RRLs and by optical emission. Mezger and Hoglund (1967) obtained an average value of T_e=5820K for the 16 sources observed in H109α line and Dieter (1967) found T_e=5200K by 39 sources in H158α. At the same time by optical measurements of HII regions indicated the temperature of 10,000K.

Goldberg (1966) was the first to suggest that the intensity of RRLs can be essentially dependent on the nonequilibrium of the population of highly exited levels (non-LTE effect). He showed that the line's absorbtion coefficient must be described by:

$$k_L = k_L^* \beta b_n ,\qquad (3)$$

where $k_L^* = 3.5 \cdot 10^{-12} (N_i N_e / T_e^{2.5})(f_{n+\Delta n, n}/n) I(\nu)$
is the absorption coefficient of local thermodynamical equilibrium (LTE) population, $f_{n+\Delta n, n}$ is the oscillator's strength (for $\Delta n=1$ $f_{n+1,n}=0.191n$), $I(\nu)$ is the normalized, $\int I(\nu)d\nu$, distribution of spectral density, b_n is the departure coefficient less than one characterising the deviation from LTE of the n-th level, and

$$\beta = [1 - (kT/h\nu)(d\ln b_n / dn)\Delta n]$$

is a value which accounts for the decrease of absorption coefficient as a consequence of non-LTE.

From expression (3) it follows that non-LTE effect may considerably change the line absorption coefficient to a vanishingly small or even to a negative value, i.e. there can be an amplification at the line frequency, a "partial maser effect" (Goldberg, 1966). For small optical depths both in the continuum and in the line, i.e., τ_c and $\tau_L \ll 1$, the expression characterising amplification in line is:

$$T_L/T_c = (T_L/T_c)^{LTE} b_n [1+(\tau_c/2)(1-\beta)] \qquad (4)$$

where $(T_L/T_c)^{LTE}$ is LTE ratio of the line and continuum temperatures. Because of $T_e \sim (T_L/T_c)^{-1}$ the increase in line intensity causes underestimation of the T_e values obtained from the RRL observations.

Subsequent theoretical and experimental studies both supported and developed Goldberg's theory. Hjellming and Churchwell (1969), considering the non-LTE effect with departure coefficiens calculated by Sejnovski and Hjellming (1969), obtained $T_e=11,000$K for Orion. Hjelming and Davies (1970) obtained $T_e=10,000$K from observations of a series of lines near 6cm (H110α, H138β, H158γ, H173δ and H186ε) from Orion and the number of HII regions. The LTE value of T_e increased with line's order. This result well agreed with the theory according to it the maser amplification effect must decrease as a function of Δn.

New RRL observations in the mm range contradicted this already accepted viewpoint. Sorochenko and Berulis (1969) and Sorochenko et al. (1969) reported H56α observations in the 8.2mm range, where maser effect is negligible and the derived T_e values do not depend on the partial maser effect, obtained $T_e=7750\pm650$K for Orion and $T_e=7500\pm1000$K for Omega.

The question of interpreting of the RRL observations and obtaining correct values of T_e, for instance, in Orion was the subject of active discussions in the 1970s. Hoang-Binh (1970) noted that the high frequency RRL measurements must give the most precise T_e values because they are only weakly affected by the partial maser effect. Placing great significance upon the high frequency observations, he obtained $T_e=8000$K for Orion. Gordon (1970) meanwhile, having made observations of H85ε, H106β, H121γ and H138δ lines and assigning the highest possible error to the H56α line's observation reaffirmed the $T_e=10,000$K value for Orion.

Shaver (1970) came to conclusion that maser effect does not have as strong an influence on T_e as suggested by Goldberg (1966); that the increase in line intensity is balanced by Stark broadening and that influence of the optical depth's influence. He found that values of T_e 8000K for Omega and Orion were correct. Nevertheless, still unexplained was the observed decrease of the RRL intensity with line's order. Soon the phenomenon was made clear. It was noted that in the high order lines Stark-broadened wings defined by Voigt line's profile were of great importance (Simpson, 1973; Berulis, Smirnov and Sorochenko, 1975; Shaver and Wilson, 1979). When a baseline selected the wings are partially c , thereby decreasing line intensities. In any given frequency range, this effect increases with the order of the transitions.

The situation with T_e and with RRL intensities respectively was finally cleared up in the last decade as the result of a series of theoretical and experimental work. New calculations using precise cross-sections of atom-electron interaction (Salem and Brocklehurst, 1979; Hoang-Binh, 1983) showed that the population of highly exited levels of hydrogen in HII regions is closer to LTE and that the maser effect is smaller than considered earlier.

Shaver (1980a) deduced that at cm wavelengths in real HII regions there is no essential change in RRL intensities due to non-LTE effects. In the majority of cases the LTE assumption describes the intensity of the RRLs, and therefore T_e, within 10-15%. Also each HII region, depending on, it's emission measure, has an optimal frequency range, ν_{opt} =0.081 $EM^{0.36}$, where the maser effect is balanced by Stark broadening and, the optical depth. Because of this T_e definition error can be reduced to 2-3% in this frequency range (Shaver, 1980b). Observations of the H66α line which is situated in optimal frequency range for Orion of ~20GHz yeilded T_e=8200∓300. Observations of the intensity ratio of the H66α and H83β radio lines correspond to the theoretical value with one σ standard definition (Wilson, Bieging and Wilson, 1979).

Now at least for hydrogen all the evidence suggests that the RRl intensity theory is fairly well developed, and that satisfact agreement with experimental data exists over a wide frequency range

2.3. The Range of RRL Investigations: How Many Distinct Levels are there in an Atom?

The RRL observations started at the range of 3-21 cm and began spreading towards both short and long wavebands. Towards shorter wavelengths, the mm range was important for a number of reasons: 1) negligible RRL Stark broadening, 2) small optical depths of HII regions and the corresponding insignificance of "maser effect" simplifing the observations' interpretation, 3) higher angular resolution for single-dish instruments. Unfortunately in mm range the RRL brightness temperature is smaller, and the equipment's sensitivity is worse than for longer waves.

The first RRL observations in mm range were carried out at 22-m radiotelescope in Pushchino. The H56α radio line (λ=8.2mm) was detected in the Omega nebula (Sorochenko et al., 1969). Soon the RRL observations shifted to the 3.5mm range: the H42α radio line was detected in Orion (Waltman et al., 1973). The shortest RRL wavelength detected at this time H30α (λ=1.3mm), was observed at 30-m radio telescope of IRAM in the CRL 618 source (Martin-Pintado et al., 1988).

The RRL investigation in the mm range helped to tackle the HII region temperature isue (sec. 2.2), to determine the population of the highly exited levels and to find the b_n values. For densities of $N_e = 10^2 - 10^4 cm^{-3}$ which are usual for HII regions, the collision processes no longer dominate level population for n<60 (λ<1cm). The b_n values noticebly

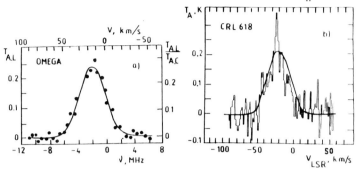

Fig.6 a)H56α line spectrum in Omega nebula (Sorochenko et al., 1969).
b)H30α spectrum in CRL618 (Martin-Pintado et al., 1988).

deviate from 1 and we can measure them. The b_n dependence on n was found to be well described by Salem and Brocklehurst's theory (1979) (Sorochenko, Rydbeck and Smirnov, 1988). The RRL research at mm range is detailed inreview by Gordon (present edition).

A very important event was the expansion of RRL investigations to a longer waves. The revised Stark broadening theory showed that the RRL are available at substantially higher atomic levels than assumed before, and the establishment of the quantum number of the limiting levels was extremely important.

The above mentioned early RRL investigations at long wavelengths were limited by a wavelength of 75 cm. The H253α line in prolonged and rarefied North America nebula (W 80) was detected in Jodrell Bank at 76m radiotelescope (Penfield, Palmer and Zuckerman, 1967). However a decade passed before any lower frequency line were detected. The lines: H274α and H275α were detected towards W49 and W51 sources with the 300-m radiotelescope in Aresibo. They were also emitted by low density HII regions (Parrish, Conklin and Pankonin, 1977).

After his theoretical analysis of the population of excited levels Shaver (1975) showed that, at low frequencies in the conditions of the rarified plasma of the ISM, the stimulated emission is important. The ISM amplifies the galactic background radiation and discrete sources at the RRL frequencies, this amplification maybe crucial for the detection of low frequency lines.

Shaver's theoretical calculations were confirmed by subsequent experimental data. H252α line of noticebly greater intensity than expected for LTE was detected towards the galactic centre with the 76-m radiotelescope in Jodrell Bank (Pedlar, Davies and Hart, 1977). Later, also in the direction of the Galactic centre were detected the H271α (λ=91.5cm) and H300α (λ=1.25m) lines (Pedlar et al., 1978). The intensity of both lines were also enhanced by stimulated emission. The good agreement between theory and experiment reqired that the RRLs originate in low density, $N_e =10 cm^{-3}$, HII region along the line of sight to the Galactic centre. The H300α line (Casse and Shaver, 1977; Pedlar et al., 1978) turned out to be the lowest known frequency for the hydrogen line; the H351α (λ=2m) towards the Galactic centre failed to be detected (Hart and Pedlar, 1980). Also attempts in Jodrell Bank failed to detect the H352α line towards W 80 and Cas A where the neutral hydrogen has large optical depth (Shaver, Pedlar and Davies, 1976).

The comprehensive RRL search programs in meter and decameter ranges carried out in USSR in the 70s were not successful either. Ariskin's et al. (1982) in Pushchino attempts to detect the H392α- H394α, H400α, and H410α (λ=3m) lines towards the Galactic centre, Cygnus, etc sources were a failure even at a line detection level of $4*10^{-4}$ of the continuum level. Similar negative results followed attempts to detect RRLs at even lower frequencies in the decameter range, undertaken in Kharkov by Konovalenko and Sodin (1979a). At a $3*10^{-4}$ upper limit they failed to detect H630α - H650α lines towards Cas A, Cygnus, and the Galactic centre.

A new stage of low frequency RRL investigations began in the 80s. Konovalenko and Sodin (1979b) with UTR-2 radiotelescope detected an absorption line in Cas A spectrum at 26.13 MHz with an opacity of $3*10^{-3}$. Based upon Shklovsky's (1956) predictions the authors identified the line as the F=5/2-3/2 hyperfine transition of nitrogen (ν =26.127MH This interpretation created difficulties since explanation of the observed line optical depth demanded that the adopted nitrogen content in ISM be increased in an order of magnitude. Blake, Crutcher and Watson (1980) offered a another explanation that the line was the 632-631 transition of carbon. It's calculated frequency fell within the measurement accuracy of the observed one. Konovalenko and Sodin (1981) confirmed this identification by detecting the similar absorption lines at the C630α and C640β frequencies.

Soon meter wavelength carbon RRLs were detected in Cas A. The C427α (λ=3.56m), C486α (λ=5.25m), C538α (λ=7.12m) and C612β (λ=3.56m) lines were observed also in absorption with the DKR-1000 radiotelescope in Pushchino (Ershov et al., 1984). In the same time more carbon RRLs, up to C732α line (λ=18m), were detected in decameter range (Konovalenko, 1984).

The meter wave RRL intensities were weaker than the decameter one, and the C400α (λ=3m) line was not detected at all (Ershov et al., 1982). Subsequently more of the carbon RRL series in the range of 26.879MHz - 87.9MHz (Anantharamaiah, Erickson and Radhakrishnan, 1985; Ershov et al., 1987) were detected in the Cas A spectrum. Meanwhile all attempts to detect the higer frequency line of C382α (λ=117MHz) were failures, even though the upper detection limit was an opacity of $6*10^{-4}$. Fig.7 shows the carbon lines profiles obtained in Pushchino and in Kharkov.

The detection of the low frequency lines is extremely interesting and it was a great surprise, It poses the questions: 1. Why at the hi est exitation levels (n>400) are the carbon lines reliably observed when the lines of more abundant hydrogen have not been detected? 2. Why are the radio lines of n=530-700 (λ=7-15m) quantum levels stronger intensive than the atomic lines of n=420-480 (λ=3.5-5m), when the n=380- 400 lines (λ=3m) have not been detected at all? 3. If the radiolines of n=732 were successfully detected and their intensity doesn't fall with increasing n, what defines the formation limit of the lines?

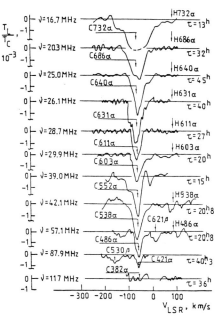

Fig.7 Profiles of the lowest frequency, meter and decameter carbon RRLs towards Cas A obtained in Kharkov and Pushchino. The abscissa is the radial velocity, the ordinate is the line contrast relative to the source's emission in the continious spectrum. On the left are the line frequencies, on the right is the signal integration time in hours. Arrows mark the calculated position of hydrogen line which yet have to be detected.

At present the answers to the mentioned questions appear to be quite clear. According to theory (Shaver, 1975) the lowest frequency intensity may be possible if the lines originate in a rarified and cool plasma. The first requirement stems from the fact that the lines would be dispersed under high densities because of Stark broadening. Every level has a limiting density of $N \sim n^{-9}$. The second requirement comes from the dependence of the RRL optical depth on temperature $\tau_L \sim N_e^2 T_e^{-5/2}$. Since Stark effects require the density N_e to be very low, the RRL must originate in cool regions only.

The condition of low N_e and T_e is fulfilled in HI-CII regions of the ISM. Hydrogen is principally neutral in the regions. Here the less abundant elements with lower ionization potentials than hydrogen are ionized by scattered interstellar UV radiation of $\lambda > 912 A$. Among these carbon is the most abundant. In this case the medium has been heated only a little bit, to $T_e < 100K$. Therefore, the carbon lines of the highest exitation levels are observed rather than the hydrogen ones (Sorochenko and Smirnov, 1987). Since for the greatest n and low temperatures the stimulated emission contribution is negligible, the lines are observed only in absorption.

The stimulated emission manifests itself at $n<500$ by decreasing the emission coefficient. Thus can be explained the decrease of corresponding lines intensities and the failures to detect RRLs of $n=380-400$ where $\tau \cong 0$. For $n<380$ both theory and experiment predicted carbon lines towards Cas A in emission (Ershov et al., 1982, 1984). This prediction was confirmed by recent observations by Payne, Anantharama and Erickson (1989), who detected C300α-C303α, C272α and C273α in emission in Cas A spectrum.

How many energy levels can there be an atom, and what defines the limit? According to Bohr's model, for n=732 the atomic dimensions are: $D = n^2 \ast 10^{-4}$ μm = 50 μm. Are there atoms of n=1000 with a size of 100 μm, or n=3000 with a size of 1mm?

Until the detections of the meter and decameter RRLs the size of the atoms had been assumed to be limited by collisions with electrons (Stark broadening), i.e. only by the electron density. The new observations showed that the highly excited atoms may originate in the very low electron density medium of N_e up to 10^{-2} – 10^{-3} cm^{-3} where Stark broadening must be small. Cross-sections for the interactions of the excited atoms with atoms in ground state increase until a limit at

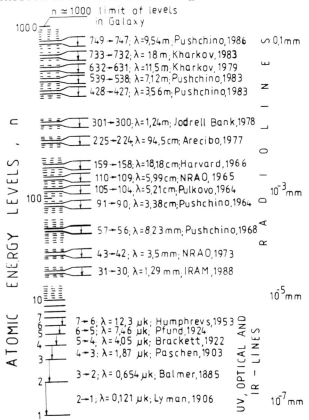

Fig.8 Hydrogen spectral lines n n-1 well known in the UV, visible and IR ranges and a few RRLs, detected from 1964-1988, which demonstrate that an atom can exist up to n=1000 exitation levels. On the right is the atomic size.

n=20-30 after which become constant. For n=1000 these cross-sections are almost 10 orders of magnitude less than the ones for interactions with electrons, and these interactions also cannot be the limiting factor on the size of the atom.

The background galactic nonthermal radiation $T_{b.gr}$ defines the maximum number of atomic exitation levels. The quantum density of the background radiation capable of inducing transitions in these highly excited atoms increases with principal quantum number n:

$$N_q = 0.6*10^{-9} \, T_{b.gr.} \, \nu = 0.1(n/100)^{4.65} \, q \, cm^{-2} Hz^{-1} s^{-1} \quad (5)$$

The induced transition's cross-sections also increases with n :

$$\sigma_{n \, n+1} = 0.5(n/100)^{-2} \, cm \quad (6)$$

Both reasons cause an abrupt ($\sim n^{5.65}$) increase in the rate of the induced transitions. As a result the life time of n~1000 level becomes so small that no distinct line is observeable (Ershov et al., 1982). Therefore atoms as quantum systems can exist in galactic conditions until n~1000, reaching a size of ~0.1mm.

Today experimental research into low frequency RRLs is aproaching the limit. Quite recently in Pushchino the C747β line was detected with the large integration time of 48 hours (Lekht, Smirnov and Sorochenko, 1989). This is the higest exitation level of an atom ever known.(further development in the field see Konovalenko, the present volume). Fig.8 shows the range of our knowledge about the atom found by RRL research.

3. RRL - AN EFFECTIVE TOOL FOR ASTROPHYSICAL RESEARCH.

RRL have proven to be effective tools for conducting astrophysical research. They are unique in the range of the transitions and in the bands which may be observed. RRLs cover more than four orders of magnitude of electromagnetic wavelengths. This great range allows us to explore objects with essentially different physical properties. Let's consider some examples.

3.1 Exploration of Physical Conditions in HII Regions.

The most precise and comprehensive electron temperature measurements of HII regions was carried out by means of RRL. To date the measurements embrace more than 300 regions of northren and southern hemispheres. The temperature of HII regions in the Galaxy have been determined to be within $T = (4-10)*10^3$ K. An important trend was

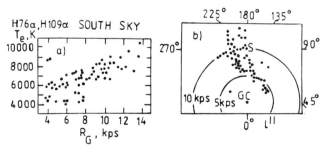

Fig.9 Gradient of electron temperature of HII regions in Galaxy (Shaver et al., 1983). a)The dependence of T_e on galocentric distance; b)the distribution of HII regions in the galactic plane according to line velocities. GC and S mark the positions of the Galactic centre, and the sun respectively.

identifieded in the T distribution: the temperature of HII regions decreases towards the Galactic centre (Churchwell and Walmsley, 1975; Churchwell, 1980). Following the most complete and accurate measurements carried out for H76α and H110α lines in the southern hemisphere the electron temperature gradient equals to 433±40 K/kpc (Shaver et al., 1983). Fig.9 gives the measurements' results.

The dependence of Stark broadening on principal number (2.1 sec.) established by RRL research makes it possible to determine the second important HII parameter - the electron density N_e.

Apart from electron temperature and density, the helium abundance, the internal motions and dynamics of HII regions - their expansion and compression - becomes available by RRL means. The latter property is of special interest for understanding of the early stages of star formation when the expansion velocity, provided it's dimensions are known, may indicate the age of star formation (Berulis and Ershov, 1983).

3.2 The Distribution of Ionized Hydrogen in the Galaxy.

RRLs were detected in not only well defined HII regions but also in many directions of the Galactic plane without HII regions. Gottesman and Gordon (1970) were the first to carry out such observations. They detected rather weak (T_L <0.05 K) but quite distinct H157α line in $l=23°.9$; $l=25°.07$ and $l=80°.09$ directions free from HII regions. Based on continuum observations at meter wavelenghts these RRLs were first considered to be emitted by cool partially ionized interstellar hydrogen, homogeneously dispersed in the Galactic plane (Cesarsky and Cesarsky, 1971). Yet subsequent observations did not confirm this

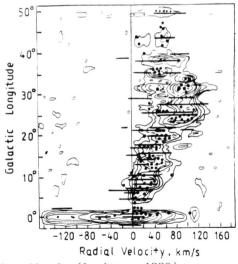

Fig.10 Longtitude-velocity diagram of recombination radio lines H272α (horizontal bars), H166α (contour lines), and H110α (dots). The width of the horizontal bars correspond to the observed half line width of H272α (Anantharamaiah, 1986)

hypothesis (Lockman, 1980).

Based on Galactic plane observations of H272α (ν =325MHz) Anantharamaiah (1986) concluded that almost all diffuse RRL emission was due to outer rarified envelopes of simple HII regions. For $l<40°$ a line of sight through the Galactic plane passes through at least one HII region envelope in any direction. The measurements suggest the envelopes to have N_e=1-10 cm^{-3}, T =3000-8000 K and dimensions 30-300pc. Fig.10 shows the longtitude-velocity diagram of H110α, H166α and H272α. As can be seen there is a good correlation between the positions of the emission of the distinct, condensed HII regions (H110α) and that of the H166α and H272α lines, indicating that the HII outer layer and their rarified envelopes are responsible for the diffuse RRL emission.

3.3 Determination of Physical Conditions Outside HII Regions.

Atoms excited up to n>400 are very sensitive indicators of rarified cool ISM physical conditions outside the HII regions. The density or $N_e T^{0.62}$, to be precise, can be determined by Stark broadening. Data obtained towards Cas A in the Perseus arm indicate N_e =0.17-0.3 cm^{-3} for a respective temperature range T =20-50K (Ershov et al., 1984). The low frequency carbon RRL make it possible to determine electron densities as low as 0.1-0.01 cm^{-3}. It's important to emphasize that the Stark broadening gives local, i.e. true, values of N_e defined through collisions. The cool, weakly ionized ISM temperature can be determined by comparing carbon RRL intensities from several transitions.

The low frequency RRL allows independent measurement of the galactic cosmic rays intensity. According to the theory the ISM is partially ionized by low energy cosmic rays. The ionization makes possible the formation of hydrogen RRLs outside HII regions. By measuring these lines, one can obtain the rate of hydrogen ionization by cosmic rays (Shaver, 1976). The detection of low frequency carbon RRLs enhances the possibility. Considering carbon as hydrogenic, we can determine these rates by the analysis of the low frequency carbon and hydrogen RRL observations, which should probably pin down many years of conjectured rates. (Sorochenko and Smirnov, 1987).

Over 25 years of RRL research quite a number of different galactic objects have been studied: HII,CII regions, planetary nebulae, s' formation regions and the rarified ISM. Observations of extragalactic sources has barely begun. Comprehensive data on physical conditions, chemical composition and construction of the evolutionary processes have been obtained. Meanwhile more new RRL properties seem to emerge as observations progress. RRLs have established themselves as a power tool for the exploration of the universe.

References

    Anantharamaiah, K.R., 1986, J.Astroph.Astr., **7**, 131
    Anantharamaiah, K.R., Erickson,W.C., and Radhakrishnan, V., 1985, Nature, **315**, 647
    Ariskin, V.I., Kolotovkina, S.A., Lekht, E.E., Rudnitskij, G.M., and Sorochenko, R.L., 1982, Astron.J.(Sov.), **57**, 38
    Berulis, I.I., and Ershov, A.A., 1983, Sov. Astr.Lett., **9**, 656
    Berulis, I.I., Smirnov, G.T., and Sorochenko, R.L., 1975, Sov. Astr. Lett., **1**, 187

Blake, D.H., Crutcher, R.M., and Watson, M.D.,1980, Nature, **287**,707
Bohr, H., 1913, Phil. Mag., **26**, 1.
Brocklehurst, M., and Seaton, M.J., M.N.R.A.S., **157**, 179
Casse, J.L., and Shaver, P.A., 1977, Astr.Ap., **61**, 805
Cesarsky, C.J., and Cesarsky, D.A., 1971, Ap.J., **169**, 293
Churchwell, E., 1971, Astr. Ap., **15**, 90
Churchwell, E., 1980, In Radio Recombination Lines. Ed. P.A. Shaver, D. Reidel, Publ.Comp., p.225
Churchwell, E., and Walmsley, C.M., 1975, Astr. Ap., **38**, 451
Davies, R.D., 1971, Ap.J., **163**, 479
Dieter, M.H., 1967, Ap.J., **150**, 435
Dravskich, Z.V., and Dravskich, A.F., 1964, Astron. Tsirk. N282
Dravskich, A.F., Dravskich, Z.V., Misezhnikov, G.S., Kolbasov, V.A. and Steinschleiger, V.B., 1965, Doklady AN USSR, **163**, 332
Dravskich, A.F., Dravskich, Z.V. and Kolbasov, V.A.; Sorochenko, R.L. and Borodzich, E.V., 1964, Papers Presented at the XII Gen. Assembly IAU, Hamburg, Trans. IAU, **12**, 360
Egorova, T.M. and Pyzkov, N.F., 1960, Izv.Glavn. Astr.Obs., **21**, 140
Ershov, A.A., Lekht, E.E., Rudnitskiy, G.M. and Sorochenko, R.L. 1982, Sov.Astr.Lett., **8**, 694
Ershov, A.A., Iljasov,Yu.P., Lekht, E.E., Smirnov,G.T., Solodkov, V.T., and Sorochenko, R.L., 1984, Sov.Astr. Lett., **10**, 348
Ershov, A.A., Lekht, E.E., Smirnov, G.T., and Sorochenko, R.L., 1987, Sov.Astr.Lett., **13**, 8
Goldberg, L., 1966, Ap.J., **144**, 1225
Goldberg, L., and Dupree, A.K., 1967, Nature, **215**, 40
Gordon, M.A., 1970, Astroph. Lett., **6**, 27
Gottesman, S.T., and Gordon, M.A., 1970, Ap.J.(Lett.), **162**, L93
Griem, H.R., 1960, Ap.J., **132**, 883
Griem, H.R., 1967, Ap.J., **143**, 547
Gudnov, V.M. and Sorochenko, R.L., 1967, Astron.J. (Sov.), **44**, 1001
Guljaev, S.A., and Sorochenko, R.L.,1974, Astron.J. (Sov.), **51**,1237
Hart, L., and Pedlar, A., 1980, M.N.R.A.S., **193**, 781
Hjellming, R.M., and Churchwell, E., 1969, Ap.Letters, **4**, 165
Hjellming, R.M., and Davies, R., 1970, Astr.Ap., **5**, 53
Hoang-Binh, D., 1970, Ap.Letters, **6**, 151
Hoang-Binh, D., 1983, Astr.Ap., **121**, L19
Höglund, B., and Mezger, P.G., 1965, Science, **150**, 339
Humphreys, C.H., 1953, Journ. of Res .Bur. of Standart, **50**, 1
Kardashev, N.S., 1959, Astron.J., **36**, 838; 1960, Sov. Astr., **3**, 813
Konovalenko, A.A., 1984, Sov.Astr.Lett., **10**, 846
Konovalenko, A.A., and Sodin, L.G., 1979a, Sov.Astr.Lett., **5**, 66
Konovalenko, A.A., and Sodin, L.G., 1979b, Nature, **283**, 360
Konovalenko, A.A., and Sodin, L.G., 1981, Nature, **294**, 135
Lekht, E.E., Smirnov, G.T., and Sorochenko, R.L., 1989, Sov.Astr. Lett., **15**, 396
Lilley, A.E., Menzel, D.H., Penfield, H., and Zuckerman, B., 1966a, Nature, **209**, 468
Lilley, A.E., Palmer, P., Penfield, H., and Zuckerman, B., 1966b, Nature, **211**, 174
Lockman, F.J., 1980, In Radio Recombination lines. Ed. P.A. Shaver,

Reidel.Publ.Comp., p.185
Lockman, F.J., and Brown, R.L., 1975, Ap.J., **201**, 134
Martin-Pintado, J., Bujarrabal, V., Bachiller, R., Gomez-Gonzalez, J., and Plamsas, P., 1988, Astron. Ap., **197**, L15
Mezger, P.G., and Höglund, B., 1967, Ap.J., **147**, 490
Minaeva, L.A., Sobelman, I.I. and Sorochenko, R.L., 1967, Astron.J. (Sov.), **44**, 995
Osterbrock, D., and Flather, E., 1959, Ap.J., **129**, 26
Palmer, P., Zuckerman, B., Penfield, H., Lilley, A.E. and Mezger, P.G., 1967, Nature, **215**, 40
Parrish, A., Conklin, E., and Pankonin, V., 1977, Astr.Ap., **58**, 319
Payne, H.E., Anantharamaiah, K.R., and Erickson, W.C., 1989, Ap.J. **341**, 890
Pedlar, A., and Davies, R.D., 1971, Nature, **231**, 49
Pedlar, A., Davies, A., and Hart, L., 1977, M.N.R.A.S., **178**, 37P
Pedlar, A., Davies, R.D., Hart, L., and Shaver, P.A., 1978, M.N.R.A.S., **182**, 473
Penfield, H., Palmer, P., and Zuckerman, B., 1967, Ap.J., **148**, L25
Salem, M., and Brocklehurst, H., 1979, Ap.J. Suppl., **39**, 633
Sejnowski, T.J., and Hjellming, R.M., 1969, Ap.J., **156**, 915
Shaver, P.A., 1970, Ap.Letters, **5**, 177
Shaver, P.A., 1975, Pramana, **5**, 1
Shaver, P.A., 1976, Astr.Ap., **49**, 149
Shaver, P.A., 1980a, Astr.Ap., **90**, 34
Shaver, P.A., 1980b, Astr.Ap., **91**, 279
Shaver, P.A., McGee, R.X., Newton, L.M., Dauns, A.C., and Pottasch. S.R., 1983, M.N.R.A.S., **204**, 53
Shaver, P.A., Pedlar, A. and Davies, R.D.,1976, M.N.R.A.S., **177**, 45
Shaver, P.A., and Wilson, T.L., 1979, Astr.Ap., **79**, 312
Shklovsky,I.S., 1956, Cosmic radio emission. (in Russia), Gostech-isdat, Moscow, p.302
Simpson, J.P., 1973, Astroph. and Space Sci., **20**, 187
Smirnov, G.T., 1985, Sov.Astr.Lett., **11**, 17
Smirnov, G.T., Sorochenko, R.L., and Pankonin, V., 1984, Astr.Ap., **135**, 116
Sorochenko, R.L., 1965, Trudy Fiz.Inst. P.N.Lebedeva Akad.nauk SSSR, **28**, 90; Proceedings P.N.Leb. Phys. Inst, **28,** New York, 1966
Sorochenko, R.L., and Berulis, I.I., 1969, Ap.Letters, **4**, 173
Sorochenko, R.L.,and Borodzich, E.V., 1965, Doklady AN USSR,**163**,603
Sorochenko, R.L., Rydbeck, G., Smirnov,G.T.,1988, Astr.Ap., **198**, 23
Sorochenko, R.L., and Smirnov, G.T., 1987, Sov.Astr. Lett., **13**, 77
Sorochenko, R.L., Puzanov, V.A., Salomonovich, A.E., and Shtein-shleger, V.B., 1969, Ap.Letters, **3**, 7
Sullivan III, W.T., 1982, Classics in Radio Astronomy. D.Reidel, Publ. Comp., p.302
Van de Hulst, H.C., 1945, Nederladsch Tijdschrift voor Naturkundl, **11**, 210 ;
Waltman, W.B., Waltman, E.B., Schwartz, P.R., and Johnston, K.J., 1973, Ap.J., **185**, L135
Wild, J., 1952, Ap.J., **115**, 206
Wilson, T.I., Bieging, J. and Wilson, W.E., 1979, Astr.Ap., **71**, 205

The morning ritual in Puschino: Walking into the lecture hall. Foreground (from right to left): Terzian, Valtc (with cart), Gordon (with shoulder strap), Bochkarev (with camera).

CAVITY QUANTUM ELECTRODYNAMICS:
RYDBERG ATOMS IN ATOMIC PHYSICS AND QUANTUM OPTICS

G. Rempe and H. Walther
Sektion Physik, Universität München,
and Max-Planck-Institut für Quantenoptik
D-8046 Garching, F.R. Germany

ABSTRACT. The progress achieved during the last years in using Rydberg atoms to investigate radiation-atom interaction is reviewed. In particular, the influence of blackbody radiation, the modification of the spontaneous emission rate in confined space, and the periodic exchange of photons between Rydberg atoms and a cavity field are discussed. In the latter case, the collapse and the revival in the dynamical behaviour of the atoms are observed. Furthermore, the generation of nonclassical radiation fields in the micromaser is addressed.

1. Introduction

The advent of the laser, and even more the development of frequency-tunable lasers have revolutionized spectroscopy. An outstanding example of the new possibilities is the spectroscopy of highly excited atomic states, the so-called Rydberg states or Rydberg atoms. These are atoms with a valence electron excited into an orbit of very high principal quantum number, i.e. far from the ionic core. The energies of these highly excited levels are well described by the Rydberg formula, because of their hydrogenic nature. The existence of hydrogen Rydberg atoms has been known from radio astronomy for many years. In space, they are produced when protons capture electrons in high orbits; radio waves are then emitted when the electron jumps to lower energy levels. Until recently, however, it was impossible to investigate Rydberg atoms in the laboratory. In particular, it is not possible to populate highly excited states in discharges, since their large collisional cross-section prevents their formation. It was the availability of tunable lasers of sufficient intensity and narrow bandwidth that first made the selective excitation of these states possible. An excellent review on the properties of Rydberg atoms is given in the book of Stebbings and Dunning [1].

Owing to their high principal quantum numbers, and also to the small energy difference between neighbouring levels, Rydberg atoms exhibit a number of classical properties. In particular, according to Bohr's correspondence principle, the transition frequency between neighbouring levels approaches the classical evolution frequency of the electron. On the other hand, these systems also represent an almost ideal testing ground for some of the most fundamental models and predictions of low energy quantum electrodynamics (QED). The reasons are the following:

(a) The dipole interaction between electromagnetic radiation and Rydberg atoms is very large, the corresponding matrix elements between neighbouring levels scale as n^2 (n is the principal quantum number). For high enough n, stimulated effects can overcome spontaneous emission even for fields with very small photon numbers. As a consequence, Rydberg atoms are very sensitive, e.g. to blackbody radiation [2-8].

(b) Because of the large wavelength of the radiation emitted in Rydberg transitions, it is possible to modify the radiative properties of the atoms by modifying the boundaries of the space in which the atoms are contained [9-19]. The consequences are the enhancement or inhibition of the rate of spontaneous emission, depending upon an external cavity being tuned on or off resonance with a transition frequency. The Lamb shift can be modified as well in such experiments [20-22].

(c) For cavities with high quality factors, the photon emitted by a Rydberg atom is stored inside the resonator long enough to be reabsorbed. In this way, it is possible to realize a single-atom maser [23-24].

(d) A single Rydberg atom inside a low-loss single-mode resonator is the experimental version of the Jaynes-Cummings model [25], which describes the interaction between a single two level atom and a single mode of the electromagnetic field. This model has been the subject of considerable attention in the past, and a number of purely quantum mechanical predictions on the dynamics of this system have been made. These include the collapse and the revivals in the dynamics of the atomic population [26-34]. Rydberg atoms for the first time offer a possibility to test these predictions.

(e) The field generated inside the cavity of the single-atom maser shows nonclassical properties [35-40]. The photon number distribution is narrower than a Poissonian distribution. Theory predicts, that under suitable conditions even a number state of the electromagnetic field can be generated [37-38]. This purely quantum mechanical state has no intensity fluctuations. The one-atom maser can therefore be used to study for the first time the interaction of atoms with a light field with fluctuations below the shot noise limit.

The rest of this paper is organized as follows. In section 2, we briefly review some important properties of Rydberg atoms. Section 3 discusses the influence of blackbody radiation on their dynamics, and section 4 shows how the rate of spontaneous emission is modified by placing the atom in an appropriate environment. In section 5, we turn to a brief discussion of the one-atom maser, while section 6 recalls some results of the quantum-mechanical Jaynes-Cummings model. Section 7 discusses the experimental observation of collapse and revival of the oscillatory energy exchange between the atom and the field. Section 8 reviews the most important properties of the field generated by the one-atom maser. Finally, section 9 contains a summary and conclusions.

2. Properties of Rydberg atoms

When a valence electron of an atom is excited into a state of high principle quantum number n, the energy of the atom can be described by the

Rydberg formula with n replaced by an effective principle quantum number n^*. In general, n^* depends on the phenomenological quantum defect δ_l of the state of angular momentum l. For low-l states, where the orbits of the classical Bohr-Sommerfeld theory are ellipses of high eccentricity, the penetration and polarization of the electron core by the valence electron lead to large quantum defects and strong departures from hydrogenic behaviour. As l increases, the orbits become more circular and the atom more hydrogenic; as a result, δ_l decreases. The radius of the Rydberg atom scales as n^{*2}. Rydberg atoms therefore are very sensitive to external electric fields. As a consequence, they ionize in rather weak fields. This opens the possibility of very effective detection.

The large Rydberg atom orbitals are characterized by natural lifetimes much longer than those of less excited atoms. The lifetimes scale as n^{*3} (l small) or as n^{*5} (l large). The rate of spontaneous emission for a transition from state n to n' is given by the Einstein coefficient

$$A_{nn'} = 16\pi^3\nu^3 e^2 <r_{nn'}>^2/3\epsilon_0 hc^3,$$

where ν is the transition frequency and $<r_{nn'}>$ the matrix element of the electric dipole operator between the initial state n and final state n'. For the case n'<<n, $<r_{nn'}>$ is small, owing to the small radial overlap of the wavefunctions for n' and n, and $A_{nn'} \propto n^{*-3}$ and n^{*-5} for small and large l, respectively. If n'≈n, the energy difference $E_n-E_{n'} \propto n^{*-3}$ and $<r_{nn'}> \propto n^{*4}$, so that $A_{nn'}$ becomes proportional to n^{*-5}. Since the matrix element $<r_{nn'}>$ (n≈n') scales as n^{*2}, this leads to rather high induced transition probabilities. Rydberg atoms therefore strongly absorb microwave radiation. As a consequence, blackbody radiation emitted e.g. by the walls of the vacuum chamber may cause strong mixing of the states.

3. Influence of blackbody radiation on Rydberg atoms

We now turn to a discussion of the scaling laws related to these blackbody induced effects. The induced transition rate due to blackbody radiation is proportional to $<r_{nn'}>^2 S_\nu$, where S_ν is the energy flux of blackbody radiation per unit bandwidth and unit surface area. At low frequencies (Rayleigh-Jeans limit) S_ν increases as ν^2. As the distance between the Rydberg states scales as n^{-3} (for simplicity, we will use n instead of n^* for the following discussion), S_ν is proportional to n^{-6}. Since $<r_{nn'}>^2 \propto n^4$, the induced transition rate scales as n^{-2}. Important in experiments is the ratio between induced and spontaneous transition rates, which scales with $n^{-2}/n^{-3}=n$ for small l, and n^3 for large l. This means that for a given atom and a given temperature, there exists an n above which the blackbody-induced rate overcomes spontaneous emission.

Blackbody radiation mainly induces transitions to nearby states. As a result the population will evolve as a function of time after pulsed laser excitation. Changes in the populations typically appear on a microsecond time scale. Experiments were done using field ionization of Rydberg atoms with an electric field pulse rising linearly in time. Since various Rydberg states ionize at different electric field strengths, the population distribution among Rydberg states can be measured in a single shot. We measured this population distribution for

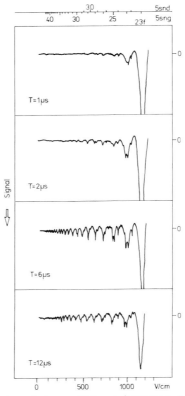

Fig.1: Blackbody-induced transitions between Rydberg states of Sr. The 23f level is excited by pulsed laser radiation. The detection of the Rydberg states is performed by an electric field increasing linearly in time and starting 1, 2, 6, and 12μs after the laser excitation. The field ionization signal at smaller field strengths results from Rydberg levels populated by blackbody radiation originating from the walls of the vacuum chamber at 300K.

various time delays between the pulsed laser excitation and the ionizing electric field pulse. It was observed that for increasing delay the population initially prepared in a Rydberg state was transferred to higher-lying states (fig.1). In these experiments, the temperature of the walls of the vacuum chamber was 300K [6].

Direct observations of the temperature dependence of the population transfer were performed in [7], where sodium atoms of a thermal beam in a cooled environment were excited to the 22d state using two continuous wave dye lasers. The interaction region was cooled to 14K. In another experiment on rubidium Rydberg atoms the environment was cooled to liquid helium temperature [8]. Fig.2 shows the experimental set-up. Two Rb atomic beams were excited stepwise by the light of three semiconductor diode lasers. The atomic beam on the left-hand side of

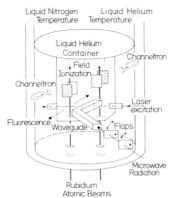

Fig.2: Experimental set-up of the 'Rydberg detector'. The two atomic beams are shielded with liquid nitrogen and liquid helium cooled surfaces. The left atomic beam is used to stabilize the frequency of the diode lasers; the right beam is used to detect microwave radiation.

fig.2 was used to stabilize the frequency of the diode lasers. Lasers 1 and 2 were stabilized using the fluorescence of the beam, whereas laser 3, which populated the Rydberg level, was stabilized by means of the signal current produced in the field ionization region.

The Rb beam on the right-hand side of fig.2 was used to study the interaction with the blackbody radiation. Fig.3 shows the signal obtained in the field ionization region of this beam when the $40P_{3/2}$ level was populated. The blackbody radiation populates the higher levels as indicated by the arrows on the left-hand side of fig.3. The strongest contribution results from the 181 GHz transition. The p-d transitions have a transition probability 100 times smaller than p-s transitions.

Changing the field strength of the ionizing field allowed the population of either the $43S_{1/2}$ or $42S_{1/2}$ level to be monitored. The contribution of the 40D and 41D levels can be neglected. The two signals are shown on the right-hand side of fig.3, as the upper and lower trace, respectively. Since the induced transition rate of the $40P_{3/2}$-$42P_{1/2}$ transition is about a factor of two larger than that of $40P_{3/2}$-$43S_{1/2}$, the field ionization signal of the $42S_{1/2}$ is correspondingly larger. The signals on the right-hand side of fig. 3 were obtained from 77K and 300K blackbody radiation, respectively. In the first case, it was emitted by the walls of the liquid nitrogen-cooled surface, and injected into the apparatus through the waveguide shown in fig. 2 (with the liquid helium-cooled flap open), while in the second case both flaps were open and the radiation was due to the 300K walls of the vacuum chamber. When comparing the results, it is of course important to take into account the different solid angles in each situation.

The evaluation of the data of fig.3 leads to a noise equivalent power of the 'Rydberg detector' of $10^{-17} WHz^{-1/2}$, a value comparable to the result of our earlier experiment [7]. But this new set-up clearly demonstrates that semiconductor lasers can be used to populate the Rydberg levels, making the Rydberg detector a more practical device.

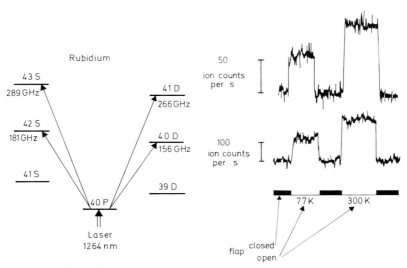

Fig.3: Left: Rubidium Rydberg states relevant for the experiment. The $40P_{3/2}$-$42S_{1/2}$ transition has the highest transition probability. Right: Field ionization signal of the $43S_{1/2}$ level (upper trace) and the $42S_{1/2}$ level (lower trace), respectively.

Aside from population transfer, blackbody radiation also affects Rydberg atoms in a more subtle way. The spectral energy density distribution of thermal radiation at 300K has its maximum at about $2 \cdot 10^{13}$Hz. This is to be compared to a typical electric dipole transition starting from an atomic ground state, which has a frequency of 10^{14}-10^{15}Hz and a transition between two Rydberg states, at about 10^{11}Hz. The blackbody radiation thus appears as a slowly varying field to a ground-state atom, and as a rapidly varying one to a Rydberg atom. This leads to a dynamic Stark shift of the Rydberg levels. An accurate evaluation of this shift was performed by Farley and Wing [41], who found that all Rydberg states experience roughly the same energy shift of about 2.4kHz at 300K.

Since the shift induced by blackbody radiation is roughly the same for all Rydberg states, it can be detected only as a change of frequency of the optical transition to the ground state. The extremely small shift was measured using Doppler-free two-photon absorption to excite the 5s-36s transition in rubidium atoms [42]. With the Ramsey method of separated fields, the spectral width of the signal at $5 \cdot 10^{14}$Hz was decreased to 40kHz. The line centre could be determined with an accuracy of 150Hz. When the temperature of the blackbody source was raised to about 500K, a shift in the line position of 1.4kHz was observed. The study of the temperature dependence also showed agreement with the predicted T^2 scaling. As pointed out in [43], the absolute temperature of the environment can be determined by measuring the blackbody-induced level shift.

4. Single atom in a cavity - modification of the spontaneous lifetime

It is well-known that the spontaneous lifetime of an excited atom is proportional to the density of modes of the electromagnetic field $\rho(\omega_k)$ about the atomic transition frequency ω_0 [44]. The spontaneous emission rate for a two-level system is increased if the atom is surrounded by a cavity tuned to the transition frequency. This was noted years ago by Purcell [9]. Conversely, the decay rate decreases when the cavity is mistuned [12]. In the case of an ideal cavity far off the atomic resonance, no mode is available to accept the emitted photon. As a consequence, spontaneous emission cannot occur. In the following, we would like to discuss this phenomenon in more detail.

Consider a cubic cavity of length L and quality factor Q. Owing to the finite Q, the cavity exhibits some losses which yield a cavity linewidth $\delta\omega = \omega/Q$. For a Lorentzian lineshape, the density of modes around a cavity mode of frequency ω_c and mode volume V_c is given by

$$\rho_c(\omega) = (1/2\pi V_c Q) \cdot [\omega_c/((\omega-\omega_c)^2 + (\omega_c/2Q)^2)] \ .$$

At resonance $\omega_0 = \omega_c$, this gives

$$\Gamma_c = \Gamma_f (Q/4\pi^2)(\lambda^3/V_c)$$

where Γ_c and Γ_f are the spontaneous emission rates in the resonator and in free space, respectively.

From this equation, we draw two important conclusions. At optical frequencies, the size of the resonator is large compared to the wavelength, resulting in a small ratio λ^3/V. However, for cavities operating near their fundamental frequencies in the microwave region, the ratio $\lambda^3/V_c \approx 1$, which typically results in high enhancements in the case of high Q resonators. We conclude that when an atom is placed inside a cavity with a single mode at its transition frequency, it radiates about Q/π^2 faster than in free space. This enhancement of the radiation rate was first pointed out by Bloembergen and Pound [45] and is the rational for using resonant cavities in lasers and masers.

Since a resonant cavity enhances spontaneous emission, it is not surprising that a nonresonant cavity depresses it. Consider for instance a cavity whose fundamental frequency is at twice the resonant frequency of the atomic transition. In this case, the radiation rate becomes

$$\Gamma_c = \Gamma_f \rho_c(\omega_0 = 2\omega)/\rho_f(\omega) = \Gamma_f/4\pi^2 Q \ .$$

Γ_c can be made arbitrarily small by making Q sufficiently large.

It is clear that cavity effects will in general also influence the Lamb shift of energy eigenstates. Since the Lamb shift itself is already fairly small for highly excited states, the expected changes turn out to be extremely small [20-22].

To demonstrate experimentally the modification of the spontaneous decay rate, it is not always necessary to use single atoms. The experiments where the spontaneous emission is inhibited can also be performed with higher densities. However, in the opposite case where an increase of the spontaneous rate is observed, a large number of excited atoms increases the field strength in the cavity and the induced transitions disturb the experiment.

The first experimental work on inhibited spontaneous emission is due to Drexhage, Kuhn and Schäfer performed in 1974 [10]. They studied the fluorescence of a dye film on a mirror and observed an alteration of the fluorescence lifetime arising from the interference of the molecular radiation with its surface image.

Inhibited spontaneous emission was observed clearly for the first time by Gabrielse and Dehmelt [13]. They studied a single electron stored in a Penning trap and observed that the cyclotron excitation showed a lifetime up to ten times larger than that calculated for a cyclotron orbit in free space. The electrodes of the trap form a cavity which decouples the cyclotron motion from the vacuum radiation field.

Recently, an experiment with Rydberg atoms was performed [14]. When the atom was placed between two parallel conducting plates a change in the absorption of infrared radiation was observed. A similar set-up was used to demonstrate inhibited spontaneous emission [15]. The atoms used in this experiment were excited to a circular orbit with n=22 and angular momentum quantum number l=21, $|m_l|$=21. The only decay channel to the state with n=21, l=20, $|m_l|$=20 was partly suppressed by the plates. This led to a longer lifetime of the excited state by a factor of 20.

Recently, experiments were done with low-lying atomic states as well. A 13-fold increase of the lifetime of the low-lying 5d state of Cesium atoms between plates separated by only one micrometer could be demonstrated [16]. In addition, shifts in the frequency of the resonance transition of Barium atoms in an optical resonator, due to radiative effects, and changes in the linewidth, due to enhanced and suppressed spontaneous emission were observed [17]. The change of the fluorescence lifetime under short-pulse laser excitation of dye molecules in a microscopic cavity was also observed [18].

The first observation of enhanced atomic spontaneous emission in a resonant cavity was published by Goy, Raimond, Gross and Haroche [19]. Their experiment was performed with Rydberg atoms of Na excited to the 23s state in a niobium superconducting cavity resonant at 340GHz. A shortening of the lifetime was observed taking advantage of the high Q value of the superconducting resonator. The cooling necessary for superconducting operation also had the advantage of totally suppressing the blackbody field contributions, a necessary requirement to test purely spontaneous emission effects in the cavity.

In spite of this enhancement of spontaneous emission, the cavity damping rate of the radiation was still much higher. The probability of reabsorbing the emitted photon either by the same atom, or by a new atom entering the cavity, could be neglected. Further improvements of the cavity Q lead to a regime where these effects become important. This allows in particular the experimental realization of a single-atom maser [23-24] and for the first time provides experimental conditions close to those required to test the Jaynes-Cummings model [25], which describes the simplest and at the same time most fundamental interaction between a single radiation field mode and a single two-level atom. This set-up will be discussed in the following.

5. The one-atom maser

The Rydberg maser experiment [23] employs an atomic beam to ensure collision-free conditions for the highly excited Rydberg states. A diagram of the vacuum chamber with atomic beam arrangement and microwave cavity is shown in fig.4. These parts are mounted inside a helium bath cryostat. In the experiment, rubidium atoms with a Maxwellian velocity distribution are used. (The velocity selector is inserted at a later stage of the experiment). The atoms enter through small apertures into the liquid helium cooled part of the apparatus. There, the atoms are excited by laser radiation to the upper maser level and enter the cavity. Behind the cavity, they are monitored by field ionization and the subsequent detection of the ejected electrons in a channeltron electron multiplier.

The $63p_{3/2}$ Rydberg state of ^{85}Rb was excited using the narrowband ($\delta\nu$=2MHz) frequency doubled ultraviolet light of a continuous wave dye laser. The cavity was manufactured from niobium. The frequency could be tuned by squeezing the cavity. The temperature could be varied from 4.3 to 0.5K, corresponding to Q factors of $1.7 \cdot 10^7$ and $3 \cdot 10^{10}$, respectively. Single photons could therefore be stored up to 200ms. The thermal background field inside the cavity is determined by the temperature of the walls. The average number of photons of blackbody radiation per mode is $n=[\exp(h\nu/kT)-1]^{-1}$, which gives about n=3.8 at 4.3K and 0.15 at 0.5K (ν=21.5GHz). The continuous wave excitation requires also a continuous detection of the Rydberg atoms. Having left the cavity, the Rb atoms move into the inhomogeneous electric field of a plate capacitor. If the maximum field strength is chosen properly, atoms in the initially prepared $63p_{3/2}$ state are selectively ionized. Transitions to other levels are thus detected by a reduction of the electron count rate.

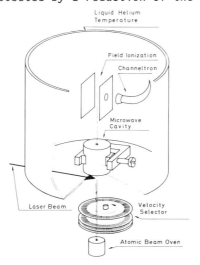

Fig.4: One-atom maser with atomic beam, velocity selector and microwave cavity. The upper part is cooled to liquid helium temperature.

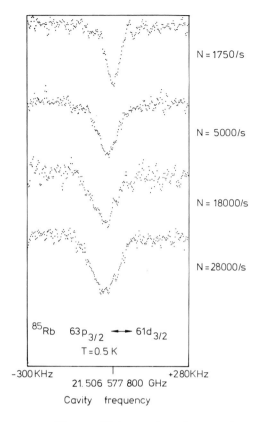

Fig.5: Maser resonances $63p_{3/2}$-$61d_{3/2}$ at a temperature of 0.5K.

To demonstrate maser operation, the cavity was tuned over the $63p_{3/2}$-$61d_{3/2}$ transition (see fig.5). An increase in flux caused power broadening and a small shift. This shift is attributed to the ac Stark effect caused predominantly by virtual transitions to the $61d_{5/2}$ level, which is only 50MHz away from the maser transition. The fact that the field ionization signal at resonance is independent of the particle flux indicates that the transition is saturated. Together with the observed power broadening, this shows that there is a multiple exchange of photons between Rydberg atoms and the cavity field.

With an average transit time of the Rydberg atoms through the cavity of 50µs, one calculates for a flux of 1750 atoms per second a probability of 0.09 of finding an atom in the cavity. According to Poisson statistics, this implies that more than 90 per cent of the events are due to a single atom. This clearly demonstrates that single atoms are able to maintain continuous oscillations of the cavity.

Since the transition is saturated, half of the atoms initially excited in the $63p_{3/2}$ state leave the cavity in the lower $61d_{3/2}$ maser

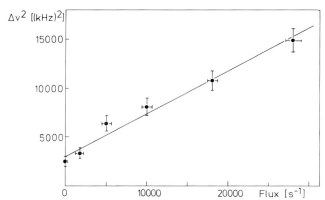

Fig.6: Squares of the halfwidth of the maser resonances versus atomic flux. The straight line shows that power broadening is present. This proves that a maser field is built up in the cavity.

level. For the average transit time of 50µs the decay to other levels can be neglected. The energy radiated by those atoms is stored in the cavity for its decay time, increasing the field strength. The average number of photons left in the cavity by the Rydberg atoms is given by

$$n_m = \tau_d(N/2)$$

where τ_d is the characteristic decay time of the cavity and N the number of Rydberg atoms entering the cavity in the upper maser level per unit time. For the highest particle flux used in our experiment $N=28\cdot 10^3$ atoms per second, one finds n=400 Photons at 0.5K ($\tau_d \approx 30$ms).

When the squares of the halfwidth $\delta\nu$ of the signal curves are plotted versus the Rydberg atom flux, a straight line is obtained as expected (fig.6). This line intersects the $(\delta\nu)^2$ axis at a finite value determined by transit time broadening.

The coupling constant between atom and radiation is large enough for a multiple exchange of photons between the cavity mode and a single Rydberg atom to occur. It follows that under the conditions of the experiment the atom performs on the average up to 3 Rabi periods during its passage through the cavity. Before discussing the experimental results of this Rabi-nutation, we review the main effects predicted by the Jaynes-Cummings model that describes the interaction of a single two-level atom with a single mode of the radiation field (see also [46-47]).

6. The Jaynes-Cummings model

For a brief discussion of the theoretical predictions of the Jaynes-Cummings model [25-34], we consider a two-level atom which enters a resonant cavity with a field of n photons. The time development of the probability $P_{e,n}$ of the atom in the excited state is then given by

$$P_{e,n}(t) = \cos^2(\Omega(n+1)^{1/2}t)$$

where Ω is the single-photon Rabi frequency, which can be determined from the dipol matrix element and the cavity mode structure.

In the realistic case, there will always be a fluctuating number of photons initially present in the cavity and the quantum Rabi solution has to be averaged over the probability distribution p(n) of having n photons in the mode at t = 0:

$$P_{e,n}(t) = \sum_{n=0}^{\infty} p(n) \cos^2(\Omega(n+1)^{1/2}t)$$

At a low atomic beam flux, the cavity contains essentially thermal photons and their number is a random quantity conforming to Bose-Einstein statistics. In this case, p(n) is given by

$$p_{th}(n) = n_{th}^n/(n_{th} + 1)^{n+1},$$

with the average number of thermal photons $n_{th}=[\exp(h\nu/kT)-1]^{-1}$. The distribution of Rabi frequencies then results in an apparent random oscillation $P_{e,th}$.

At higher atomic beam fluxes, the number of photons produced by stimulated emission in the cavity will increase and the statistics change. For a Poissonian distribution representing a coherent field:

$$p_c(n) = \exp(-<n>) \cdot <n>^n/n!$$

As first shown by Cummings, the Poisson spread in n gives dephasing of the Rabi oscillations, therefore $P_{e,c}$ first exhibits a collapse [26]. This is described in the resonant case by the approximate envelope $\exp(-\Omega^2 t^2/2)$ and is independent of the average photon number <n> [30]. After the collapse, there is a range of interaction times for which $P_{e,c}$ is independent of time, and this even in the absence of a decay mechanism in the model [26-34]. Later, $P_{e,c}$ exhibits recorrelations (revivals) and starts oscillating again in a complex way. As has been shown by Eberly and co-workers, the recurrences occur at times given by [30]

$$t = kT_R \quad (k = 1,2,3,...), \quad \text{with } \Omega T_R = 2\pi n^{1/2}$$

The revival in the coherent field is a pure quantum feature and has no classical counterpart. Eberly and co-workers have shown that the physical reason for collapse and revival is the Poisson photon number distribution leading to a "granularity" of the quantized radiation field present even at high average photon numbers.

This can be understood by noting that only a classical field can have a well defined amplitude and phase. However, in quantum mechanics, Heisenberg's uncertainty relation imposes an upper limit on the product of the variances of two canonically conjugate variables like the number of photons and the phase of the field. The minimum uncertainty state of the radiation field corresponds to a Poissonian number distribution. This represents a coherent wave in the classical limit. In this case, the phenomena of collapse and revival of the Rabi-nutation are observed.

The inversion also collapses and revives in the case of a chaotic Bose-Einstein field [33]. Here, the photon number spread is larger than for a coherent state. As a consequence the collapse time is much shorter. In addition, the revivals completely overlap and interfere to produce a very irregular time evolution. A classical thermal field represented by a Gaussian distribution of amplitudes also shows collapse, but revivals are absent. Therefore, the revival can be considered as a clear quantum feature, but the collapse is less clear-cut as a quantum effect.

We conclude this section by noting that in the case of Raman type two photon processes the Rabi frequency turns out to be $2\Omega n$ rather than $2\Omega(n+1)^{1/2}$, enabling the sum over the photon numbers in $P_{e,c}$ to be carried out in simple closed form. In this case, the inversion revives perfectly with a completely periodic sequence [34].

7. Single atom inside a resonant cavity - oscillatory regime

The experimental set-up described in section 5 is suitable to test the Jaynes-Cummings model. An important requirement is that the atoms of the beam have a homogeneous velocity so that it is possible to observe the Rabi-nutation induced by the cavity field directly. This is not possible with the broad Maxwellian velocity distribution. A Fizeau-type velocity selector is therefore inserted between the atomic beam oven and the cavity, so that a fixed atom-field interaction time is obtained [24]. Changing the selected velocity leads to a different interaction time and leaves the atom in another phase of the Rabi cycle when it reaches the detector.

The experimental results obtained for the $63p_{3/2}$-$61d_{5/2}$ transition are shown in fig. 7-9. In the figures the ratio between the field ionization signals on and off resonance are plotted versus the interaction time of the atoms in the cavity. The solid curve in fig.7 was calculated using the Jaynes-Cummings model, which is in very good agreement with the experiment. The total uncertainty in the velocity of the atoms is 4% in this measurement. The error in the signal follows from the statistics of the ionization signal and amounts to 4%. The measurement is made with the cavity at 2.5K and $Q=2.7 \cdot 10^8$ (τ_d=2ms). There are on the average 2 thermal photons in the cavity. The number of maser photons is small compared with the number of blackbody photons.

The experimental result shown in fig. 7 is obtained with very low atomic beam flux (N = 500s^{-1} and n_m = 0.5; n_m is the number of photons accumulated in the cavity). When the atomic beam flux is increased, more photons are piled up in the cavity. Measurements with N = 2000s^{-1} (n_m = 2) and N = 3000s^{-1} (n_m = 3) are shown in figs. 8 and 9. The maximum of $P_e(t)$ at 70μs flattens with increasing photon number, thus demonstrating the collaps of the Rabi nutation induced by the resonant maser field. Figs. 8 and 9 together show that for atom-field interaction times between 50μs and about 130μs $P_e(t)$ does not change as a function of time. Nevertheless, at about 150μs, $P_e(t)$ starts to oscillate again (fig. 9), thus showing the revival predicted by the Jaynes-Cummings model. The variation of the Rabi nutation dynamics with increasing atomic beam fluxes and thus with increasing photon numbers in the cavity generated by stimulated emission is obvious from figs. 7-9.

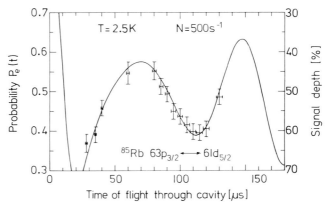

Fig.7: Measured probability of finding the atom in the upper maser level with the cavity tuned to the $63p_{3/2}$-$61d_{5/2}$ transition. The flux of Rydberg atoms is $N=500s^{-1}$. The solid line represents the theoretical prediction.

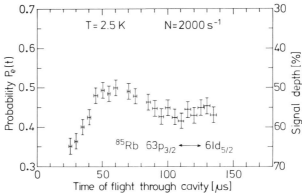

Fig.8: Same as fig. 7, but the flux of Rydberg atoms is $N=2000s^{-1}$.

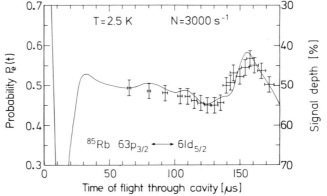

Fig.9: Same as fig. 7, but the flux of Rydberg atoms is $N=3000s^{-1}$.

8. Maser radiation below the shot noise limit

The results discussed in section 7 represent the change of the Rabi-nutation with increasing flux of atoms. This means that the photon statistics also change starting with a Bose-Einstein distribution. Now we will discuss the photon statistics in the maser cavity, which under appropriate conditions turn out to be nonclassical [35-40]. The maser field therefore has to be described within the framework of quantum mechanics.

There are two approaches to the quantum theory of the one-atom maser. Filipowicz et al. [35] describe the device via a microscopic approach. On the other hand, Lugiato et al. [36] show that the standard macroscopic quantum laser theory leads to the same steady state photon statistics. The special features of the one atom maser were not emphasized in the standard laser theory because the broadening due to spontaneous decay obscured the Rabi cycling of the atoms [48]. When similar averages in the microscopic theory associated with inhomogeneous (interaction time) broadening are performed, equivalent results are obtained.

In the one-atom maser, the photon statistics depend on the normalized interaction time $\theta = \Omega t(\tau_d/t_a)^{1/2}$, where t is the interaction time of the atoms and the field, τ_d the photon damping time of the cavity, and t_a the average time between two atoms entering the cavity after each other. At threshold $\theta=1$, the photon statistics are first strongly super-Poissonian (i.e. $\sigma^2=(<n^2>-<n>^2)/<n> \gg 1$). There are further super-Poissonian regions around $\theta \approx 2\pi k$ (k=1,2,...). These peaks become less pronounced for increasing θ. Between these peaks, the field becomes sub-Poissonian. Since the experimental results are in good agreement with the Jaynes-Cummings theory it can also be expected that the photon statistics of the micromaser field is sub-Poissonian for most interaction times θ.

For a cavity with a quality factor of $Q=3 \cdot 10^{10}$, the lifetime of the photons in the resonator amounts to 200ms. This time is then long compared to the buildup time of the micromaser field; furthermore, no thermal field is present anymore since the cavity is operated at temperatures lower than 0.5K. Under these circumstances, fluctuations are suppressed in the experiment and it is possible to obtain a pure number state of the field in the cavity [37-38]. The Rydberg atoms which were injected into the cavity in the upper state are probed for excitation when they leave the cavity. When all the atoms are counted which are in the lower state (and which therefore have emitted a photon), the total number of photons in the cavity is known, assuming that there were no photons in the cavity at the beginning of the experiment.

This means that the state of the field is reduced to a pure number state by the determination of the number of atoms that have emitted a photon. The photon number of the field is always fixed and known after the experiment. Before the experiment starts, however, only the probability to obtain a certain photon number n is known, depending on the number of interacting atoms m. This probability distribution must not be confused with the photon statistics of the maser field. In this case, the photon number distribution of the field is always a δ-function. Therefore, the one-atom maser is a new tool for studying the influence of repeated measurements on a single quantum system. New dynamics and instabilities are expected [49] which are not apparent when conventional ensemble average predictions are considered. Experiments to measure the highly nonclassical maser field have just been completed in our laboratory and will be presented elsewhere [50].

9. Conclusion

We have reviewed progress towards developing Rydberg atoms as tools to test basic principles of quantum electrodynamics. Their unique properties lead to a situation where it is for the first time possible to study light-matter interaction in confined space such as: switching-off the vacuum, enhancing spontaneous emission, modifying the Lamb shift and testing the Jaynes-Cummings model. Further areas of investigation not mentioned here, such as the classical/quantum chaos [51-53], present further evidence for the increasing importance of Rydberg atoms in fundamental research. In addition, the one-atom maser can be used as a new probe of complementarity in quantum mechanics [54]. Rydberg atoms were also used to demonstrate for the first time maser operation on a two-photon transition [55]. The tremendous experimental progress of the last years also promises exciting and surprising results for the future.

References

[1] Stebbings, R.F., and Dunning, F.B., (ed.), Rydberg States of Atoms and Molecules, (Cambridge University Press).
[2] Gallagher, T.F., and Cooke, W.E., 1979, Phys. Rev. Lett., **42**,835.
[3] Haroche,S., Fabre,C., Goy,M., Gross,M. and Raimond,J., 1979, Laser Spectroscopy IV, ed. H. Walther, and K.W. Rothe, (Berlin: Springer)
[4] Beiting, E.J., Hildebrandt, G.F., Kellert, F.G., Foltz, G.W., Smith K.A.,Dunning, F.B., and Stebbings, R.F., 1979, J.Chem.Phys.,**70**,3351
[5] Koch, P.R., Hieronymus, H., Van Raan, F.J., and Raith, W., 1980, Phys. Lett. A **75**,273.
[6] Rempe, G., 1981, Thesis, University of Munich (unpublished).
[7] Figger, H., Leuchs, G., Straubinger, R., and Walther, H., 1980, Optics Commun., **33**,37.
[8] Rempe, G., 1986, Thesis, University of Munich (unpublished).
[9] Purcell, E.M., 1946, Phys. Rev., **69**,681.
[10] Drexhage, K.H., 1974, Progress in Optics, Vol. 12, ed. E. Wolf (Amsterdam: North-Holland), p. 165.
[11] Milonni, P.W., and Knight, P.L., 1973, Opt. Commun. **9**,119.
[12] Kleppner, D., 1981, Phys. Rev. Lett., **47**,233.
[13] Gabrielse, G., and Dehmelt, H., 1985, Phys. Rev. Lett., **55**,67.
[14] Vaidyanathan, A., Spencer, W., and Kleppner, D., 1981, Phys. Rev. Lett., **47**,1592.
[15] Hulet,R.G., Hilfer,E.S., Kleppner,D., 1985, Phys.Rev.Lett. **55**,2137
[16] Jhe, W., Anderson, A., Hinds, E.A., Meschede, D., Moi, L., Haroche, S., 1987, Phys. Rev. Lett. **58**,666
[17] Heinzen,D.J., Childs,J.J., Thomas,J.F., Feld,M.S., 1987, Phys. Rev. Lett **58**,1320. Heinzen,D.J., Feld,M.S., 1987, Phys.Rev.Lett. **59**,2623
[18] DeMartini, F., Innocenti, G., Jacobovitz, G.R., Mataloni, P., 1987, Phys. Rev. Lett. **59**,2955
[19] Goy, P., Raimond, J.D., Gross, M., and Haroche, S., 1983, Phys.

Rev. Lett., **50**,1903.
[20] Dobiasch, P., and Walther, H., 1985, Ann. Phys. Fr. **10**,825.
[21] Lütken, C.A., and Ravndal, F., 1985, Phys. Rev. A **31**,2082.
[22] Barton, G., 1987, Proc. R. Soc. London, A**410**,141 and 175.
[23] Meschede,D., Walther,H., and Müller,G., 1985, Phys.Rev.Lett.,**54**,551
[24] Rempe, G., Walther, H., and Klein, N., 1987, Phys.Rev.Lett. **58**,353.
[25] Jaynes, E.T., and Cummings, F.W., 1963, Proc. IEEE, **51**,89.
[26] Cummings, F.W., 1965, Phys. Rev. A, **140**,1051.
[27] Stenholm, S., 1973, Phys. Rep., **6**,1.
[28] Von Foerster, T., 1975, J. Phys. A, **8**,95.
[29] Meystre, P., Geneux, E., Quattropani, A., and Faist, A., 1975, Nuovo Cim. B**25**,521.
[30] Eberly, J.H., Narozhny, N.B., and Sanchez-Mondragon, J.J., 1980, Phys. Rev. Lett., **44**,1323.
[31] Knight, P.L., and Milonni, P.W., 1980, Phys. Rep., **66**,21.
[32] Yoo, H.-I., and Eberly, J.H., 1985, Phys. Rep. **118**,239.
[33] Knight, P.L., and Radmore, P.M., 1982, Phys. Lett. **90A**,342.
[34] Knight, P.L., 1986, Physica Scripta, T**12**,51.
[35] Filipowicz, P., Javanainen, J., and Meystre, P., 1986, Phys. Rev. A **34**,3077.
[36] Lugiato,L.A., Scully,M.O., and Walther,H., 1987, Phys.Rev. A **36**,740
[37] Filipowicz, P., Javanainen, J., and Meystre, P., 1986, J. Opt. Soc. Am. B **3**,906
[38] Krause, J., Scully, M.O., and Walther, H., 1987, Phys.Rev.A **36**,4547
[39] Krause, J., Scully, M.O., Walther, T., and Walther, H., 1989, Phys. Rev. A **39**,1915.
[40] Meystre, P., Rempe, G., and Walther, H., 1988, Opt. Lett., **13**,1078.
[41] Farley, J.W., and Wing, W.H., 1981, Phys. Rev. A **23**,2397.
[42] Hollberg, L., and Hall, J.L., 1984, Phys. Rev. Lett. **53**,230.
[43] Gallagher, T.F., Sandner, W., Safinya, K.A., and Cook, W.E., 1981, Phys. Rev. A **23**,2065.
[44] Weisskopf, V., and Wigner, E., 1930, Z. Phys., **63**,54.
[45] Bloembergen, N., and Pound, R.V., 1954, Phys. Rev., **95**,8.
[46] Haroche, S., and Raimond, J.M., 1985, Advances in Atomic and Molecular Physics, Vol. 20, ed. B. Bates and B. Bederson (New York: Academic Press), p. 347.
[47] Gallas, J.A.C., Leuchs, G., Walther, H., and Figger, H., 1984, Advances in Atomic and Molecular Physics, Vol. 20, ed. B. Bates and B. Bederson (New York: Academic Press), p. 412.
[48] Sargent, M., Scully, M.O., and Lamb Jr., W.E., 1974, Laser Physics, (Reading: Addison Wesley).
[49] Meystre, P., and Wright, E.M., 1988, Phys. Rev. A **37**,2524
[50] Rempe, G., Schmidt-Kaler, F., and Walther, H., to be published
[51] Casati, G., Chirikov, B.V., and Shepelyanski, D.L., 1984, Phys. Rev. Lett., **53**,2525.
[52] Bayfield, J.E., and Pinnaduwage, L.A., 1985, Phys.Rev.Lett., **54**,313
[53] Blümel, R., Graham, R., Sirko, L., Smilanski, U., Walther, H., and Yamada, K., 1989, Phys. Rev. Lett. **62**,341.
[54] Scully, M.O., and Walther, H., 1989, Phys. Rev. A **39**,5229.
[55] Brune, M., Raimond, J.M., Goy, P., Davidovich, L., and Haroche, S., 1987, Phys. Rev. Lett., **59**,1899.

REVIEW OF THE POPULATIONS OF HIGHLY-EXCITED STATES OF ATOMS IN LOW-DENSITY PLASMAS

S. A. GULYAEV
Astronomical Department
University of the Urals
620083 Sverdlovsk
USSR

ABSTRACT. Brief survey of the problem is discriebed. Results obtained with the use of analytical and numerical methods and the methods proper are reviewed. The influence of the nonideal effects on the highly-excited level populations is considered.

1. Introduction

1.1. EQUATION OF STATISTICAL BALANCE

The population of the atomic energy level with principal quantum number n in ideal plasma under thermodynamic equilibrium is given by the Saha-Boltzmann equation

$$N_n^{TE} = N_+ \frac{g_n}{S N_e} \exp \frac{E_n}{kT} \qquad (1)$$

$$S = 2\left(\frac{2\pi m kT}{h^2}\right)^{3/2} \frac{1}{N_e} \qquad (2)$$

Here N_e is the electron density and N_+ is the ionic density for the element X. T is the electron temperature, g_n is the statistical weight and E_n is the ionization potential of the level n. In this equation S corresponds to the statistical weight of the free electron.

The astrophysical plasma usually forms an open system, generally far from thermodynamic equilibrium. Radiation is lost from the plasma, and energy is injected into it from external sources both as radiation and as thermal kinetic energy of the free particles.

The time development or equilibrium of the populations of the atoms in such a plasma is described by the equation of statistical balance:

$$\frac{d}{dt} N_n = -N_n \Gamma_n + \sum_m N_m \Gamma_{mn} + N_+ R_n \qquad (3)$$

Here Γ_n (c^{-1}) is the total loss rate coefficient from level n, including the rate coefficients from level n to other levels due to spontaneous (ΣA_{nm}), induced radiative ($\Sigma B_{nm} I_\nu$) and collisional (ΣC_{nm}) processes, and, also the rate coefficients of photo- and collisional ionization (P_{ni} and C_{ni}). The rate of population of level n is determined by collisional and radiative (spontaneous and induced) transitions from other levels (Γ_{nm}) and by recombination on the level n. The rate coefficient for recombination processes (R_n) is equal to the sum of photorecombination ($N_e \alpha^r_{in}$) and three-body recombination ($N_e C_{in}$) rate coefficients. In the case of non-hydrogenic atoms, it is necessary to add the rate coefficient of dielectronic recombination ($N_e \alpha^d_{in}$).

The radiation field I_ν in an astrophysical context might be both thermal and nonthermal. Its origin depends on the concrete astrophysical model of an object and is typically due to stellar radiation or an adjacent H II region or the general galactic background or an adjacent supernova remnant. The radiation field is usually described by diluted blackbody radiation at temperature $T_r(\nu)$ and dilution W_ν or by a superposition of such radiation fields.

The largest cross-sections of collisional excitation and de-excitation by electrons are those for which n=n' and l=l'+1 because they are zero energy difference transitions of long-range dipole type. The cross-sections for such collisions in hydrogen have been calculated by Pengelly and Seaton (1964). These cross-sections are so large that a very good approximation is obtained by assuming statistical relative populations for the l-states. This assumption has been justified by Brocklehurst (1971) and Dupree (1972) who demonstrated the great effectiveness of elastic collisions in redistributing the angular momentum for highly excited states (n > 40).

Neutral particles interact much less strongly than charged particles and may be neglected in practically all astrophysical situations of interest here. Even in H I regions, where the hydrogen density could be greater than the electron density by a factor of 2500, neutral particles can be ignored, because the rate coefficients for the H(1s) - H(n) collisions are smaller by a factor of 10^{11} than the corresponding electron-hydrogen rates (Flannery (1970)).

The ion (proton) collisions must be taken into account in inducing transitions between neighbouring highly-excited states (Hoang-Binh, this book).

To find the level populations one needs to solve the set of equations (3) with the conditions that the plasma remains neutral and that the number of nuclei of the element remains constant.

The main simplification of equation (3) is connected with the assumption of the time-independence of the level populations. The relaxation time of N_n to its stationary value is approximately $1/\Gamma_n$. The analysis of rate coefficients shows that the relaxation of the excited levels is much more rapid than evolutionary processes in astrophysics. Therefore we can rewrite (3) as

$$\Gamma_{nm} N_m = N_+ R_m \qquad (4)$$

neighbour collisions, the statistical balance equation for the highly-excited levels may be replaced by a differential equation:

$$\text{Rad} + \text{Rec} + C_{ni}(1-b_n) + C_{nn+1}\frac{d^2 b_n}{d n^2} + (C_{nn+1}-C_{nn-1})\frac{db_n}{dn} = 0 \quad (7)$$

Here, the terms Rad and Rec include the rate coefficients of the spontaneous emission and recombination.

In the first studies of radio recombination lines, beginning with the pionier work of Kardashev (1959), the computation of line intensities neglected the differences in b_n for highly-excited levels because they vary by only a few per cent from unity. However, Goldberg (1966) showed that very slight differential departures of the equilibrium populations of the lower and upper levels of a transition may lead to population inversion and amplification of the lines by stimulated emission.

Goldberg's work stimulated a new wave of investigations. These studies tried to achieve maximum precision in calculations and to take into account, if possible, a maximum number of factors which determine the level populations.

Further investigations were carried out in both analytical and computational directions.

2. Analytical Methods

The differential equation method has limitations in that collisional transitions between more distant levels are of importance, and the consistent matching of the low level solution is difficult. The technique has been extended by Hoang-Binh (1968) and Dyson (1969) to include second and third neighbour collisions. This technique relies on the slow variation of the populations.

The next approach allows inclusion all collisional transitions between highly-excited levels.

Belyaev and Budker (1956) proposed and Pitaevkii (1962), Gurevich and Pitaevskii (1964) developed "diffusion approximation". In their approach the movement of electrons between excited levels is described as a diffusion in the space of quantum levels.

Neglecting radiative processes, they defined the flux of electrons $j(n)$ in the space of quantum numbers n as

$$j(n) = \sum_{k' \geq 0} \sum_{k \geq 0} (N_{n+k'+1} C_{n+k'+1, n-k} - N_{n-k} C_{n-k, n+k'+1}) \quad (8)$$

The first term describes the flux of electrons from levels $n'>n$ to levels $n''<n$ (the flux directed to the ground state), the second term corresponds to the flux of electrons from levels $n'<n$ to all discret levels with $n''>n$ (the flux directed to the continuum). In b_n-factors

Here the index notation and summation convention are used, and Γ_n is denoted as Γ_{nn}.

There is one such equation for each value of n, for n from the ground level n_o to ∞. n_o is the effective ground level and may correspond to the principal quantum number other than 1. It is connected with the fact that the rates of radiative excitation and ionization from the ground state depend upon the radiation field of the plasma itself. Equation for the level population is thus coupled with the problem of radiation transfer. A frequently used solution to this problem is to simplify the transfer equation by assuming that all radiation can escape freely (case A, $n_o = 1$), or that radiation due to transitions to the ground state is reabsorbed, all other radiation escaping freely (case B, $n_o = 2$).

The solution of the equation (4) is the set of values N_n. It may be expressed in terms of b_n-factors, where

$$b_n = N_n / N_n^{TE} \qquad (5)$$

1.2. BRIEF HISTORY

Before discovery of radio recombination lines the populations of highly-excited levels were studied in the context of calculating optical spectra for both laboratory and astrophysical plasma. The determination of the spectrum produced by the processes of radiative capture and cascade for an assembly of electrons and protons was first made by Plaskett (1928). Solutions were obtained for an atom with a finite number of levels, assuming statistical populations for the l-states. In this assumption Baker and Menzel (1938) obtained solutions for an infinite number of levels by using an iterative scheme. Seaton (1959) obtained simular solutions using the cascade matrix technique and deduced the assimptotic expression for b_n in the limit of very high n :

$$b_n = \frac{3 \ln n - 4.84}{3 \ln n - 1.84} \qquad (6)$$

The great advantage of the cascade matrix technique is that elements of this matrix involve only atomic constants and can be calculated once and for all. Seaton (1959) calculated cascade matrix elements for hydrogen. Pengelly (1964) calculated this matrix, taking l-values into account. Detailed calculations have been made by Clarke (1965) and Brocklehurst (1971). The analytical expression for cascade matrix element in the case of very high n was obtained by Beigman and Mikhalchi (1969) in Kramer's approximation.

The first calculations of b_n for very high n, in which collisional processes were taken into account, were carried out by Seaton (1964), primarily to illustrate the dependence of the departure coefficients on the electron density. In his b_n calculations, Seaton included only collisional transitions for which $\Delta n=1$ and, in order to calculate the cascade term, asummed that $b_n = 1$ for all higher levels.

Finally Seaton (1964) showed that by simplifying of the cascade term, neglecting of the radiation field and including of only first

putting $b_{n+k} \approx b_n + \frac{db_n}{dn} k$ and taking into account only terms which are proportional to k^2, we have

$$j(n) = \frac{N_+}{S} e^{E_n/kT} \frac{db_n}{dn} g_n \sum_k k^2 C_{n,n+k} \qquad (9)$$

Then, introducing the continuous variable $\varepsilon = 1/n^2$ we obtain $j(\varepsilon) \propto \frac{db(\varepsilon)}{d\varepsilon}$. The diffusion equation in the variable ε may be written as

$$2 \varepsilon^{3/2} \frac{d}{d\varepsilon} j(\varepsilon) = N_e N_+ C_{i\varepsilon} - N(\varepsilon) C_{\varepsilon i} \qquad (10)$$

where $C_{i\varepsilon}$ and $C_{\varepsilon i}$ are the rate coefficients for collisional recombination and ionization, respectively.

This equation resembles, in principle, Seaton's second-order differential equation with one essential difference that $j(n)$ consists of collisional transitions between all levels.

Two boundary conditions are necessary to find the unique solution of the second order differential equation. In the limit $n \to \infty$, the function $b(\varepsilon)$ transforms in the corresponding distribution function of free electrons. If the distribution function is Maxwellian: $\lim_{\varepsilon \to 0} b(\varepsilon) = 1$. The second boundary condition may be chosen as the numerical solution for lower levels.

A homogeneous diffusion equation corresponds to this approximation when the flux j is constant. The solution in this approximation is (Vainshtein et al. (1981))

$$1 - b_n \propto \varepsilon^{5/2} \propto n^{-5} \qquad (11)$$

where $C_{nn+k} \propto n^4 k^{-3}$ (as in the Born approximation) and $g_n = 2n^2$. The solution of Seaton's equation can be expressed as $1 - b_n \propto n^{-S}$, where S converges to 7 as n increases to infinity. The difference seems to be connected with the allowance for collisional transitions with $\Delta n > 1$ in the diffuse approximation.

Beigman (1975) included the radiative term and replaced Seaton's differential equation with the integral-differential equation. He found its analytical solution in the cases of pure radiative or collisional populations of levels, and formulated boundary conditions which allows the consistently including the solution for low levels. He has shown that the common case defies analitical solution. (See also paper by Beigman and Gaisinsky (1982) and review by Beigman (1988)).

3. Numerical Solutions

3.1. HYDROGENIC ATOMS

Progress in calculating atomic level populations went in the direction of allowing for more and more numbers of reactions in plasma and more accurate definitions of their rate coefficients. External conditions became more various, the range of the physical parameters of the plasma widened. Numerical methods were changed from simple iterative schemes to the matrix condensation technique first suggested by Burgess and Summers (1969).

Let us shortly consider this evolution in the historical aspect.

Sejnowski and Hjellming (1969) used an iterative procedure to solve a truncated set of equations. Collisions between adjacent levels were included for $\Delta n=1$ to 20.

Brocklehurst published detailed results obtained by using a matrix condensation technique. In a series of works, he considered the most essential cases of interest in H II and H I region physics: the case of very high levels in H II regions (Brocklehurst (1970)); the very important case of H and He^+ levels with n<40, where the full allowance for collisional redistribution of angular momentum and energy was made (Brocklehurst (1971)); and finally, the case of an H I region with the effect of thermal continuum from a neighbouring H II region (Brocklehurst (1973)).

In 1972 Dupree had already calculated the highly-excited level populations in H I region with strong thermal continuum radiation from neighbouring ionized region. But the collisional cross-sections used were much too large (Percival and Seaton (1972)) and thus the populations calculated were rather approximate.

Brocklehurst and Seaton (1972) improved the solution by Brocklehurst (1970) for H II regions, adding the thermal radiation field of the nebula itself.

Ahmad (1974) has included hitherto neglected $|\Delta n|>1$ radiatively induced transitions in the balance equation.

Shaver (1975) presented calculations of the atomic populations in the cold interstellar cloud which take into account both thermal and nonthermal external radiation fields. He included in the set of 1000 equations the bound-bound transitions (collisional and induced) only between adjacent levels. For the "left" boundary condition he used the pure radiative solution for n = 10.

Burgess and Summers (1976) have tabulated b_n for a wide range of physical conditions in H I and H II regions. They demonstrated the influence of radiation and free particle temperature and density on the excited level populations of hydrogen. Proton collisions, the 3 K background and extremely diluted stellar radiation fields were taken into account. The population structure of the lower levels and their influence on the populations of intermediate levels were analysed very carefully.

Brocklehurst and Salem (1977) have proposed a common computer programm for the numerical solution of equation (4). They have given the most complete and precise tabulation of b_n and related factors, covering a wide range of physical parameters (Salem and Brocklehurst (1979)).

These two results - Burgess and Summers (1976) and Salem and Brocklehurst (1979) - essentially agree with each other. Some differences between them may be due to the use of different expressions for collision cross-sections. The latter uses the cross-sections given

possibility of the additional process, of the dielectronic recombination. The importance of dielectronic recombination in astrophysical plasmas was well-established theoretically by Burgess (1964), Goldberg and Dupree (1967).

Optimal conditions for dielectronic recombination are determined by the placement of the resonance transition in the next stage of ionization. When the electrons in the thermal plasma have energies comparable to this transition, recombination through doubly-excited levels can occur at a rate greater than direct radiative recombination. As a consequence, large overpopulations of high levels can arise. Doubly-excited states are usually highly-energetic; hence, the dielectronic recombination become effective under high temperatures.

In 1969 the b_n-curves allowing for dielectronic recombination were obtained for calcium by Burgess and Summers, for carbon by Dupree and by Gayet et al. under T = 10 000 K. Shaver (1976 a,b) calculated b_n's for neutral He, C, Be, Mg, Ca, for conditions of H II regions and hot interstellar medium.

Another type of dielectronic recombination for carbon has been put forward by Watson et al. (1980). They suggested that recombination to a high n state occurs simultaneously with the fine structure excitation of the C^+ core. The recombination occurs directly to a low l quantum state and is stabilized against the autoionization by rapid l-changing collisions. The excitation energy of the core is $\Delta E/k=92$ K, so the process is effective at temperatures near this value, which is assumed to be typical for cold interstellar clouds.

Calculated b_n's depend critically on the assumed population of the $^2P_{1/2}$ and $^2P_{3/2}$ fine structure states of C^+. Both electrons and neutrals can influence the statistical equilibrium of these states. The degree to which the transition is thermolized is also a function of the degree of carbon depletion onto grain surfaces. Authors estimated, that for undepleted carbon the ratio $[N(C^+\ ^2P_{3/2})]/[N(C^+\ ^2P_{1/2})]$ is 0.1 of the thermal equilibrium value if $N_e = 0.1$ cm^{-3}. The calculations were performed for this value (the so called "sub-thermal" case) and for the thermal equilibrium value.

Walmsley and Watson (1982) have calculated departure coefficients and related functions for the sub-thermal and thermal cases in the range of T and N_e. In their approach b_n are calculated as the superposition of $b^*_{3/2}$ and $b^*_{1/2}$ factors, computed separately for level n with $^2P_{3/2}$ and $^2P_{1/2}$ cores. $b^*_{1/2}$ is the population computed without the dielectronic recombination process. The relation between the b_n and b^*_j is

$$b_n = \frac{b^*_{1/2}(n) + 2 b^*_{3/2}(n) \exp(-\Delta E/kT)}{1 + 2 \exp(-\Delta E/kT) R} \tag{13}$$

where R measures the deviation of the ratio of the population of $^2P_{3/2}$ to $^2P_{1/2}$ carbon ions in the gas from the ratio in thermodinamical equilibrium.

by Gee et al. (1976), while the former uses a combination of expressions given by Seaton (1962), Van Regemorter (1962), Percival and Richards (1970) and Banks et al. (1973). Both of them have used the same method of the matrix condensation.

Ungerechts and Walmsley (1979) performed calculations for H I regions at low temperatures using a different computational technique. They studied the influence of the 3 K background radiation and of the infrared radiation field from a nearby H II region upon the level populations and found these effects are small. Their numerical experiments show explicitly the importance of including collisions with large Δn when the temperature is low.

Beigman et al. (1980) and Hoang-Binh (1983) have taken into account the role of stellar radiation in the photoexcitation of high levels in H II regions. Van Blerkom (1969) and Seaton (1969) had already considered the influence of the absorption of Lyman line photons on the level populations of atomic hydrogen, assuming that the ionizing star radiates as a black body at temperature T_*. They found that, for $T_* >$ 50 000 K, photoexcitation does not significantly affect the Balmer decrement, and is thus unimportant. But Beigman et al. (1980) and Hoang-Binh (1983) have shown that, if the effective temperature of the exciting star is suitable ($T_* <$ 40 000 K) and the Lyman jump in its spectrum is step-like, the contribution of radiative excitation in the level population may be significant. However, Sorochenko et al.(1988) and Gordon (1989) have demonstrated in their experiments the ineffectiveness of direct photoexcitation of high levels for Orion nebula.

Shinohara (1984) presented calculations of b_n factors of hydrogen atoms at temperatures ranging from 2500 K to 80 000 K.

3.2. HELIUM AND HEAVY ELEMENTS

Radio recombination lines were observed not only for neutral hydrogen, but for the ionized helium and for heavier neutral elements (Gulyaev and Sorochenko (1983)). He^+ recombination lines were observed in H II regions with temperatures of about 10 000 K. At this temperature the effect of Coulomb interaction with plasma particles are small enough for highly-excited states of He^+ (Pengelly and Seaton (1964)). Weisheit and Walmsley (1977) have shown that, if strong external radiation is absent, the departure coefficients $b_n(N_e,Z)$ for the hydrogenic system of nuclear charge Z and those for atomic hydrogen $b_n(N_e,1)$ satisfy approximately the relationship

$$b_n(N_e,Z) = b_n(N_e/Z^6,1) \qquad (12)$$

Due to the more rapid radiative decay, the departure of He^+ level populations from LTE is more essential than that for hydrogen. To the extent of the accuracy of the scaling approximation, this relationship is valid at all densities, and for all n, Z and T satisfying the requirement $n^2 T > 10^6 Z^2$ (K).

Physical processes for complex neutral atoms in highly-excited states are analogous to those for the hydrogen atom, except for any given n and N_i a certain positive probability w_n (depending on n and N_i) that the bound level n is realized. This is equivalent to a reduction of the statistical weight of the discrete level n. Then, for the actual (effective) statistical weight g_n of the level n one can write (Gundel(1970,1971))

$$g_n = 2n^2 \, w_n(N_i) \qquad (17)$$

Here the realization probability is

$$w_n(N_i) = \int_0^{\beta_{cr}} P(\beta) \, d\beta \qquad (18)$$

where $P(\beta)$ is the microfield distribution function. In case $\beta_{cr} \gg 1$ the realisation probability w_n is very close to unity. When $\beta_{cr} \ll 1$, w_n is very close to zero. Using the approximation of the Holtsmark distribution function we have

$$w_n \approx (4/9\pi) \, \beta_{cr}^3 \propto n^{-12} \quad , \quad \beta_{cr} \ll 1 \qquad (19)$$

A critical value of n can be defined as one corresponding to the transition region with $\beta_{cr} = 1$:

$$n_{cr} \approx 4000 \, N_i^{-1/6} \qquad (20)$$

When $n=n_{cr}$, the realisation probability $w_n = 0.1$.

Let us find the range of n where ions behave quasistatically. Using the usual criterion (Gulyaev and Sholin (1986)) for an atom in state i = n - 1 we get

$$n > 25 \, (T/A)^{1/4} N^{-1/6} \qquad (21)$$

where A is the ion mass in the atomic mass units. Consequently, in the transition region ($n \simeq n_{cr}$) ions may be treated quasistatically if temperature is not very high ($T<10^7$ K).

To include the nonideal effects in the balance equation Gulyaev (1987) chose the boundary condition $\lim_{n \to \infty} b_n = 0$, which imitate formally the decreasing of the level realization probability with n, and solved analitically the differential form of the equation. Gulyaev and Nefedov (1989) calculated the set of the equations with the boundary condition $b_n = 0$ when $n \geq n_{cr}$. They showed that the solutions in these cases exist and found that effects of amplification of the carbon decametric lines and transition from emission to absorption at $n \simeq 400$ may be interpreted without the use of the dielectronic recombination mechanism.

The following approach may be used for nonideal effects to be taken into account more consistently.

The Saha-Boltzmann equation (1) with reduced statistical weights g_n gives the level populations N_n^* which may be considered as the "nonideal" equilibrium populations. N_n^* is practically equal to N_n^{TE}, when $n \ll n_{cr}$. The renormalised departure coefficient is $b_n = N_n / N_n^*$.

4. The Role of Nonideal Effects in Plasma

Two extreme regimes are usually used to describe interactions between atom and charged particles - the impact regime and the quasistatic regime. The behavior of ion perturbers can span the entire range from impact to quasistatic depending on the physical conditions and the type of plasma. The electrons can, to a good approximation, be treated with the impact theory. The impact interactions are included in the collision terms of the balance equation.

To allow for the influence of quasistatic ions on highly-excited level populations, Gulyaev (1986,1987) proposed to use the "non-realisation model" (see review by Fortov and Yakubov (1984)). In this model one assumes that all bound levels lying within an energy ΔE of the unperturbed atom's ionization continuum are destroyed by interaction with particles in the plasma. Critical analysis of the other approaches has been given recently by Hummer and Mihalas (1988).

Unsold (1948) considered two models, one with a spatially uniform electric field, and the other with the field arising from a single ion at a distance r from the atomic nucleus. In both cases the potential experienced by the atomic electron, which is composed of the sum of the Coulomb potential due to the ionic core and the perturbing electric field, exhibits a saddle point above which no bound state can exist. Thus a state α with ionization potential E_α in the unperturbed atom can be bound only if the field strength F is less than the critical value $F_{cr}(\alpha)$.

For a uniform electric field the critical value is

$$F_{cr}(n) = e/(16\, a_0^2 n^4) \qquad (14)$$

The critical value of the reduced field strength β_{cr} can be found from the equation (Gulyaev (1986))

$$3(\frac{4\pi}{3})^{1/3} N_i^{1/3} \beta_{cr}^{1/2}(n,i) = \frac{1}{2a_0 n^2} - \frac{3}{2} a_0 n(n_1-n_2)(\frac{4\pi}{3})^{2/3} N_i^{2/3} \beta_{cr}(n,i) \qquad (15)$$

The reduced field strength is $\beta = F/F_0$, where $F_0 = e(4\pi N_i/3)^{2/3}$ is the normal Holtsmark field. In (15) the Stark splitting of the level is taken into account. The electric quantum number is $i = n_1 - n_2$, where n_1 and n_2 are the parabolic quantum numbers. The lowest value of β_{cr} corresponds to the state with the maximum value of i : $i = n-1$. For the central (unshifted) Stark components with $i=0$:

$$\beta_{cr} = (1/6a_0)^2 (3/4\pi)^{2/3} n^{-4} N_i^{-2/3} \qquad (16)$$

Due to the electron impacts, the mixing of states with small values of i with those of large i is sufficiently rapid that the higher critical fields for those states is irrelevant. So all closely coupled states will ionize together with the one having the lowest threshold. However this effect is not very sensible because, as it follows from (15), the minimum value of β_{cr} is only 15% lower than $\beta_{cr}(n,0)$.

Since the strength of the ion microfield in plasma is a quantity statistically distributed around the mean value F_0, there exists for

Note, that results of Goldberg (1966) are entirely valid for b_n in this normalization. Gulyaev and Nefedov (1990) rewrote the balance equation for these b and solved the set of the equations numerically with the boundary condition $b_n =1$ when $w_n < 10^{-10}$ ($n \geq 5n_{cr}$). In Fig. 1 the functions b_n and $dlnb_n/dn$ calculated with allowence for the nonideal effects (Gulyaev and Nefedov (1990)) are represented by solid lines. The dotted lines show solutions obtained by Salem and Brocklehurst (1979). The calculations predict the descent of the b_n function with n and existence of the negative branch of its derivative.

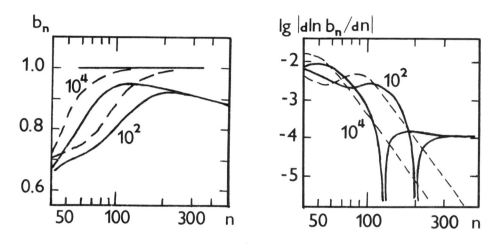

Figure 1. Computed b_n factors and $|d \ln b_n/dn|$ for n in range 40 - 500, computed with (full curves) and without (dashed curves) consideration of the nonideal effects in plasma. Curves are computed for an electron densities 10^2 and 10^4 cm^{-3} and temperature 10^4 K.

5. Conclusion

Investigations of radio recombination lines from the very beginning have required knowing the populations of highly-excited levels in plasma. To a great extent it is these investigations that stimulated the working up of the new fine computational methods. To realize these methods, the computing programms taking into account all basic reactions in the recombining plasma were created. At present there is a confidence in atomic data; the approaches to consideration of the dielectronic recombination mechanism and nonideal effects in plasma, such as a lowering of the ionization potential and the merging of the energy levels of a Rydberg atom, are outlined. As a consequence now there are all possibilities that, using observations of radio recombination lines available, to solve reliably the problem of definition of the physical conditions in astrophysical objects.

Acknowledgements

I acknowledge useful discussions with I.L.Beigman and R.L.Sorochenko.

References

Ahmad, I. A. (1974), Ap. J. 194, 503.
Baker, J. G. and Menzel, D. H. (1938), Ap. J. 88, 52.
Banks, D., Percival, I.C. and Richards, D. (1973), Ap. Letters 14, 161.
Beigman, I. L. (1975),Preprint No.167, Lebedev Phys.Inst., Moscow.
Beigman, I. L. (1988), Proc. Lebedev Phys. Inst., v. 179, Nova Science Publishers, p. 211.
Beigman, I. L. and Gaisinsky, I. M. (1982),J.Quant.Spectrosc.Rad. Transf. 28, 441 .
Beigman, I.L., Gaisinsky, I.M.,Smirnov, G.T. and Sorochenko, R.L. (1980), Preprint No. 141, Lebedev Phys. Inst., Moscow.
Beigman, I. L. and Mikhalchi, E. D. (1969), J.Quant.Spectrosc.Rad. Transf. 9, 1365.
Belyaev, S. T. and Budker, G.I.(1956) 'Many-Quantum Recombination in Ionized Gases', in M. A. Leontovich (ed.), Plasma Physics and Controlled Thermonuclear Reactions Problem, v. 3, Moscow, p. 41.
Brocklehurst, M. (1970), M.N.R.A.S. 148, 417.
Brocklehurst, M. (1971), M.N.R.A.S. 153, 471.
Brocklehurst (1973), Ap. Letters 14, 81.
Brocklehurst, M. and Salem, M. (1977), Computer Phys. Comm. 13, 39.
Brocklehurst, M. and Seaton, M. J. (1972), M.N.R.A.S. 157, 179.
Burgess, A. (1964), Ap. J. 139, 776.
Burgess, A. and Summers, H. P. (1969), Ap. J. 157, 1007.
Burgess, A. and Summers, H. P. (1976), M.N.R.A.S. 174, 345.
Clarke, W. (1965), in Stars and Stellar Systems 7, B. Middlehurst and L. Aller (eds.), Nebulae and Interstellar Matter, Chicago, Univ. of Chicago Press, 1968, p. 504.
Dupree, A. K. (1969), Ap. J. 158, 491.
Dupree, A. K. (1972), Ap. J. 173, 293.
Dyson, J. E. (1969), Ap. J. 155, 47.
Flannery, M. R. (1970), Ap. J. 161, L41.
Fortov, V. E. and Yakubov, I. T.(1984) Physics of Nonideal Plasmas, Chernogolovka.
Gayet, R., Hoang-Binh, D., Joly, F. and McCarroll,R.(1969),Astr.Ap. 1, 365.
Gee, C. S., Percival, I. C., Lodge, J. G. and Richards, D. (1976), M.N.R.A.S. 175, 209.
Goldberg, L. (1966), Ap. J. 144, 1225.
Goldberg, L. and Dupree, A. K. (1967), Nature 215, 41.
Gordon, M. (1989), Ap. J. 337, 782.

Gulyaev, S. A. (1986) 'Populations of Highly-Excited Levels of Hydrogen-Like Atoms in Atmospheres of Hot Stars', in T. Nugis and I. Pustil'nik (eds.), Wolf-Rayet Stars and Related Objects, Proc.of the Workshop held in Elva,14-17 Oct.1986,Tallin,1988,p.245
Gulyaev, S. A. (1987), Astr. Circ. USSR, No. 1516.
Gulyaev, S. A. and Nefedov, S. A. (1989), Astr. Nachr. 310, 403.
Gulyaev, S. A. and Nefedov, S. A. (1990), Astr. Nachr., submitted.
Gulyaev, S. A. and Sholin, G. V. (1986), Astron. Zh. 63, 50.(Soviet Astron.).
Gulyaev, S. A. and Sorochenko R. L. (1983), Catalogue of Radio Recombination Lines. Preprints Nos.145,146 and 168. Lebedev Phys. Inst., Moscow.
Gundel, H. (1970), Beitr. Plasmaphys. 10, 455.
Gundel, H. (1971), Beitr. Plasmaphys. 11, 1.
Gurevich, A. V. and Pitaevskii, L. P. (1964), Zh.Eksp.Teor.Fiz. 46, 1281 (Soviet Phys. - JETP).
Hoang-Binh, D. (1968), Ap. Letters 2, 231.
Hoang-Binh, D. (1983), Astr. Ap. 121, L19.
Hummer, D. G. and Mihalas, D. (1988), Ap. J. 331, 794.
Kardashev, N. S. (1959), Astron.Zh. 36, 838.(Soviet Astron. 3,813).
Pengelly, R. M. (1964), M.N.R.A.S. 127, 145.
Pengelly, R. M. and Seaton, M. J. (1964), M.N.R.A.S. 127, 165.
Percival, I. C. and Richards, D. (1970), J. Phys. B 3, 1035.
Percival, I C. and Seaton, M. J. (1972), Ap. Letters 11, 31.
Pitaevskii, L. P. (1962), Zh. Eksp. Teor. Fiz. 42, 1326 (Soviet Phys. - JETP 15, 919).
Plaskett, H. H. (1928), Publ. Dom. Ap. Obs. Victoria 4, 187.
Salem, M. and Brocklehurst, M. (1979), Ap. J. Suppl. 39, 633.
Sejnowski, T. J. and Hjellming, R. M. (1969), Ap. J. 156, 915.
Seaton, M. J. (1959), M.N.R.A.S. 119, 81.
Seaton, M. J. (1962), Proc. Phys. Soc. 79, 1105.
Seaton, M. J. (1964), M.N.R.A.S. 127, 117.
Seaton, M. J. (1969), M.N.R.A.S. 145, 91.
Shaver, P. A. (1975), Pramana 5, 1.
Shaver, P. A. (1976 a), Astr. Ap. 46, 127.
Shaver, P. A. (1976 b), Astr. Ap. 47, 49.
Shinohara, M. (1984), Ann. Tokyo Astr. Obs. 19, 639.
Sorochenko, R. L., Rydbeck, G. and Smirnov, G. T. (1988), Astr. Ap. 198, 233.
Ungerechts, H. and Walmsley, C. M. (1979), Astr. Ap. 80, 325.
Unsold, A. (1948), Zs. Ap. 24, 355.
Vainshtein, L. A., Sobel'man, I. I. and Yukov, E. E. (1981) Atom Excitation and Spectral Line Broadening, Springer.
Van Blerkom, D. (1969), M.N.R.A.S. 145, 75.
Van Regemorter, H. (1962) Ap. J. 136, 906.
Walmsley, C. M. and Watson, W. D. (1982), Ap. J. 260, 317.
Watson, W. D., Western, L. R. and Christensen, R. B. (1980), Ap. J. 240, 956.
Weisheit, J. and Walmsley, C. M. (1977), Astr. Ap. 61, 141.

THE BROADENING OF RADIO RECOMBINATION LINES BY ION COLLISIONS: NEW THEORETICAL RESULTS

D. HOANG-BINH
Laboratoire Atomes & Molécules
LAM, Observatoire de Paris
92195 Meudon Cedex
France

ABSTRACT. We give new theoretical data for calculating the ion impact broadening of radio recombination α, β, and γ lines. A comparison with other broadening mechanisms is presented.

1. Introduction

Carbon radio recombination lines (RRLs) have been detected in several dark clouds in the Galaxy. Omont and Encrenaz (1977) have pointed out that contrary to the case of hydrogen RRLs observed in HII regions, ion impact, besides the Doppler effect, may be the most important process for line broadening in several circumstances. However, the variation of the line width Δw (in angular frequency unit, 2π Hz) with Δn ($\Delta n = m-n$, m and n being the principal quantum numbers of the upper and lower states of the transition m $\rightarrow$ n respectively) has not been explored in sufficient detail. Omont and Encrenaz have used a Δn^2 rule in their work on the carbon RRLs, without discussing its derivation. This Δn dependence is important, because the analysis of the widths of α and higher order lines, observed at the same frequency (e.g. C167α, C210β and C240γ near ν =1400 MHz), may yield interesting information on the physical conditions in the emitting source. Furthermore, the formula given by Omont and Encrenaz contains the observed line width, the presence of which in a theoretical formula does not seem justified and restricts its general use. This has incited us to present a more detailed analysis. In the following, except where otherwise stated, cgs units will be used.

2. Theory

The ion impact width $\Delta w_i(2\pi$ Hz) is given by (Griem 1967)

$$\Delta w_i = (4\pi/3v)(h/2\pi\, m_e Z)^2\, N_i\, \ln(\rho_{max}/\rho_{min})\, F(n, \Delta n), \tag{1}$$

where v is the relative ion-emitter velocity, h is the Planck constant, m_e is the electron mass, Ze is the emitter charge, N_i is the ion density; ρ_{max}, ρ_{min} are respectively the maximum and minimum impact parameters; and $F(n, \Delta n)$ is a quantity defined below.

For radio-frequency H nα lines, n≈100, Griem has given a simple estimate of $F(n, 1)$, and he has also noted that $F(n, \Delta n)$ is proportional to Δn^2. Using his results, Omont and Encrenaz have derived a formula, in which Δw is proportional to Δn^2. However, in a previous paper (Hoang-Binh et al. 1987) we have found that $F(n, \Delta n)$ is not proportional to Δn^2 for 5≤n≤20, and this may be also the case for higher values of n. Furthermore, Δw is not proportional to $F(n, \Delta n)$, because ρ_{min} depends also on $F(n, \Delta n)$. We will give here a more accurate treatment.

We have generally, for a transition $n+\Delta n \rightarrow n$,

$$F(n, \Delta n) = \sum_{l,m} |<n+\Delta n, l+1, m|z|n, l, m>|^2\, f(n, l, \Delta n), \tag{2}$$

where l and m are respectively the orbital and magnetic quantum numbers, and

$$\begin{aligned}f(n, l, \Delta n) =\ & (9/4)n^2(n^2-l^2-l-1) + (9/4)(n+\Delta n)^2(n^2-l^2+2n\Delta n-3l-3+\Delta n^2) \\ & -\{\,[\,2l/(2l+1)\,]\, R(n,l;n,l-1)\, R(n+\Delta n, l+1; n+\Delta n, l) \\ & + [\,(2l+4)/(2l+3)\,]\, R(n, l+1; n, l)\, R(n+\Delta n, l+2; n+\Delta n, l+1) \\ & \cdot R(n, l+1; n+\Delta n, l+2)\,\} / R(n, l; n+\Delta n, l+1)\,.\end{aligned} \tag{3}$$

The factors $|<n+\Delta n, l+1, m|z|n, l, m>|^2$ are normalized, $\sum |<n+\Delta n, l+1, m|z|n, l, m>|^2 = 1$ (Hoang-Binh et al. 1987), and $R(n, l;\, n', l')$ is the dipole radial integral for the transition $(n, l) \rightarrow (n', l')$ (Green et al. 1959).

3. Results

3.1. CALCULATION OF $F(n, \Delta n)$

As we can see above, the fundamental quantity in the theory of ion impact broadening is $F(n, \Delta n)$, the evaluation of which requires a very large number of dipole radial integrals.

Accurrate values of R(n,l; n',l') have been obtained using a method proposed by Hoang-Binh (1986). We have calculated F(n, Δn), for various values of n, and for Δn= 1,2,3.

Table 1. Values of F(n,Δn) and R(n,Δn) = F(n,Δn) /F(n,1)

n	F(n,1)	F(n,2)	F(n,3)	R(n,2)	R(n,3)
5	1.591E+02	1.177E+03	3.294E+03	7.39	20.70
6	2.260E+02	1.610E+03	4.399E+03	7.12	19.46
7	3.050E+02	2.111E+03	5.665E+03	6.92	18.57
8	3.962E+02	2.681E+03	7.090E+03	6.77	17.90
9	4.997E+02	3.320E+03	8.677E+03	6.64	17.36
10	6.156E+02	4.027E+03	1.042E+04	6.54	16.93
11	7.440E+02	4.803E+03	1.233E+04	6.46	16.58
12	8.850E+02	5.648E+03	1.440E+04	6.38	16.27
13	1.039E+03	6.563E+03	1.663E+04	6.32	16.01
14	1.205E+03	7.546E+03	1.902E+04	6.26	15.79
15	1.384E+03	8.599E+03	2.158E+04	6.21	15.59
16	1.576E+03	9.722E+03	2.429E+04	6.17	15.42
17	1.781E+03	1.091E+04	2.717E+04	6.13	15.26
18	1.998E+03	1.218E+04	3.021E+04	6.09	15.12
19	2.229E+03	1.351E+04	3.341E+04	6.06	14.99
20	2.473E+03	1.491E+04	3.677E+04	6.03	14.87
50	1.600E+04	8.968E+04	2.134E+05	5.61	13.34
100	6.637E+04	3.569E+05	8.361E+05	5.38	12.60
200	2.764E+05	1.435E+06	3.324E+06	5.19	12.03
300	6.388E+05	3.248E+06	7.482E+06	5.08	11.71
500	1.856E+06	9.135E+06	2.088E+07	4.92	11.25
800	5.172E+06	2.393E+07	5.403E+07	4.63	10.45

As can be seen in Table 1, F(n, Δn) is not proportional to Δn^2, even for very large values of n ($\approx$800).

According to our results, we may write, for n in the range 50 - 1000, with an accurracy of $\approx$ 10%,

$$F(n,1) = (9/4) n^2 [3/2 + 0.271 \ln(2n/3) + 1.92/n^{0.422}], \qquad (4)$$

$$F(n,2) = (5.49/n^{0.5} + 4.83) F(n,1), \qquad (5)$$

$$F(n,3) = (17.9/n^{0.5} + 10.81) F(n,1) . \tag{6}$$

3.2. LINE WIDTHS

Let the ion-emitter reduced mass be M' (g) and μ_i (a.m.u.). The average of (1/v) is $(2M'/\pi kT)^{1/2}$, where T (°K) is the kinetic temperature. Denoting the proton mass by M (g), the ion impact width of a line $(n+\Delta n, n)$ is then

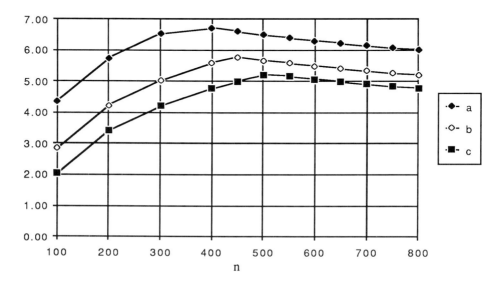

Figure 1. $\ln(\rho_{max}/\rho_{min})$ as a function of n, for T=20 K, $N_i = N_e = 0.3$ cm^{-3}; a, b, c refer to Cnα, Cnβ, Cnγ lines respectively. The colliding ion is C$^+$.

$$\Delta w_i(n, \Delta n) = (4/3) (2\pi M/k)^{1/2} (h/2\pi m_e Z)^2 (\mu_i/T)^{1/2} N_i F(n, \Delta n)$$
$$\cdot [1/2 + \ln(\rho_{max}/\rho_{min})], \tag{7}$$

$$\rho_{max} = \min [(\pi/2)^{1/2} \lambda/2\pi, \rho_D], \tag{8}$$
$$\rho_{min} = 8.301 \cdot 10^{-5} F(n, \Delta n)^{1/2} (\mu_i/T)^{1/2}, \tag{9}$$
$$\rho_D = 6.90 [T/(N_i + N_e)]^{1/2}, \tag{10}$$

where k is the Boltzmann constant, N_e is the electron density, λ and ρ_D are respectively the line wavelength and the Debye radius. In (7), the term 1/2 in the square brackets accounts for the strong collision term. It should be noted that the impact approximation breaks down if $1/2 \geq \ln(\rho_{max}/\rho_{min})$. This is generally not the case for RRLs, as can be seen on Figure 1.

Finally, the full width at half-maximum (FWHM), $\Delta v_i (n,\Delta n) = \Delta w_i (n, \Delta n)/\pi$, is given by

$$\Delta v_i(n,\Delta n) = 1.569 \cdot 10^{-4} Z^{-2} (\mu_i/T)^{1/2} N_i F(n,\Delta n)[1/2 + \ln(\rho_{max}/\rho_{min})]. \qquad (11)$$

Figure 2 shows the ratios of line widths of carbon RRLs, broadened by collisions with C^+

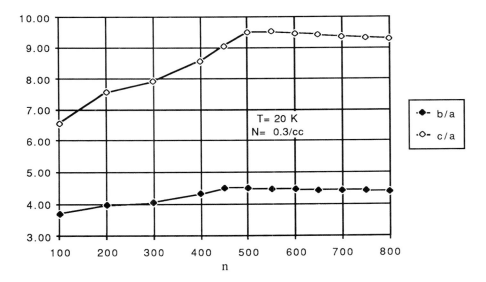

Figure 2. Ratios $\Delta v(n,2)/\Delta v(n,1)$ (referred to as b/a) and $\Delta v(n,3)/\Delta v(n,1)$ (referred to as c/a) of carbon line widths, as functions of principal quantum number n. cc stands for cm^3.

as functions of n, for T=20 K, $N_i = N = 0.3$ cm^{-3}. The formulae given above may be used for impact broadening of RRLs by various ions, such as Mg^+, C^+, Si^+, etc.; provided that due account is taken of the difference in the reduced mass. Thus, for carbon RRLs broadened by collisions with C^+, $\mu_i=6$.

4. Comparison with other broadening mechanisms.

For a line of frequency ν_0, emitted by an atom of mass M_E (g), and μ (a.m.u.), the normalized Doppler profile is

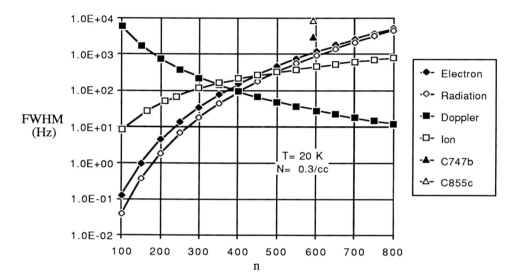

Figure 3. $Cn\alpha$ line widths due to different broadening mechanisms. cc stands for cm^3. C747b, and C855c refer to the ion widths of C747ß, C855γ respectively. It is assumed that N_e and N_i are equal to the number density N of C^+.

$$\Phi_\nu = \exp(-x^2)/\Delta_D\sqrt{\pi},$$

where $\Delta_D = \nu_0 \sqrt{(2kT/M_E c^2)}$, $x = (\nu-\nu_0)/\Delta_D$, c being the velocity of light. Thus the Doppler FWHM is

$$\Delta\nu_D = 7.1364 \ 10^{-7} \ \nu_0 \ \sqrt{(T/\mu)} \ . \tag{12}$$

Broadening by electrons has been extensively investigated (see e.g. Griem 1967, Ershov et al. 1984). According to Ershov et al. (1984), the electron width $\Delta v_e(n,1)$ and the radiation width $\Delta v_r(n,1)$ are given by

$$\Delta v_e(n,1) = 1.16 \cdot 10^{-3} \, N_e \, (n/100)^{5.1} \, (T/100)^{0.62} \text{ kHz,} \tag{13}$$

$$\Delta v_r(n,1) = 3.82 \cdot 10^{-5} \, (n/100)^{5.63} \text{ kHz.} \tag{14}$$

Figure 3 plots the line widths due to ion (this paper), electron, radiation (Ershov et al. 1984), and Doppler broadenings, for one set of physical parameters typical of dark clouds. It can be seen that ion (C^+) broadening dominates over electron broadening for n < 450; but, due to a stronger n-dependence, electron broadening becomes more important for decameter Cnα lines. In the region where Doppler broadening dominates, ion broadening may be important for higher order lines or/and in the case of higher ion densities. The widths of the C747ß and C 855γ lines, which have about the same frequency as the C593α line are also shown for comparison. Formulae presented in this paper should be preferred to that in Omont and Encrenaz (1977) and Griem (1967).

5. Discussion

It is most interesting to observe simultaneously Cnα and higher order lines at about the same frequency; e.g. C167α, C210ß and C240γ near 1400 MHz. In effect; since they have the same Doppler width and refer to about the same region (because the antenna beam width varies with frequency), any differences in the observed widths would reflect the n and Δn dependences of pressure broadening. A theoretical knowledge of the latter would allow us to estimate the Doppler and Stark widths separately; hence to deduce the temperature and density of the emitting region.

References

Ershov, A.A., Ilyasov, Yu. P., Lekht,E.E., Smirnov, G.T., Slodkov ,V.T., Sorochenko, R.L. 1984, Pisma A.J., **10**, 833.
Griem, H.R. 1967, Ap. J., **148**, 547.
Hoang-Binh, D. 1986, in *Europhysics Conference Abstracts* , **18 th EGAS Marburg,** ed. G. Thomas (Geneva), p. 335.
Hoang-Binh, D., Brault, P., Picart ,J., Tran-Minh, N., Vallée, O. 1987, Astr. and Ap.,**181**, 134 .
Omont, A., Encrenaz, P. 1977, Astr. and Ap., **56**, 447.

High resolution radio recombination line observations

P.R. ROELFSEMA
Space Research Groningen
Postbus 800
9700 AV Groningen
The Netherlands

Abstract. Over the last decade very sensitive observations of radio recombination line emission using high angular resolution synthesis telescopes have become available. As a result it has now become possible to image the physical parameters deduced from radio recombination lines of individual sources. In the case of HII regions this work has resulted in detailed maps of radial velocities, the electron temperature and the abundance of singly ionized helium (Y^+). Furthermore the influence on radio recombination line emission towards HII regions of variations in Y^+, of pressure broadening and of the continuum optical depth has been demonstrated unequivocally. Other new results due to the high angular resolution are a better understanding of the properties of the emission from the partially ionized medium adjacent to HII regions (CII and H^o regions). Finally the new observational possibilities show great promise for the study of the physics and kinematics of e.g. extragalactic objects.

1 Introduction

Since their first prediction (Kardashev, 1959) and subsequent discovery (Dravskikh and Dravskikh, 1964; Sorochenko and Borozich, 1964) radio recombination lines have been used extensively as probes of the physical conditions in ionized interstellar material. The most obvious advantage of radio recombination lines lies in the fact that, contrary to optical emission, radio emission is not attenuated by matter along the line of sight. Thus we can be fairly sure that the line emission as observed is not in some way distorted by intervening material. This means that observed line parameters like intensity and velocity can directly be interpreted in terms of the properties of the emitting gas without having to apply an undoubtedly model dependent correction for extinction.

In the sixties and seventies an extensive database of single dish radio recombination line observations was built up. A number of projects studying individual HII regions in many transitions were undertaken, but also several large radio recombination line surveys were done. Unfortunately all this work has been constrained by the attainable spatial resolution. In virtually all single dish radio recombination line observations of HII regions the beam size is at best comparable to the angular size of the object under study (a notable exception is e.g. Orion). Thus corrections had to be made for the non-uniformity of HII regions, their assumed spherical or cylindrical morphology, their kinematical properties etc. etc. All this single dish work led to refinements in the theory of radio recombination line emission, but especially to refinements in the analysis and interpretation of radio recombination line data (Shaver, e.g. 1980).

As a result presently the theory of radio recombination line emission is very well understood. Non-LTE effects (e.g. pressure broadening , stimulated emission) can be recognized in observations, and can be traced back to the physical properties of the plasma under study. Thus with the advent of very stable and sensitive radio interferometers (WSRT, VLA) we should now be able to study radio recombination line emission with very high angular resolution and understand the observed line properties.

A first series of interferometric radio recombination line observations was carried out by van Gorkom (1980) using the Westerbork array. These observations already indicated some of the possibilities; variations in density within HII regions were observed, pressure broadening effects were seen, and even the suggestion of narrow line H° emission was found. Since these first observations important improvements in sensitivity and stability of the interferometers have come about. Furthermore from these this pioneering work we have learned how to properly calibrate and analyze interferometric radio recombination line observations.

1.1 WHAT CAN BE LEARNED FROM RADIO RECOMBINATION LINES

A number of different physical properties can be derived from radio recombination line observations. Firstly the physical parameters of the line-emitting plasma can be determined. The electron temperature T_e can be derived from the observation of a single transition (e.g. Shaver, 1980). By combining observations of two or more transitions also a relatively good determination of the r.m.s. electron density $n_{e,rms}$ can be obtained.

Secondly kinematical information can be obtained through observations of radio recombination lines . The central velocity of the line can be used to investigate systematic motions of the gas and the width of a line (e.g. the Gaussian FWHM ΔV_G) gives information about turbulence in the plasma under study.

Thirdly by comparing H and He lines information can be obtained regarding abundance of singly ionized helium Y^+ and the ionization structure of the gas. Thus we obtain some information about e.g. the source of the ionizing radiation or the presence or absence of dust absorbing parts of the ionizing radiation *in* an HII region.

When in addition IR or optical emission lines are observed, a direct determination of the external extinction is obtained. Thus we can also study the molecular material along the line of sight towards a radio recombination line emitting region.

1.2 HII REGIONS AND THEIR ENVIRONS

The first candidates for RRL line observations are HII regions. In our simplest model such objects consists of spherically symmetric balls of ionized hydrogen at a temperature T_e of ~10000 K. Thus line shapes and intensities from such objects can easily be understood. A somewhat more advanced model of an HII region also incorporates ionized helium , partially ionized hydrogen (H°), ionized carbon and neutral material. Such a model is shown in figure 1. In this figure the possible locations of the different RRL emitting regions are shown. For simplicity still the simple spherical configuration is used. Clearly a real-life HII region will have a much more complex morphology.

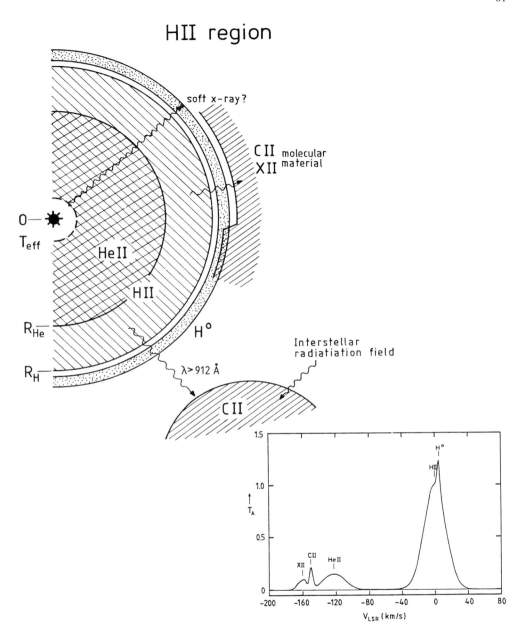

Figure 1: The basic HII region and its surroundings. The inset shows a complete radio recombination line spectrum observed towards such an HII region.

The basic HII region is the result of ionization of gas around a young stellar object characterized by an effective temperature T_{eff}. The material in the immediate vicinity of the ionizing source has been used to form the star, and thus in general an HII region will have a central cavity. The stellar radiation will ionize the surrounding material thus forming the HII region proper. Temperatures in different HII regions are found to range from $T_e \sim 5000$ to 15000 K. The size of the HII region (the HII region Strömgren radius R_H) is dependent on the properties of the ionizing radiation (i.e. T_{eff}, and absorbtion by dust inside the HII region) and the density of the gas. If the ionizing radiation contains photons capable of ionizing He (i.e. $\lambda \leq 504$Å) then also a HeII region will be formed with a HeII region Strömgren radius R_{He}. It is important to note that in principle $R_{He} \leq R_H$.

Beyond this region of predominantly ionized material the gas is neutral, and for an important part in molecular form. However photons with wavelengths longer than the Lyman-α limit of 912Å(i.e. not capable of ionizing hydrogen) will escape from the HII region. Such photons can ionize other atomic species in the neutral gas; carbon ($\lambda \leq 1088$Å), magnesium ($\lambda \leq 1632$Å), sulphur ($\lambda \leq 1204$Å) etc. Thus these species may exist in ionized form, and thus emit RRL's, *outside* of the HII region proper. These regions will have temperatures of several tens to a few hundred degrees K. Obviously such e.g. CII regions need not be adjoining to HII regions. The general interstellar radiation field will also ionize isolated clouds, thus heavy element RRL emission can also be expected from such clouds.

The third region of interest for RRL research is a region where hydrogen is only partially ionized ($\frac{n(H^+)}{n(H)} \leq 10^{-3}$), the so called H° region. Such low ionization can be caused by leakage of soft X-ray photons from the HII region (Pankonin et.al., 1977; Krügel and Tenorio-Tagle, 1978) or by the interaction of a weak D-type shock front with the neutral material (Hill, 1977). To date it is not clear whether such a region is confined to a thin shell surrounding the HII region proper or whether H° regions also exist e.g. associated with isolated dark clouds. Also the question whether H° and CII regions are coexistent or physically separate has not been answered satisfactorily yet.

Note that high density clumps of neutral materal may exist *within* an HII region. Such clumps can also contain CII, XII and H° regions.

1.2.1 Which radio recombination lines From an HII region as described in the previous section a number of different lines can be observed. The HII region proper will give rise to relatively broad RR lines of ionized hydrogen and ionized helium (typical FWHM of $\Delta V_G \sim 20$ to 30 km s^{-1} for LTE conditions). From the cooler neutral gas narrow carbon and heavy element lines can be expected with widths of several km s^{-1}. Similar widths of a few km s^{-1} are expected for lines from the H° region. The inset of figure 1 shows such a typical radio recombination line spectrum. Note that near the CII line a blend of heavy element lines appears, the so-called XII line.

2 High angular resolution observations

In the following sections some of the results of high resolution observations of radio recombination lines from ionized material shall be described. The data reported on were obtained using synthesis telescopes (WSRT and VLA) achieving a spatial resolution of typically a

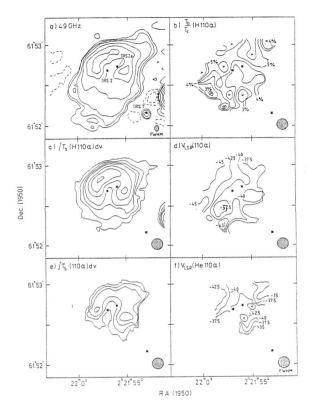

Figure 2: 110α observations of W3A (from Roelfsema, 1987).
a) 4.9 GHz continuum.
b) Peak line to continuum ratio T_l/T_c for the H110α line in percent.
c) Intensity distribution of the H110α transition.
d) Radial velocities V_{lsr} of the H110α emission.
e) Intensity distribution of the He110α transition.
f) Radial velocities V_{lsr} of the He110α emission.
The resolution of the observations is indicated by the hatched ellipses in all panels. The positions of the infrared sources IRS2, 2a and 7 are indicated by stars.

few arcseconds. With a velocity resolution of ∼2/kms, the r.m.s. noise in these observations is typically only a few mJy/beam (corresponding to a few degrees K).

Since the radio recombination line radiation (that is, lines in the *radio* regime!) itself usually is very weak as compared to the radio continuum emission towards e.g. HII regions, special care is required to properly calibrate the shape of the receiver pass band . Usually this is done by observing a very strong point like calibrator source several times during an observing run. Variations seen in the spectrum towards such a source are then assumed to be entirely due to the pass band of the receiver system.

Fortunately the pass band shapes, both at Westerbork and at the VLA, are found to be very stable; observing a band pass calibrator need only be done once every 2–3 hours. However, the total amount of time spent on the pass band calibrator may be a significant fraction of the observing time. After observing a pass band calibrator (with a flux density of S_{bp}Jy) for time t_{bp}, the r.m.s. noise in the pass band calibration data will be σ_{bp}. Thus the *lowest* possible line to continuum ratio that is detectable in this calibration spectrum is σ_{bp}/S_{bp}. Since this spectrum is used to calibrate the pass band of the observation, σ_{bp}/S_{bp} will also be the limit of the line to continuum ratio in the data if the pass band calibrator is observed for too short a time.

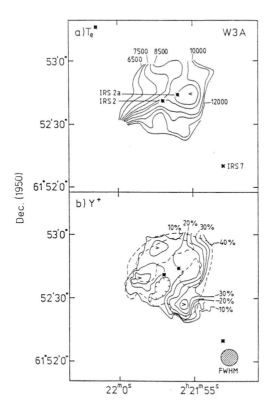

Figure 3: **a)** The LTE electron temperature distribution T_e^* for W3A as derived from the H76α line intensity and the 14.7 GHz continuum. The contour levels are indicated in °K.
b) The apparent abundance of singly ionized helium in W3A as derived from the He110α and H110α line intensities. The 4.9 GHz continuum map contours at 150 mJy beam^{-1} and 300 mJy beam^{-1} are superposed as dashed lines.
The hatched ellipses indicate the resolution. The stars indicate the positions of the infrared sources IRS2, 2a and 7. (from Roelfsema, 1987)

2.1 HII REGIONS

The obvious result of high angular resolution observations is that line parameters (intensity, width, central velocity) can be mapped across an HII region. As an example W3A as observed in the H and He 110α lines is shown in figure 2. At a resolution of $\sim 8''$ the shell of the HII region is clearly resolved in the line emission.

2.1.1 Electron temperature From the data presented in figure 2 the LTE electron temperature T_e^* can easily be calculated assuming low optical depth and a constant abundance of singly ionized helium Y^+. A distribution of T_e^* in W3A is shown in figure 3a. This figure suggests that the electron temperature within W3A fluctuates dramatically; from $\sim 6000°$K in the east to about 12000°K in the west of the shell. However, as is shown in figure 3b the abundance of singly ionized helium also varies greatly. In fact when figure 3a is corrected for the variations in Y^+, T_e is found to be virtually constant over the entire source at 8000±750°K. Indeed, in many of the compact HII regions observed at high angular resolution, it is found that T_e is roughly constant. Different HII regions have different temperatures, but each region is close to isothermal.

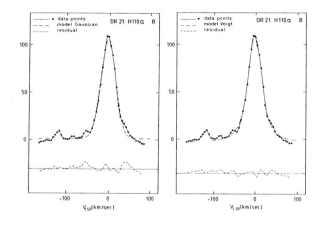

Figure 4: An H110α line profile observed towards component B of DR21 showing pressure broadening due to high density ($n_e \geq 10^4$ cm^{-3}) clumps (from Roelfsema et. al., 1989).
The left panel shows a Gaussian fit, the right panel a Voigt fit superposed on the observed profile. The residuals after subtraction of these fitted profiles are indicated by the dashed lines.

2.1.2 Pressure broadening Already in 1980 van Gorkom concluded from the comparison of theoretical and observed line to continuum ratios that pressure broadening must play an important role in RRL emission from compact HII regions. In very high density clumps the collisions of the electrons in high quantum states with neighboring atoms will result in a redistribution of the line energy to the wings of the line. As a result the radio recombination line will have a lower amplitude than under LTE and become broader. The profile will obtain the shape of a Voigt function. The resolution of the observations of van Gorkom unfortunately was not enough to actually see broader lines. In figure 4 an H110α line observed towards DR21 is shown. The line profile is clearly non-Gaussian. The left panel shows the residuals (dashed line) after subtraction of a Gaussian model profile. From this residual it is clear that both at the center *and* in the wings a Gaussian underestimates the true line profile. Indeed a Voigt profile (left panel) gives a significantly better fit.

Detecting the change in line shape due to pressure broadening really shows the importance of angular resolution. Profiles as shown in figure 4 have only been found on a very small angular scale; towards DR21 only in a single resolution element. Observing such a source with a lower spatial resolution will effectively average the low amplitude Voigt profile with higher amplitude Gaussian profiles due to the lower density gas surrounding the high density clump. As a result the Voigt profile itself will be drowned and the only remaining observable effect of pressure broadening will be a depressed line to continuum ratio $\frac{L}{C}$ of the average profile.

2.1.3 The influence of the continuum optical depth In several compact HII regions it is found that the optical depth of the continuum (τ_c) does have a significant effect. In W3B the H110α lines show a slight decrease in line to continuum ratio $\frac{L}{C}$ towards higher density regions: typically 3–3.5% is observed as compared to a theoretical LTE $\frac{L}{C}$ of ∼4%. This decrease of $\frac{L}{C}$ is consistent with τ_c ∼0.3 at 4.9 GHz as is observed (Roelfsema, 1987). A more extreme case is found in Sgr B2 where the H76α line towards one of the most compact components is depressed due to a large continuum optical depth of τ_c ∼0.8 (Roelfsema et. al., 1987b). Thus even at high frequencies τ_c may still be of importance. As is the case for pressure broadening, the lines that are affected by a high τ_c are weak as compared to those that are not affected. Therefore observing the influence of τ_c is only possible with sufficient angular resolution.

2.1.4 Kinematics Studying the kinematics of HII regions is very well possible using radio recombination line observations. Clearly the velocities of HII regions can be used to study e.g. galactic dynamics. With sufficient resolution also the internal dynamics of HII region can be investigated. As can be seen from figure 2d quite detailed velocity information can be derived from high resolution radio recombination line data. Unfortunately, interpreting these velocities in terms of what the systematic motions within an HII region are, is less straightforward. For W3A we can only say that the velocity of the gas along the shell seems to be more negative then the velocity of the gas near the central region (-42 km s^{-1} v.s. -39 km s^{-1}), suggesting that the shell is expanding towards the observer. However, this picture is complicated by the velocity gradient (from -40 km s^{-1} to -45 km s^{-1}) that is present along most of the north east rim of the source.

The velocities are very important when comparing e.g. H and He lines. If these lines originate from the same volume of gas, they should be at the same velocity (or, when observed within one spectrum as shown in figure 1 they should be separated by exactly 122.1 km s^{-1}). Clearly the good correspondence of the H and He velocities as seen in figures 2d and 2f indicates that the He110α emission observed towards W3A originates from the same volume of space as does the H110α emission.

2.2 HeII REGIONS

Our knowledge on HeII regions has also increased due to sensitive high angular resolution radio recombination line observations. The He radio recombination line emission can now be mapped towards HII regions without too much effort. From the observations it has become clear that the ionized He need not be distributed uniformly over an HII region. Clumps with a relatively high abundance of singly ionized helium Y$^+$ are found in several sources (up to an extreme Y$^+$ of ~35% as observed in W3A, see figure 3). Again the angular resolution is of the utmost importance, the high Y$^+$ clumps in W3A typically have sizes of ~10″. A more extensive discussion on helium radio recombination lines by W.M. Goss can be found elsewhere in this volume.

2.3 H° REGIONS

Observing narrow lines from the partially ionized medium (PIM) near HII regions at high frequencies is probably one of the most challenging experiments in radio recombination line research. The line is narrow (typically a few km s^{-1}), weak ($\frac{L}{C}$ of order 1%) and superposed on a relatively strong line (the ratio of HII to H° line amplitude is found to be anywhere in the range 1–10) and an even stronger radio continuum. At lower frequencies these problems are less severe because stimulated emission becomes a more effective emission mechanism, and the lines from the HII region proper become very weak.

When Pankonin reviewed the available observations regarding H° regions during the previous radio recombination line conference (Pankonin, 1980), his list was disappointingly

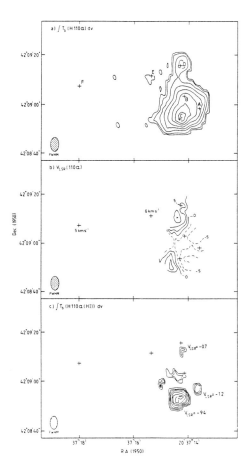

Figure 5: H110α line emission towards DR21 (from Roelfsema et.al., 1989).

a) H110α line intensity.

b) velocities of the H110α lines. Contours are drawn at −7.5 through 15 km s^{-1} with 2.5 km s^{-1} intervals.

c) H110α(HI) line intensity. The velocities of the narrow lines observed towards the different continuum components are indicated.

Crosses indicate the positions of the main continuum components. The hatched ellipses indicate the FWHP.

short; only NGC2024, Mon R2 and (probably) W3 showed narrow line emission. In all cases the data consisted of spectra of only a few different transitions, with no spatial information. Since then this list has grown considerably; e.g. DR21 (see figures 5 and 6), K3-50 (figure 7), W49 and others were added to the list, and the narrow line emission from W3 was confirmed (Roelfsema, 1980).

More important than the mere detection of narrow lines towards HII regions, has been the improvement in spatial resolution. Many of these H° regions have now been mapped successfully. An example of this is shown in figure 5 from Roelfsema et.al. (1989).

2.3.1 The extent of H° regions   The three panels of figure 5 show the H110α line intensity, velocity and narrow line intensity observed towards DR21. A striking feature of the narrow line emission is the fact that it is significantly less extended than the HII emission. The major part of the H110α(HI) emission is observed towards the southern parts of the source.

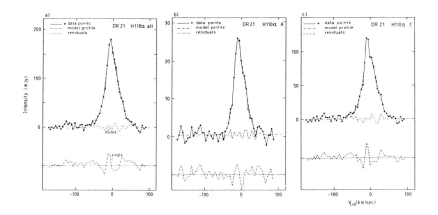

Figure 6: H110α and narrow H110α(HI) lines observed towards DR21 (from Roelfsema et.al., 1989). The dotted lines indicate the residuals after subtraction of a single Gaussian profile (marked single) or a double Gaussian profile (marked double).

Note also that the velocity of the narrow line emission changes considerably over the source.

Due to the variations in the properties of the narrow line emission the global H110α profile of DR21 (figure 6a, equivalent to the profile as it would be observed with a single dish telescope) shows a fairly weak H110α(HI) line. The average narrow line $\frac{L}{C}$ is only ~0.7% while towards component C, the region with the strongest narrow line emission, $\frac{L}{C}$ is found to be at least twice as high. Similar data obtained towards other sources also suggests that the narrow line emitting H° regions have a filling factor $f_{H°} < 1$, with typical values of $f_{H°}=0.2$–0.6. In DR21 the most intense H110α(HI) emission is found in a region of high H_2CO optical depth. This correspondence suggests a close relation between the molecular material and the H° region. However, the velocities of the H_2CO and H110α(HI) lines differ by up to 5 km s^{-1} and thus seem to contradict such a relation.

2.3.2 Kinematics of H° regions   Apart from a more accurate determination of $\frac{L}{C}$ for narrow line emission, also better velocities are obtained with high resolution observations. The narrow line velocities ($V_{H°}$) in general are different from the velocity of both the CII region (V_{CII}) and the HII region (V_{HII}) by a few km s^{-1}. These velocity differences imply that in general H°, CII and HII regions are *not* coextensive.

The present data also provide spatial information on the H° velocities. Mapping of the H110α(HI) line velocities towards DR21 shows velocity gradients in the narrow line emission similar to those observed in the emission from the HII region proper (see figure 5b and 5c). This similarity is consistent with the suggestion that the H° region is a (partial) shell surrounding, and being accelerated by, the expanding HII region (see figure 1). With such a model the DR21 H° region lies on the near side of the HII region. Thus the narrow line emission can be explained as being mainly due to stimulated emission.

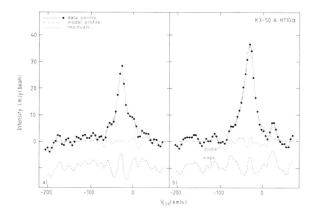

Figure 7: H110α and narrow H110α(HI) lines observed towards two locations at opposite sides of the center of K3-50A. The velocity of the H° line shifts from -5 km s^{-1} to +5 km s^{-1} w.r.t. the HII line. The dotted lines indicate the residuals after subtraction of a single Gaussian profile (marked single) or a double Gaussian profile (marked double).

A counter example showing that $V_{H°}$ and V_{HII} are not always different is found in K3-50A (Roelfsema et.al., 1988). In this HII regions velocity gradients are found both in the broad HII and in the narrow H° line emission. However, contrary to DR21 where the difference between $V_{H°}$ and V_{HII} is roughly constant, in K3-50A $V_{H°} - V_{HII}$ varies from -5 km s^{-1} to +5 km s^{-1} (see figure 7). Such different velocity gradients are not consistent with the simple concentric shell model for the HII and H° regions, in this case a much more complicated geometry is definitely required by the observations.

2.3.3 The structure and ionization of H° regions   Clearly with such high resolution narrow line data the validity of models regarding their origin can now be tested in much more detail. Both the values of $\frac{L}{C}$ and the variations of $\frac{L}{C}$ with frequency are consistent with the model where the H° region is ionized by X-rays from the central star of the HII region as proposed by Krügel and Tenorio-Tagle (1978). Since this model does not require the H° region to be part of e.g. the outer edge of the HII region, it does not constrain $V_{H°}$. The model of Hill (1977) explaining the H° region as being due to a weak D-type shock in the outer edge of the HII region does suggest a velocity structure as found in DR21. However, the Hill model predicts values of $\frac{L}{C}$ which are lower than observed at high resolution.

2.4 CII REGIONS

Observations of carbon radio recombination lines have also been very successful over the last decade. These lines, although weak like narrow lines from H° regions, are less difficult to observe. This is mainly due to the fact that they are not blended with other lines as the H° lines are. Carbon lines have been observed towards a number of sources, in transitions at frequencies from several tens to of MHz up to 20 GHz. Both the low and the high frequency end of the spectrum will be treated extensively by others, therefore here only a few comments will be made on observations done in the 1–10 GHz range.

One of the main issues in CII radio recombination line observations has been the question of the emission mechanism. Before CII regions could be mapped in detail, the only way to approach this question has been by observing many different transitions. The observed line intensities were then compared with a model HII-CII complex (e.g. Hoangh-Binh and

Walmsley, 1974; Pankonin et.el., 1977; Qaiyum and Ansari, 1983). With high resolution data a much simpler approach has become available. When CII lines are observed at a frequency where the HII region is optically thick, obviously the emission cannot come from the far side of the HII region. In this case at positions towards the HII region the equation of transfer *must* contain a stimulated emission term. This immediatly constrains the emission measure of those parts of the CII region which lie at the near side of the HII region. In the case of W3A this approach was used successfully to determine the properties of the CII region on the basis of C166, 110 and 76α line observations (Roelfsema et.al., 1987a). As a result it was found that the W3A CII region at the near side of the source must be a thin sheet ($l \sim 0.01 pc$ along the line of sight, which is $\sim$5% of the HII region radius) with an electron density between 30 and 100 cm^{-3}.

High resolution CII line observations have also shown correlations between the CII intensity and e.g. H$^\circ$ intensity, H$_2$CO column density and leakage of high energy uv-photons from HII regions.

2.5 OTHER SOURCES

Two other types of objects are worth mentioning with regards to high resolution radio recombination line observations; planetary nebulae and extragalactic sources. Both can be mapped, albeit that in both classes to date only a single object has been mapped. Of the planetary nebulae only NGC 7027 is large enough to be mapped, and of the extragalactic sources only M82 seems to have the right properties for radio recombination line studies. However, these sources do show the possibilities of using radio recombination lines to study objects that are otherwise invisible. Both objects are very much obscured by dust. Thus physical parameters of the ionized material as derived from optical observations suffer greatly. The radio recombination line data allow us to determine T_e and n_e with a very high accuracy. Furthermore high resolution line data have been used to study the kinematics of both objects (Roelfsema, 1987).

Recent VLA H92α and H110α observations of M82 have shown once more the importance of high resolution. As is the case in e.g. H$^\circ$ regions it was found that the line emitting region is considerably smaller than the radio continuum emitting disk; here the filling factor can be as small as 0.1. Thus the true $\frac{L}{C}$ of the lines is much larger than what would be observed using a single dish instrument. If this low filling factor is characteristic for extragalactic sources, many of the previously undetected sources may be detectable when observed at high angular resolution.

3 What is next?

Clearly the high sensitivity and the improved resolution of synthesis instruments has increased our knowledge of HII regions and the adjacent partially ionized and neutral material. The question of course is, what more is required?

Firstly, to completely understand a source, it has to be observed not only at high angular resolution but also in many different transitions of many different atomic species. This is illustrated by the results obtained from the extensive body of observations of the HII complex W3 (Roelfsema, 1980; Roelfsema and Goss, in prep). For several HII complexes

(DR21, K3-50 and others) still more observations are required. Also with new wavelengths that are becoming accessible (e.g. the low frequency 327 MHz bands of the WSRT and the VLA, and the very sensitive 8.4 GHz band at the VLA) the number of transitions observed for these sources should be increased.

Also the body of HII regions that have been mapped is still fairly small. Here also new instrumentation will help out: the AT will open the southern hemisphere which to date has been virtually inaccessible except for single dish work. This telescope will also allow us to study e.g. HII regions in the Magellanic clouds, 30 Doradus etc.

The study of CII and H° regions is still in its infancy, here clearly many more observations are required. To definitely distinguish between the different possible ionization mechanisms again the comparison of different transitions is necessary. Again here the 327 MHz and 8.4 GHz bands are bound to be of great importance. Possibly for these regions even a 327 MHz galactic plane survey should be attempted.

Finally also the surveys of extra galactic radio recombination lines as done in the seventies should be repeated, utilizing the high resolution instruments at their most sensitive frequencies (e.g. 8.4 GHz at the VLA).

References

Dravskikh, Z.V., and Dravskikh, A.F.: 1964, *Astron. Tsirk.*, **158**, 2
Hill, J.K.: 1977, *Astrophys. J.* **212**, 692
Hoang-Binh, D., Walmsley, C.M.: 1974, *Astron. Astrophys.* **35**, 49
Gorkom, J. van: 1980, Ph.D. Thesis, University of Groningen, The Netherlands.
Kardashev, N.S.: 1959, *Soviet Astr.*,**3**, 813
Krügel, E., Tenorio-Tagle, G.: 1978, *Astron. Astrophys.* **70**, 51
Pankonin, V., Walmsley, C.M., Wilson, T.L, Thomasson, P.: 1977, *Astron. Astrophys.* **57**, 341
Pankonin, V.: 1980, in Proc. Workshop on Radio Recombination lines, ed. P.A. Shaver, Reidel, Dordrecht.
Quaiyum, A., Ansari, S.M.R.: 1983, *Monthly Notices Roy. Astr. Soc.* **205**, 719
Roelfsema, P.R., Goss, W.M., Wilson, T.L.: 1987a, *Astron. Astrophys.* **174**, 232
Roelfsema, P.R., Goss, W.M., Whiteoak, J.B., Gardner, F.F., Pankonin, V.: 1987b, *Astron. Astrophys.* **175**, 219
Roelfsema, P.R.: 1987, Ph.D. Thesis, University of Groningen, The Netherlands.
Roelfsema, P.R., Goss, W.M., Geballe, T.R.: 1988, *Astron. Astrophys.* **207**, 132
Roelfsema, P.R., Goss, W.M., Geballe, T.R.: 1989, *Astron. Astrophys.* **222**, 247
Shaver, P.A.: 1980, *Astron. Astrophys.* **91**, 279
Sorochenko, R.L., Borozich, E.V.: 1964 *Reports of the Sov. Academy of Science*, **162**, 3, 603

RADIO RECOMBINATION LINES FROM COMPACT HII REGIONS

GUIDO GARAY
Departamento de Astronomía, Universidad de Chile
Casilla 36-D, Santiago, Chile

ABSTRACT. We compile measurements made with high angular resolution of the H76α line and radio continuum emission from over thirty compact HII regions. The aim was to investigate global trends exhibited by the observed and derived parameters and to provide explanations for the tendencies in terms of the physical conditions of the ionized gas. The line widths of compact HII regions are considerably broader than those of diffuse regions. We conclude that the large widths are due to turbulent motions produced by the progressive ionization of a magnetized and inhomogeneous medium around a newly formed star. The line to continuum ratios are spread over a much wider range than that span by extended regions. We show that the large ratios are due to line enhancement by stimulated emission, while the small ones can be attributed to line suppression in regions with large continuum opacities.

1. INTRODUCTION

Compact HII regions are small (L $\leq$ 0.1 pc) and dense ($N_e \geq$ 5x10^3 cm^{-3}) regions of ionized gas surrounding and excited by newly formed massive stars (Mezger 1968). Because they are deeply embedded in the dense molecular cloud from which they formed, compact HII regions are invisible at ultraviolet and optical wavelengths. Further, since they are usually found in groups and/or associated with more diffuse regions, their observations at radio wavelengths, which are not absorved by dust, require the use of telescopes providing high angular resolution. For these reasons, until recently, the number of studied compact HII regions was small. The advent of filled-aperture instruments, such as the Westerbork Synthesis Radio Telescope (WSRT) and the Very Large Array (VLA), providing high angular resolution, sensitivity, and spectral resolution has nevertheless opened up the field for investigation.

Observations of radio recombination lines from compact HII regions show that their line parameters are distinctly different from those of more diffuse regions. The full line widths at half maximum from compact regions are considerable larger than those from the extended ones (Garay and Rodríguez 1983). At a given transition, the line to continuum ratio shows marked variations from one compact region to another. For instance, van Gorkom et al. (1980) reported variations in the H109α line to continuum ratio of about an order of magnitude; while for diffuse regions the spread in the observed ratio is typically less than a factor of two. In this paper we give a qualitative discussion of the factors that determine the values of the line parameters from compact HII regions. To achieve this goal we first summarize the results of radio recombination line observations made with high angular resolution and then discuss the derivation of physical parameters. We finally investigate correlations between observed and derived parameters and compare them with the theoretical predictions of simple models.

2. THE SAMPLE AND PARAMETERS OF COMPACT HII REGIONS

The sample analyzed in this paper consists of 33 compact HII regions, given in Table 1, for which the H76α line emission has been observed with high angular resolution. This line was chosen because is the most commonly observed with interferometric instruments and hence is the one for which enough information is available for statistical analysis. The data were mainly taken from Garay, Reid and Moran (1985), Garay (1985), Roelfsema (1987), and Wood and Churchwell (1989).

The essential data obtained from observations of the radio continuum emission of an HII region are its radio spectrum and its angular size. By modelling the radio spectrum as due to free-free emission from an isothermal and homogeneous region of ionized gas, the electron temperature and emission measure, EM, can be determined from the optically thick and thin portions of the spectrum, respectively (e.g., Gordon 1988). In addition, if the distance to the HII region is known, its physical size, L, and its average rms electron density, N_e, can be derived. Nebular parameters of the compact HII regions derived from radio continuum observations are given in columns 2-5 of Table 1.

There is a strong correlation between the electron densities and linear sizes of the compact HII regions in our sample, as shown in Figure 1. The values plotted are those cited in each study, which in most cases were derived assuming the HII regions to be homogeneous in density and temperature and having gaussian shapes. A least squares fit for $\log <N_e^2>^{1/2}$ = A log L + B, gave A = -1.0 ± 0.2 and B = 3.5 ± 0.2. A similar power-law relation between molecular hydrogen density and cloud size has been found for molecular clouds (Larson 1981; Myers 1983). We will not discuss here the meaning of the above relation, but mention that the constraint between the electron densities and sizes is of considerable relevance when studying the effects of stimulated emission in the line strength. Since $N_e \propto L^{-1}$, then EM $\propto N_e$; hence models of HII regions with high emission measure and low electron density, which should show most markedly the effects of stimulated emission, are unrealistic.

The parameters obtained from observations of radio recombination lines (RRL) are the line center velocity, the ratio of the emission in the line to that in the continuum, and the line width. Observed parameters of the H76α line emission from the compact HII regions in our sample are given in columns 6 and 7 of Table 1. In the following section we review these line parameters and discuss the physical conditions and mechanisms leading to the observed values.

TABLE 1. PARAMETERS OF COMPACT HII REGIONS.

Source	Radio continuum parameters				H76α line parameters		Ref.
	L (pc)	EM/10^6 (pc cm^{-6})	$N_e/10^3$ (cm^{-3})	τ_C a)	Δv (km s^{-1})	T_L/T_C	
W3(OH)	0.015	500.	130.	0.75	42.1	0.068	1
Sgr B2 M	0.39	84.	12.	0.09	34.3	0.089	1
Sgr B2 S	0.07	180.	41.	0.25	37.2	0.102	1
W49S(K)	0.22	160.	22.	0.31	39.9	0.120	1
W51 d	0.20	120.	20.	0.24	27.0	0.183	1
W51 e	0.33	47.	10.	0.10	25.4	0.204	1
Sgr B2 1	0.24	16.	7.	0.035	26.7	0.196	2
Sgr B2 2	0.18	47.	13.	0.119	39.3	0.152	2
Sgr B2 3	0.39	98.	13.	0.096	31.9	0.084	2
Sgr B2 4	0.07	520.	90.	0.706	36.1	0.098	2
Sgr B2 5	0.57	22.	5.	0.034	32.9	0.120	2
Sgr B2 6	0.17	21.	8.	0.056	31.6	0.196	2
DR21 A	0.05	35.	27.	0.06	30.1	0.133	3
DR21 B	0.06	51.	30.	0.09	30.6	0.127	3
DR21 C	0.07	61.	30.	0.10	31.0	0.136	3
DR21 D	0.05	25.	23.	0.04	25.7	0.147	3
DR21 E	0.08	5.	8.	0.008	43.0	0.063	3
DR21 F	0.1	3.	5.	0.005	50.	0.055	3
DR21 ALL	0.2	20.	10.	0.03	32.2	0.119	3
W3 A	0.40	29.	8.4	0.015	28.4	0.089	3
W3 B	0.19	53.	17.	0.029	27.5	0.138	3
W3 C	0.082	32.	20.	0.017	26.7	0.141	3
G5.89-0.39A	0.044	2445.	237.	2.67	49.0	0.046	4
G5.89-0.39B	0.044	2445.	237.	2.67	56.7	0.031	4
G10.62-0.38	0.048	1160.	156.	1.27	34.3	0.115	4
G11.94-0.62	0.043	159.	61.	0.17	23.9	0.185	4
G29.96-0.02	0.109	211.	44.	0.23	29.9	0.183	4
G30.54+0.02	0.167	52.	18.	0.06	38.4	0.134	4
G35.20-1.74	0.039	289.	86.	0.32	31.8	0.142	4
G45.12+0.13	0.119	1517.	113.	1.66	47.8	0.084	4
G34.3+0.2	0.068	770.	130.	0.78	55.5	0.054	5
G45.07+0.13	0.040	530.	160.	0.57	48.1	0.059	5
GM 24	0.04	83.	38.	0.10	42.8	0.121	6

a) 15 GHz continuum optical depth.
References: 1 Garay, Reid, and Moran (1985); 2 Garay (1985);
3 Roelfsema (1987); 4 Wood and Churchwell (1989); 5 Garay, Rodríguez, and van Gorkom (1986); 6 Roth et al. (1988).

3. DISCUSSION

3.1. Line Center Velocity

The line center velocity of RRL offers the possibility to determine the relative motions of compact HII regions with respect to the molecular clouds in which they are embedded. For instance, from a comparison of the center velocities of RRL and of molecular lines, van Gorkom et al. (1980) found, in the K3-50 region, two young regions of ionized gas at the edge of the cloud in the so called blister stage of evolution. Garay, Reid and Moran (1985) found that the OH masers associated with compact HII regions have velocities that are redshifted from the velocity of the ionized gas, suggesting that the masers are part of a remnant envelope which is still collapsing toward the newly formed star.

With the advent of radio synthesis instruments, allowing to resolve the source spatial structure, it is now possible to map the line center velocity of RRL across compact HII regions and therefore to study the kinematics of the gas within them. Large gradients in the line center velocity have been found across the ultra compact regions G34.3+0.2 (Garay, Rodríguez and van Gorkom 1986) and W3(OH) (Keto et al. 1990), which are attributed, respectively, to rotation and to champagne-like flows in a medium with density gradients. In a similar way, from maps of the velocity structure of recombination lines, Garay et al. (1986) and Wood et al.(1990) have presented evidence for the expansion of young HII regions into the cooler interstellar medium.

An interesting and novel result, first reported by Berulis and Ershow (1983), is that the center velocity of RRL from W3(OH) increases with the frequency of the transition. Berulis and Ershow (1983) suggested that the shift in the velocity centroid is due to optical depth effects in a dense and expanding shell of ionized gas. Alternatively, the observed trend could be explained as the result of champagne flows in a medium with gradients in its density (Welch and Marr 1987; Wilson et al. 1987; Keto et al. 1990).

3.2 Linewidths

The half power widths of the H76α line from the regions in Table 1 range from 24 to 57 km/s, and have an average value of 36 ± 9 km s^{-1}. These widths are considerable broader than the width produced by the thermal motions of the particles in an ionized gas, which for an electron temperature of 10^4 K is ~21.4 km s^{-1}. Hence other mechanism(s) must broaden the line. Possibilities include : i) pressure broadening; ii) large scale systematic motions; and iii) turbulence. In the following we consider the relative importance of each of these mechanisms in explaining the observed line profiles.

3.2.1. Pressure broadening. Electron impacts provide the major contribution to the pressure broadening. The FWHM linewidth of the Lorentzian profile is given by (Brocklehurst and Seaton 1972)

$$\left(\frac{\Delta v_I}{km\ s^{-1}}\right) = 4.29 \left(\frac{n}{100}\right)^{7.4} \left(\frac{10^4\ K}{T_e}\right)^{0.1} \left(\frac{N_e}{10^4\ cm^{-3}}\right). \quad (1)$$

For the HII regions in Table 1, Δv_I ranges from 0.3 to 14 km s^{-1}. Figure 2 plots the observed line widths versus the electron densities derived from the radio continuum observations. The continuous curve corresponds to a theoretical curve of line width versus density computed assuming that the line is broadened by electron impacts and has a Doppler width of 25 km s^{-1}, the average value observed in diffuse HII regions (Garay and Rodríguez 1983). The term Doppler width is used here to describe the broadening of an spectral line by all possible mechanisms besides pressure broadening. It is clear from Figure 2 that the line widths of compact HII regions are considerable larger that those of diffuse regions and that pressure broadening is negligible at the frequency of the H76α line. Therefore, other mechanism is responsible for the line broadening in young HII regions.

3.2.2. Large scale systematic motions. As discussed in section 3.1, recent VLA observations have revealed large gradients in the line center velocity across compact HII regions. Gradients of 23 and 12 km s^{-1} were found across G34.3+0.2 (Garay et al. 1986) and W3(OH) (Keto et al. 1990), respectively. Large scale motions might thus be an important source of broadening of recombination lines. However, in most of the cases where RRL have been mapped with high angular resolution, the width of the line emission within a radio beam is still considerable larger than the thermal width and remains approximately constant across the source. Hence, other effect besides large scale motions must account for the line broadening observed in compact HII regions.

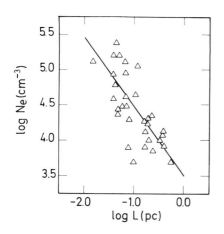
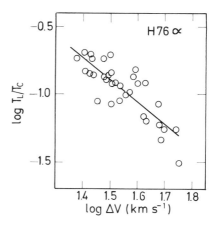

Figure 1. Plot of rms electron density versus linear diameter for compact HII regions. The line is a least squares fit to the data.

Figure 3. Plot of H76α line to continuum ratio versus line width for compact HII regions. The line is a least squares fit to the data.

3.2.3. Turbulence. The velocity fields of small scale eddies within a radio beam could be an important source of broadening of recombination lines in compact HII regions. Turbulence acts similarly to the microvelocities of the thermal motions of the gas producing a Gaussian line with a FWHM width given by

$$\Delta v = 21.4 \left\{ \left(\frac{T_e}{10^4 \text{ K}} \right) + 0.87 \left(\frac{\Delta v_{tur}}{20 \text{ km s}^{-1}} \right)^2 \right\}^{1/2} \text{ km s}^{-1}, \qquad (2)$$

where Δv_{tur} is the FWHM turbulent width. For the objects in our sample, using the electron temperatures cited in the literature, we find an average value for Δv_{tur} of 30 ± 10 km/s. Possible sources of turbulence in compact HII regions are the expansion of dense, small scale eddies of gas into the less dense surrounding medium and champagne flows, namely the expansion of newly ionized gas in a medium having density gradients (Tenorio-Tagle 1979). The typical velocity of the streaming gas is ~10 km s^{-1}, although, in the champagne model, flow velocities of up to 30 km s^{-1} may be reached in the low density medium (Yorke et al. 1984). Thus, turbulence due to the presence of density gradients is likely to be an important source of the broadening of the line profiles, even though it may not completely account for the observed large widths. Another possibility is that the turbulence may have a magnetic origin. Disturbances of a magnetized cloud of ionized gas will generate a spectrum of hydromagnetic waves travelling with velocities of about the Alfven speed, $V_A = B/(4\pi\rho)^{1/2}$, where B is the magnetic field strength and ρ the density of the cloud. The energy for the turbulence may originate from the expansion of the HII region in a magnetized cloud (Arons and Max 1975). Assuming that the non-thermal kinetic energy density is equal to the magnetic energy density, the FWHM turbulent width is given by (Myers and Goodman 1988a).

$$\Delta v_{mag} = 9.4 \left(\frac{B}{\text{mGauss}} \right) \left(\frac{10^5 \text{ cm}^{-3}}{N_e} \right)^{1/2} \text{ km s}^{-1}. \qquad (3)$$

To estimate Δv_{mag}, and since magnetic fields have not been directly observed in compact HII regions, we use a magnetic field strength of ~ 1 mGauss at densities of ~10^5cm^{-3} as inferred from observations of molecular clouds (Myers and Goodman 1988b), finding Δv_{mag} ~ 10 km s^{-1}. Broadening due to Alfvenic turbulence is thus comparable in magnitude to that produced by turbulent motions in a medium with density fluctuations; however, it may not be large enough to entirely account for the observed large widths. On the other hand, during the early stages of evolution of an HII region it is possible to generate super-Alfvenic fluid motions which could then constitute the main source of turbulence.

We conclude that turbulence, possibly generated by two or more mechanisms operating concurrently, is the dominant factor producing the large line widths observed in compact HII regions. We suggest that the turbulent motions might be originated by the progressive ionization of a magnetized medium containing widespread density fluctuations and/or large density gradients. Since turbulence is dissipated rapidly (t~10^4 years) and the conditions needed to sustain it are likely to be found only close to the exciting star, we expect turbulent broadening to be most important during the early stages of evolution of an HII region.

3.3 Line to Continuum Ratio

The H76α peak line to continuum ratio among the compact HII regions in our sample ranges from 0.03 to 0.21. For more diffuse regions such ratio is usually measured to be ~0.10. Possible reasons for the large extent observed are : 1) lines have constant integrated intensity but exhibit an ample range of widths; ii) wide range of electron temperatures; iii) amplification of the line by stimulated emission; and iv) large continuum optical depths.

Assuming that the integrated intensity under the line is constant, for large line widths the power will be distributed over a large range of frequencies and consequently the peak T_L/T_C ratio will be decreased, roughly as $1/\Delta v$. Figure 3, which plots the observed peak line to continuum ratio versus linewidth for the compact HII regions in our sample, shows that in fact the peak T_L/T_C ratio decreases as the line width increases. However a least squares fit for log T_L/T_C = C log Δv + D gave a slope of -1.6±0.2, suggesting that models assuming that the total power under the line is constant can not entirely explain the observations. We conclude that part, but not all, of the wide range in the observed peak T_L/T_C ratio among compact HII regions is due to differences in their linewidths. Therefore, a mechanism of either line enhancement or line supression must be at stake in these young regions.

Stimulated effects could produce an enhancement of the line emission and consequently an increase in the line to continuum ratio. To illustrate the importance of this effect we have made calculations of T_L/T_C for simple models of HII regions. For an isothermal, homogeneous nebula, T_L/T_C is given by (Lockman and Brown 1978)

$$\frac{T_L}{T_C} = \frac{\tau_C + b\tau_L^*}{\tau_C + b\beta\tau_L^*} \frac{1 - e^{-(b\beta\tau_L^* + \tau_C)}}{1 - e^{-\tau_C}} - 1, \qquad (4)$$

where τ_C is the continuum optical depth, τ_L^* is the peak line optical depth assuming LTE, and b and β are the usual departure coefficients from LTE (Brocklehurst 1970). In general the line to continuum ratio is a function, through τ_C, τ_L^*, b and β, of the nebular parameters electron temperature, electron density, and path length and of the width of the recombination line. However, as discussed in section 2., for the compact HII regions there is a relationship between N_e and L and hence the T_L/T_C ratio can be expressed in terms of three independent variables. Figure 4 shows how does the H76α peak line to continuum ratio varies with continuum optical depth for LTE and non-LTE models. Curves are shown for regions with electron temperatures of 8000 and 12000 K and a Doppler width of 36 km s^{-1}. Pressure broadening was taken into account using the approximation of van Gorkom et al. (1980). The electron densities,

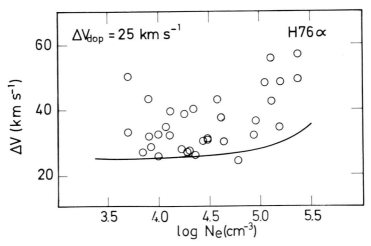

Figure 2. Plot of H76α line width versus electron density for compact HII regions. The curve indicates a theoretical run of Δv versus N_e computed assuming that the line is broadened by electron impacts and has a Doppler width of 25 km s^{-1}.

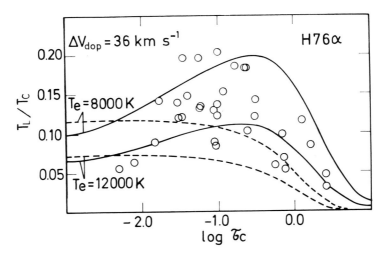

Figure 4. Theoretical line to continuum ratios for homogeneous, isothermal model HII regions. Dashed and continuous curves represent computations made assuming LTE and non-LTE conditions, respectively. All calculations assume pressure broadening. The Doppler width of the line is 36 km s^{-1}. The circles indicate observed peak H76α line to continuum ratio versus 15 GHz continuum optical depth for compact HII regions.

needed to compute the widths of the Lorentz profiles as well as b and β, were determined from τ_C using the relation $N_e = 3.2 \times 10^3 \, (L/pc)^{-1.0}$ cm^{-3} as derived from the continuum observations. For small (≤ 0.01) values of τ_C, the peak line to continuum ratio is approximately constant and equal for both models. As τ_C increases, for the non-LTE model the peak T_L/T_C ratio increases due to stimulated emission. Finally, for large (≥ 1) values of τ_C, T_L/T_C decreases sharply due to both pressure broadening and to the depression of the line in full thermodynamic equilibrium conditions.

Figure 4 also plots the observed peak line to continuum ratio versus the derived continuum optical depth of the compact HII regions in our sample. Two conclusions can be derived from this figure. First, the observed run of the peak T_L/T_C ratio with continuum optical depth is best fitted by the non-LTE model suggesting that stimulated emission significantly increases the intensity of RRL from compact HII regions. A similar conclusion was previously reached by van Gorkom (1980). Second, the compact HII regions having the largest continuum opacities exhibit the smaller line to continuum ratios, result which is most easily explained as due to line suppression in full thermodynamic equilibrium. Of course, by using the simple model we have ignored a variety of complications such as for instance that compact HII regions might be far from homogeneous, displaying fluctuations in density and temperature as well as exhibiting irregular shapes. However, the qualitative conclusions that one derive from Figure 4 should be correct.

4. CONCLUSIONS

The main results obtained from a compilation and analysis of the H76α line emission data, obtained with high angular resolution, from over thirty compact HII regions are as follows.

The observed widths are invariable larger than those of more diffuse regions. We interpret this result as evidence that young HII regions are not yet in equilibrium with the surrounding medium but in a dynamical stage were turbulent motions are appreciable. We suggest that the large line widths can be accounted by turbulence in a magnetized medium having density inhomogeneities or density gradients.

The H76α peak line to continuum ratios differ by as much as a factor of 7, while for diffuse HII regions the spread is usually less than a factor of 2. We conclude that the large extent observed is the result of the combination of three effects : i) broad range of line widths; ii) large optical depths; and iii) stimulated emission. The first two effects produce a decrease, with respect to the average value, of the peak line intensity, while the third produces an enhancement of the line emission.

The run between peak line to continuum ratio versus continuum opacity shows that non-LTE effects are of considerable importance within compact HII regions, as previously suggested by van Gorkom (1980).

Future observations, with high angular resolution, of RRL from compact HII regions will certainly provide significantly more information on the physical conditions and dynamics of the gas near recently formed massive stars.

ACKNOWLEDGMENTS

The author is the grateful recipient of an Henri Chrétien Award. This work was supported in part by the Chilean FONDECYT through grant 0502.

REFERENCES

Arons, J., and Max, C.E. 1975, Ap.J.(Letters), 196, L77.
Berulis, I.I., and Ershow, A.A. 1983, Sov. Astron. Lett., 9, 341.
Brocklehurst, M. 1970, M.N.R.A.S., 148, 417.
Brocklehurst, M., and Seaton, M.J. 1972, M.N.R.A.S., 157, 179.
Garay, G., and Rodríguez, L.F. 1983, Ap. J., 266, 263.
Garay, G., Reid, M.J., and Moran,J.M. 1985, Ap. J., 289, 681.
Garay, G., Rodríguez, L.F., and van Gorkom, J.H. 1986, Ap. J., 309, 553.
Garay, G. 1986, ESO-IRAM-Onsala Workshop on (Sub)Millimeter Astronomy,
    eds. P.A. Shaver and K. Kjar, ESO Conf. Workshop Proc. N° 22, p. 517.
Gordon, M.A. 1988, Galactic and Extragalactic Radio Astronomy, eds.
    G.L. Verschuur and K.I. Kellerman (New York:Springer-Verlag), p. 37.
Gorkom, J.H. van 1980, Ph.D. Thesis, University of Groningen, Holland.
Gorkom, J.H. van, Goss, W.M., Shaver, P.A., Schwarz, U.J., and Harten,
    R.H. 1980, Astr. Ap., 89, 150.
Keto, E.R., Reid, M.J., Welch, W,J., Carral, P., and Ho, P.T.P. 1990,
    Ap. J., submitted.
Larson, R.B. 1981, M.N.R.A.S., 194, 809.
Lockman, F.J., and Brown, R.L. 1978, Ap. J., 222, 153.
Mezger, P.G. 1968, Interstellar Ionized Hydrogen, ed. Y. Terzian (New
    York:Benjamin), p. 33.
Myers, P.C. 1983, Ap. J., 270, 105.
Myers, P.C. and Goodman, A.A. 1988a, Ap. J., 329, 392.
_____. 1988b, Ap. J. (Letters), 326, L27.
Roelfsema, P.R. 1987, Ph.D. Thesis, University of Groningen, Holland.
Roth, M., Tapia, M., Gómez, Y., and Rodríguez, L.F. 1988, Rev. Mexicana
    Astron. Astrofís., 16, 3.
Tenorio-Tagle, G. 1979, Astr. Ap., 80, 110.
Welch, W.J., and Marr, J. 1987, Ap. J. (Letters), 317,L21.
Wilson, T.L., Mauersberger, R., Brand, J., and Gardner, F.F. 1987, Astr.
    Ap., 186, L5.
Wood, D.O.S., and Churchwell, E. 1989, Ap. J. Suppl., 69, 831.
Wood, D.O.S., et al. 1990, in preparation.
Yorke, H.W., Tenorio-Tagle, G., and Bodenheimer, P. 1984, Astr. Ap.,
    138, 325.

RADIO RECOMBINATION LINE EMISSION FROM ULTRACOMPACT HII REGIONS

E. CHURCHWELL
Astronomy Dept.
475 N. Charter St.
Madison, WI 53706
USA

C. M. WALMSLEY
Max-Planck-Inst.
Auf dem Hügel 69
5300 Bonn 1
West Germany

D. O. S. WOOD
Center for Astrophysics
60 Garden St.
Cambridge, MA 02138
USA

H. STEPPE
IRAM
Avenida Divina
Pastora 7
18012 Granada
Spain

ABSTRACT. We report observations of H30α, H42α, and H76α line emission toward a selection of 12 ultracompact (UC) HII regions. These same UC HII regions have also been imaged with the VLA. Continuum and radio recombination line (RRL) models have been calculated and compared. For the highest density nebulae, single component models do not seem to be consistent with both the continuum and RRL observations. Possisble correlations of line velocities and line widths are considered. Evidence for electron impact broadening is shown.

1. Introduction

High resolution studies of ultracompact (UC) HII regions have recently become possible because of the NRAO Very Large Array (VLA), the IRAM 30m telescope, and the Nobeyama 45m telescope. UC HII regions are typically only a few arc seconds in diameter (usually <10"), have emission measures >10^7 pc cm^{-6}, and are optically thick at wavelengths longward of ~6 cm and shortward of ~5 μm. When observed with resolutions commensurate with their angular diameters, they are among the brightest objects in the Galaxy.

UC HII regions are manifestations of newly formed OB stars that are still embedded in their natal molecular cloud. As such, they are important tracers of massive star formation in the Galaxy. They also provide opportunities to study the impact a newly formed massive star has on its environment. One can study the circumstellar ionized gas via radio free-free continuum emission, radio recombination line (RRL) emission, infrared recombination line emission, and infrared (IR) fine structure lines. The surrounding molecular gas can be observed via rotational and vibrational transitions of a variety of molecular species; since both quiescent and shocked molecular gas exist near these objects and the radiation densities are high, the chemistry is likely to be complicated. The cocoon of circumstellar dust can be studied at IR wavelengths.

In this paper, we report new high spatial resolution observations of H76α, H42α, and H30α line emission toward 11 UC HII regions. It is important to note that RRL emission is confined to a relatively narrow frequency range in UC HII regions. At $\nu \leq 5$ GHz, the nebulae are usually optically thick, so RRLs cannot be observed. At $\nu \geq 230$ GHz, the continuum begins to be dominated by thermal dust emission, which must be separated out to obtain meaningful line-to-continuum ratios. The RRLs reported here span most of the frequency range noted above.

2. Observations

2.1. VLA OBSERVATIONS

The H76α data were obtained using the VLA in the D-array. The spectral resolution was 7.97 km s^{-1} and the velocity range observed was ±123.6 km s^{-1} about the LSR velocity of the source. The HPBW achieved depends on the source declination; they are given for each source in Table 1 along with the pixel size used to image each source. The VLA was sensitive to objects with diameters ≤ 110" in this configuration. The flux density scale was calibrated by observing 3C286 and assuming that its flux density at 14.690 GHz is 3.50 Jy (Baars *et al.* 1977).

2.2. IRAM 30m OBSERVATIONS

The H42α and H30α data were obtained simultaneously with the IRAM 30m telescope in Dec. 1988. During the same period, we also obtained molecular line observations of CS (2-1 and 5-4), CH_3CN (6-5 and 12-11), CO (1-0 and 2-1), and ^{13}CO (1-0 and 2-1). A more detailed description and analysis of these data is being prepared for publication. The weather was clear and optical depths at 230 GHz at the zenith were < 0.1 during the entire observing period. Both lines were observed simultaneously using cooled SIS receivers. The average system temperatures at an elevation of ~30° were typically ~600 K and ~800 K at 3mm and 1.3mm, respectively. The HPBW was 27" and 12" at 3mm and 1.3mm, respectively. The continuum temperatures at 3mm and 1.3mm were obtained both from the average ON-OFF differences and from continuum scans through the sources after careful pointing had been performed. Pointing was done at 3mm (where free-free emission dominates) toward each source before the line observations were made; pointing errors were ~2". The positions found by Wood and Churchwell (1989; hereafter WC) with a resolution of 0.4" were assumed.

TABLE 1

ANTENNA PARAMETERS FOR H76α OBSERVATIONS

Source	Observation Date	Peak Position[†] α (1950) (h min sec)	δ (1950) (° ′ ″)	Image Pixel Size (arc sec)	Synthesized Beam Major Axis (arc sec)	Minor Axis (arc sec)	Position Angle (degrees)
G5.89-0.39	23-MAR-87	17 57 26.717	-24 03 57.20	1.0 x 1.0	6.87	3.65	2.74
G10.62-0.38	23-MAR-87	18 07 30.670	-19 56 29.42	1.0 x 1.0	5.22	4.18	29.00
G11.94-0.62	23-MAR-87	18 11 04.373	-18 54 19.57	1.0 x 1.0	4.36	4.06	40.19
G23.96+0.15	01-AUG-88	18 31 42.560	-07 57 10.90	1.3 x 1.3	4.50	4.18	-37.73
G29.96-0.02	23-MAR-87	18 43 27.073	-02 42 36.42	1.0 x 1.0	4.59	4.32	-38.84
G31.41+0.31	01-AUG-88	18 44 59.353	-01 16 04.50	1.3 x 1.3	4.38	4.35	-44.32
G33.92+0.11	01-AUG-88	18 50 17.460	00 51 45.67	1.3 x 1.3	5.57	5.28	-49.65
G34.26+0.15A	01-AUG-88	18 50 46.179	01 11 12.80	1.3 x 1.3	4.49	4.21	-38.61
G35.20-1.74	01-AUG-88	18 59 14.060	01 09 03.19	1.3 x 1.3	6.32	6.08	-47.34
G43.89-0.78	01-AUG-88	19 12 02.813	09 17 19.00	1.3 x 1.3	5.48	4.57	-62.80
G45.12+0.13	01-AUG-88	19 11 06.225	10 48 25.80	1.3 x 1.3	5.77	5.19	-52.09
G45.45+0.06	01-AUG-88	19 12 00.070	11 03 59.50	1.3 x 1.3	5.23	4.41	-69.00

[†] For highest resolution, positions found from the 2cm observations are given; from WC, Tables 11-15.

3. The Data

The measured line and continuum parameters are given in Table 2. We point out that the data for the H76α entries have been integrated over the central 3x3 pixels of the line and continuum images; this is only slightly less than a resolution element. See Table 1 for the

TABLE 2

Observed Parameters[†]

Source	Line	V_{lsr} (kms^{-1})	ΔV (kms^{-1})	$\int T_L dv$ $(Kkms^{-1})$	T_L (K)	T_C (K)	$L_n^{\ddagger}$ (GHz^{-1})	T_e^* (K)	Notes
G5.89-0.39	H30α	11.1 ± 3.0	61.1 ± 1.3	35.5 ± 0.4	0.55	1.02 ± 0.16	0.233	19220	1
	H42α	6.7 ± 0.4	59.4 ± 1.0	90.8 ± 0.9	1.44	2.05 ± 0.10	0.82	6278	
	H76α	4.9 ± 1.1	74.8 ± 2.2	1956 ± 76	24.6 ± 0.9	457 ± 46	0.366	8093	2
G10.62-0.38	H30α	0.40 ± 0.40	29.4 ± 1.0	20.9 ± 0.5	0.67	1.07 ± 0.09	0.270	30930	
	H42α	0.17 ± 0.26	28.0 ± 0.6	51.1 ± 1.0	1.71	1.25 ± 0.10	1.596	6718	
	H76α	0.72 ± 0.61	35.3 ± 0.9	1212 ± 170	32.2 ± 1.2	272 ± 27	0.806	7845	
G11.94-0.62	H30α	44.0 ± 1.3	25.8 ± 2.8	4.8 ± 0.5	0.17	0.19 ± 0.02	0.386	25680	
	H42α	40.4 ± 1.0	25.5 ± 2.3	10.1 ± 0.8	0.37	0.29 ± 0.02	1.489	7652	
	H76α	41.4 ± 1.7	25.5 ± 2.5	1452 ± 69	16.0 ± 2.2	68.4 ± 6.8	1.59	5829	
G23.96+0.15	H76α	73.7 ± 1.6	27.8 ± 2.3	1159 ± 47	11.8 ± 1.4	40.9 ± 4.1	1.96	4564	
G29.96-0.02	H30α	98.2 ± 0.6	27.2 ± 1.6	9.8 ± 0.5	0.34	0.38 ± 0.04	0.386	24590	
	H42α	97.5 ± 0.4	28.4 ± 1.2	30.1 ± 1.0	1.0	0.72 ± 0.07	1.62	6564	
	H76α	97.1 ± 0.7	30.9 ± 1.0	3167 ± 50	28.9 ± 1.3	137 ± 14	1.44	5416	
G31.41+0.31	H76α	102.9 ± 1.6	23.3 ± 2.4	256 ± 43	10.3 ± 1.5	36.4 ± 3.6	1.93	5365	
G33.92+0.11	H30α	96.3 ± 4.9	40.3 ± 9.4	2.8 ± 0.7	0.07	0.17 ± 0.01	0.178	33700	
	H42α	104.1 ± 1.2	31.7 ± 2.5	4.6 ± 0.3	0.14	0.22 ± 0.2	0.743	11250	
	H76α	101.4 ± 2.6	32.5 ± 4.1	679 ± 41	5.89 ± 1.01	45.6 ± 4.6	0.879	7815	
G34.26+0.15	H30α	–	–	–	–	1.71 ± 0.14	–	–	3
	H42α	51.4 ± 0.6	57.5 ± 1.6	78.1 ± 1.6	1.28	1.97 ± 0.16	0.758	6853	
	H76α	38.2 ± 1.2	57.2 ± 2.0	1858 ± 100	30.5 ± 1.4	426 ± 43	0.487	7979	
G35.20-1.74	H30α	48.0 ± 0.5	29.0 ± 1.1	11.2 ± 0.4	0.36	0.43 ± 0.04	0.361	24640	
	H42α	48.4 ± 0.3	27.9 ± 0.6	34.1 ± 0.7	1.15	0.92 ± 0.08	1.46	7239	
	H76α	45.0 ± 1.0	29.3 ± 1.6	784 ± 66	25.1 ± 1.8	194 ± 19	0.881	8513	
G43.89-0.78	H76α	53.7 ± 1.9	27.7 ± 2.8	211 ± 35	7.15 ± 1.01	41.7 ± 4.2	1.17	7048	
G45.12+0.13	H30α	61.4 ± 0.4	40.7 ± 1.1	20.0 ± 0.4	0.46	0.63 ± 0.05	0.315	20890	
	H42α	60.3 ± 0.5	44.0 ± 1.2	51.8 ± 1.2	1.11	1.25 ± 0.10	1.04	6614	
	H76α	58.4 ± 1.4	46.8 ± 2.3	1404 ± 107	28.2 ± 1.8	371 ± 37	0.517	8982	
G45.45+0.06	H30α	56.3 ± 1.0	34.0 ± 2.4	6.5 ± 0.4	0.18	0.14 ± 0.01	0.554	15300	
	H42α	55.9 ± 0.9	31.3 ± 2.0	14.4 ± 0.8	0.43	0.40 ± 0.03	1.24	7508	
	H76α	54.6 ± 1.5	27.8 ± 2.3	243 ± 32	8.19 ± 0.94	64.6 ± 6.5	0.863	9052	

Notes:
1. The H30α line is blended with one or more weaker and narrower molecular lines.
2. The H76α line is non-Gaussian (see profile of WC, Figure 83).
3. The H30α line is strongly blended with molecular lines. Hence the recombination line parameters cannot be derived with reasonable accuracy.

[†] Errors reported for T_C are VLA systematic errors, $\approx 10\%$. Other errors are statistical errors of fit reported at the 2σ level. The H30α and H42α lines were fit with a different routine which does not report T_L errors.

[‡] $L_n \equiv \left(\frac{100}{\nu}\right) \cdot \left(\frac{T_L}{T_C}\right)$ where ν is in GHz and n is the principal quantum number.

pixel sizes.

In Figures 1 and 2, we show the spectra of all three lines for G5.89-0.39 and G34.26+0.15. These figures illustrate that the signal-to-noise is large (typically > 20 for

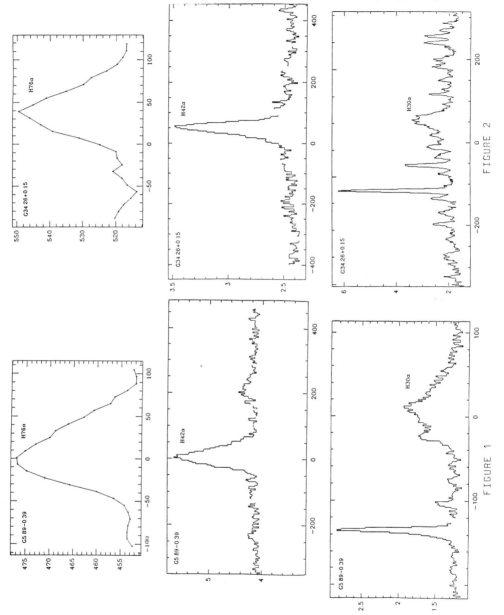

all the lines, except for H30α toward G34.26+0.15). The line profile toward G5.89-0.39 is not gaussian; it appears to have two components. Finally, the H30α spectrum toward G34.26+0.15 illustrates another problem with mm-RRL observations toward UC HII regions, namely blending with weak molecular line emission (especially at 1.3 mm and shorter). Three sources have been observed with the VLA for which we were not able to obtain corresponding 3 mm and 1.3 mm observations. We report the H76α data for these sources in Table 2, but do not attempt to model them.

4. Discussion

4.1. CONTINUUM MODELS

Using the observed images and high resolution radio flux density distributions reported by by WC, we have calculated continuum models which reproduce to first order the **observed brightness distributions** and accurately fit **the radio continuum spectra**. Numerical solutions of the radiative transfer were obtained for the specific geometry of each source (inferred from the images of WC). These models establish rms electron densities $<n_e>$, average electron temperatures T_e, and emission measures EM_c. The EM_c are used to constrain the nonLTE RRL models, described below. The continuum models show that all the sources under discussion have extremely high EM_c values; ranging from 4.8×10^7 to 2.9×10^9 pc cm^{-6}.

4.2. NonLTE RRL MODELS

We use the departure coefficients b_n and corrections for stimulated emission $(1-\beta_n)$ given by Walmsley (1989). For a given set of assumed n_e, T_e, and EM_c values we calculate the following quantities:

where
$$L_n = [100/\nu(GHz)](T_L/T_C) = Y_n L_n^* \quad (1)$$

and
$$Y_n = b_n\{f(\tau_c) + (1-\beta_n)[1- f(\tau_c)]\} \quad (2)$$

and
$$f(\tau_c) = \tau_c/(e^{\tau_c}-1) \quad (3)$$

$$L_n^* = [100/\nu(GHz)](T_L^*/T_C) \quad (4)$$

L_n^* is the value of L_n if the line is emitted in LTE at the assumed n_e and T_e values. L_n is the calculated value of the line-to-continuum ratio as defined in eq. (1), corrected both for electron impact broadening and for stimulated emission. We assume a relative helium abundance by number of 0.1. The continuum optical depth is τ_c and Y_n is the correction factor for nonLTE effects. In the above equations n is the lower principal quantum number. The electron density and temperature were varied until the calculated values of L_{42} and L_{76} agreed with the observed values, with the one constraint that EM_c found from the continuum observations remain unchanged. Such models are, of course, over simplified because there are clearly density inhomogenieties apparent in the high resolution continuum images and probably a range of temperatures as well in these nebulae. Basically, the H42α line determines T_e and the H76α line determines the density via its sensitivity to electron impact broadening and stimulated emission. We do not attempt to fit the H30α line-to-continuum ratio in our nonLTE model calculations because of serious contamination of the continuum emission by thermal dust emission. One can see from the

T_e^* values given in Table 2 for the H30α line, that it gives systematically higher values than the other lines. This is because at λ1.3 mm, thermal dust emission is an important contributor to the continuum emission. Thus, our RRL model results depend only on two lines. We show in Figure 3 the calculated values of L_n and L_n^* along with the observed values L_n(obs) for G34.26+0.15 to illustrate the behavior of these parameters with principle quantum number. It also illustrates how well the model fits the observations and shows that a third line, such as H66α, is needed to further constrain the models.

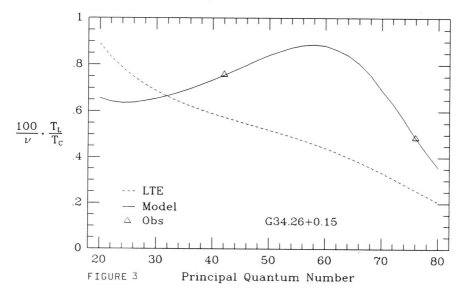

FIGURE 3

We can summarize the results of our continuum and RRL models as follows. Generally, the RRL models require local densities about an order of magnitude larger than the rms densities inferred from the continuum models. This has the effect of producing broad H76α lines due to electron impact broadening and a rising continuum spectrum between 2 cm and 3mm due to large optical depths. Neither effect is apparent in the data to the extent implied by the RRL models. For two sources, G11.94-0.62 and G33.92+0.11, the rms and local densities differ by less than a factor of 4, the nebulae are optically thin at 2 cm and shorter wavelengths, and the model results seem reasonably compatible with the observatons (line and continuum). The continuum and RRL models are still in a preliminary state, so any general conclusions based on them are tentative. Nonetheless, it appears that single component models usually are not consistent with the whole range of continuum and line observations when n_e(rms) and n_e(local) differ by factors larger than about 5.

4.3. CORRELATIONS

In Figure 4 the velocity differences H76α - H42α and H30α - H42α are plotted against the H42α velocity for 8 of the sources listed in Table 2. This plot illustrates that the systematic shift of velocity with principal quantum number found by Welch and Marr (1987) toward W3(OH) is not a general phenomenon that occurs in all UC HII regions. In our sample, the velocity systematically increases with frequency in G5.89-0.39 and perhaps in

G34.26+0.15, although only two lines are used in the latter source due to severe blending of the H30α line with molecular lines. In both G45.12+0.12 and G45.45_0.06, the line velocities systematically increase with frequency, but the differences are all within the uncertainties. With the exception of G5.89-0.39 and perhaps G34.26+0.15, we find no general trend with frequency in our sample.

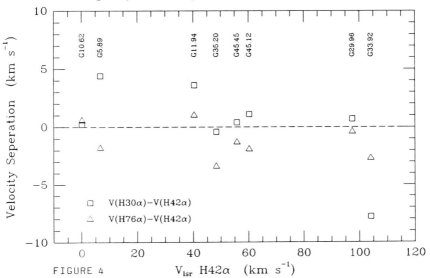

FIGURE 4

In Figure 5, we plot the velocity differences NH$_3$-H42α, CS-H42α, and CH$_3$CN-H42α for eleven UC HII regions. The CS and CH$_3$CN data were obtained with the IRAM

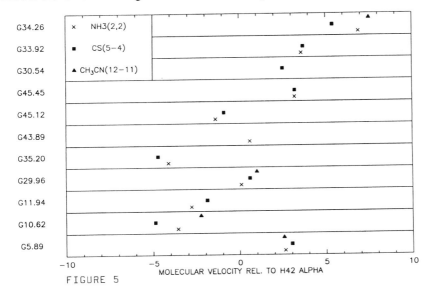

FIGURE 5

30m telescope during the same period that the H42α observations were obtained; they also have about the same HPBW. The NH$_3$ (1,1) and (2,2) data were taken from Churchwell *et al.* (1990); the HPBW is 40". There is also no systematic trends apparent here. Two sources have essentially zero differences, 6 have positive velocity differences, and 4 have negative velocity differences. No molecule is systematically red-shifted or blue-shifted relative to the others. Since the lines are all in emission, one does not know whether the molecular gas lies mostly between us and the HII region or mostly behind the HII region. Therefore, one cannot determine if the molecular and ionized gas have systematic relative velocities of approach or recession from these data.

Finally, in Figure 6 we plot the RRL linewidths against emission measure. In principle, this should identify the sources in which electron impact broadening is important. When electron impact broadening is important, the H76α linewidth should be greater than that of the mm lines. In Figure 8, it is apparent that for log(EM)≥8.5 the H76α linewidths are systematically larger than the mm wavelength lines. For log(EM)>9, the widths of even the 42α and 30α lines sharply increase with EM. Some of this increase may be due to systematic flows within the HII region. For example, G5.89-0.39 has a very powerful bipolar molecular outflow associated with it. We now have evidence from high resolution RRL imaging that the outflow can be identified even in the HII region. Therefore, some fraction of the linewidths in this source is probably due to ordered motions, but these cannot explain the fact that the H76α line is systematically broader than that of the mm wavelength lines. Thus, the evidence is strong that some of the increase in linewidths with EM is due to impact broadening.

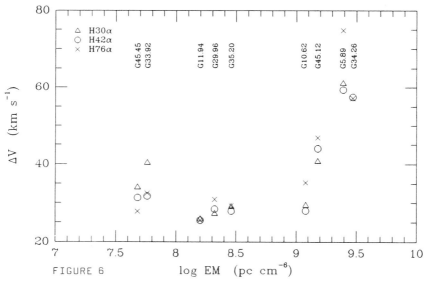

FIGURE 6

5. Summary

We have reported new high resolution observations of H76α, H42α, and H30α toward 11 UC HII regions. Preliminary single component line and continuum models have been calculated for those nebulae for which all three lines have been observed and for which high resolution continuum images are available. For all but two nebulae (G11.94-0.62 and

G33.92+0.11) the RRL models require electron densities greater than 5 times that of the rms densities derived from the continuum models. For these nebulae, the single component models predict H76α linewidths that are significantly larger than the observed linewidths. Also, the RRL models which require the largest electron densities predict increasing flux densities between 2 cm and 3mm, an effect that is not supported by the observations. We, therefore, conclude that single component (i.e. single density and temperature) models are usually not consistent with both the continuum and line observations in the densest UC HII regions.

Possible correlations of RRL velocities with principal quantum number and of relative motions of molecular and ionized gas have been examined and found not to show systematic trends. In particular, the trend of changing line velocity with principal quantum number found by Welch and Marr (1987) in W3(OH) is not found in our sample of UC HII regions.

In an attempt to determine if electron impact broadening is important in UC HII regions at 2 cm and shorter, the linewidths were plotted against emission measure for all three lines (H76α, H42α, and H30α). It was found that for emission measures greater than $\sim 10^{8.5}$ pc cm^{-6}, the H76α line width is systematically larger than that of the other lines. At $EM_c > 10^9$ pc cm^{-6}, the linewidths of even the mm lines increase dramatically. In some of these sources, the linewidths may be enhanced by ordered motions such as bipolar outflows or flows around shocks, but the systematically larger H76α linewidths make a strong case that electron impact broadening is operative in these high EM sources.

Much more work needs to be done on mapping RRL emission with high spatial resolution in UC HII regions to identify possible bipolar outflows and to test the stellar wind supported bow shock models proposed by Van Buren *et al.* (1990) and Mac Low *et al.* (1990). Such observations would make possible more realistic models of HII regions that conform both to the continuum and the RRL data.

REFERENCES

Baars, J. W. M., Genzel, R., Pauliny-Toth, I. I. K., Witzel, A. (1977) "The Absolute Spectrum of Cas A: An Accurate Flux Density Scale and a Set of Secondary Calibrators", Ast. Ap., **61**, 99.
Churchwell, E., Walmsley, C. M., Cesaroni, R. (1990) "A Survey of Ammonia and Water Vapor Emission from Ultracompact HII Regions", Ast. Ap., in press.
Mac Low, M.-M., Van Buren, D., Wood, D. O. S., Churchwell, E. (1990) "On the Formation of Ultracompact HII Regions", in preparation.
Van Buren, D., Mac Low, M.-M., Wood, D. O. S., Churchwell, E. (1990) "Cometary Compact HII Regions are Stellar Wind Bow Shocks", Ap. J., to appear in the Apr. 20 issue.
Walmsley, C. M. (1989) "Level Populations for Millimeter Recombination Lines", Ast. Ap., in press.
Welch, W. J., Marr, J. (1987) "Study of the H42α (86 GHz) Recombination Line in W3(OH)", Ap. J., **317**, L21.
Wood, D. O. S., Churchwell, E. (1989) "The Morphologies and Physical Properties of Ultracompact HII Regions", Ap. J. Suppl., **69**, 831.

Theoreticians Walmsley (left) and Hoan-Binh (right) debating the importance of Lyman Line Excitation.

RADIO RECOMBINATION LINES AT MILLIMETER WAVELENGTHS IN HII REGIONS

M. A. GORDON
National Radio Astronomy Observatory
949 North Cherry Avenue
Tucson, Arizona 85721-0655
USA

ABSTRACT. Radio recombination lines from HII regions involve much smaller principal quantum numbers at millimeter than at centimeter wavelengths. In this regime pressure broadening and partial maser effects are small and can usually be neglected. Free of these complications, millimeter-wave lines have been used to determine the electron temperatures of HII regions and to explore the atomic population processes for principal quantum numbers between 25 and 50. In general, observations appear to fit the transfer theory suggested by Goldberg (1966) with typical departure coefficients such as those calculated by Salem and Brocklehurst (1979).

This review is intended to cover the salient features of radio recombination lines (RRLs) at millimeter wavelengths but excludes discussion of the mm-wave maser lines from the star MWC349, which is described in a separate article in this volume.

1. Overview

When RRLs were detected from HII regions, the electron temperatures derived from them were significantly lower than those derived from optical emission lines. The RRL intensities relative to the underlying free-free continuum emission constitute a partition function between bound and free electrons, which is usually parameterized by an *LTE electron temperature*. Goldberg (1966) suggested that the intensities of the RRLs were sensitive to slight variations in the level populations—departures from Local Thermodynamic Equilibrium (LTE). In his theory two competing effects occur: a weakening of the lines due to an underpopulation of the atomic levels relative to LTE, and an amplification of the lines due to a slight overpopulation of upper relative to lower atomic levels. In addition, pressure broadening due to an inadiabatic linear Stark effect (Griem 1967) broadens the lines, redistributes radiation from the the line core to the wings where it is indistinguishable from the continuum, also reducing the line intensities relative to the continuum. The net effect of these processes is to produce an electron temperature different than the kinetic temperature of the free electrons.

Qualitatively, Fig. 1 illustrates the behavior of RRLs from 0.5 to 500 GHz relative to the background free-free continuum emission. At decimeter wavelengths, the Stark broadening the RRLs, redistributing the radiation from the line core to the wings where

Fig.1–Observed line to continuum ratios (*crosses*) for Hα RRLs observed in NGC 1976 (Orion Nebula) and thermodynamic equilibrium values (*filled circles*) for an electron temperature of 8500K plotted against frequency. The filled circles also mark the rest frequencies of the RRLs. Data reported by Lockman and Brown (1975), Gordon (1989), and Gordon and Walmsley (1990).

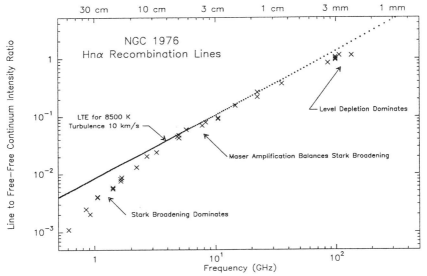

it becomes indistinguishable from the background of free-free continuum, and causing the measureable line intensities to fall below the LTE values. At centimeter wavelengths, although the line amplification from the partial maser effect exceeds the line weakening from level underpopulation, the resultant increase in the line intensity is offset by the line broadening from the Stark effect. Happily, these compensating mechanisms allow simple LTE formulas to describe the line intensities (Simpson 1973, Shaver 1980) as can be seen by the good agreement of observations and the LTE line in Fig. 1. At millimeter wavelengths, the line intensities fall below the LTE curve because the underpopulation of the quantum levels dominates the partial maser effect and because, at these quantum levels, the small "size" of the atom renders it immune to Stark broadening.

Quantitatively, in the mm-wave region, the intensities I_L of RRLs may be described by simple equations because the opacities in the line, τ_L, and continuum, τ_c, are small.

$$\int I_L \, d\nu = b_n \left(1 - \frac{\tau_c}{2}\gamma\right) \int I_L^* \, d\nu \qquad \tau_c, \tau_L \ll 1 \qquad (1)$$

where

$$\gamma \equiv 1 - \frac{d \ln b_n}{dn}(m-n)\frac{kT_e}{h\nu}. \qquad (2)$$

Here I_L is the line intensity is the observed line intensity, I_L^* is the line intensity which would be observed under LTE conditions, ν is the line frequency, b_n is the departure coefficient for the lower principal quantum level n, m is the quantum number for the upper level, k is Boltzmann's constant, T_e is the electron temperature of the emitting gas, and h

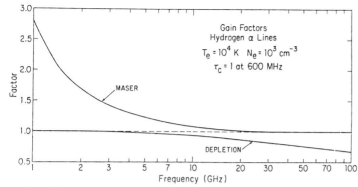

Fig. 2–The behavior of the first and second terms of Eq. 1 as a function of frequency for α-type RRLs under HII region conditions. From Gordon(1989)

is Planck's constant.

In HII regions, the first term in Eq. 1, b_n, accounts for line weakening due to a population underabundance with respect to LTE values. The second term, $(1 - \tau_c\gamma/2)$, accounts for line enhancement due to the gradient of population between the upper and lower quantum levels—the "partial maser" effect. Fig. 2 illustrates these effects for α-type lines as a function of principal quantum number. While exceptions can occur in situations where excitation by stellar Lyman lines and, in multi-electron atoms, where dielectronic recombination become important, Observations show that the mm-wave RRLs from HII regions generally behave as shown in Figs. 1 and 2.

A common use of RRLs is to determine the $<T_e>$ and $<N_e>$ of HII regions, where the values are the average of the gas included in the telescope beam. In general,

$$T_e^* \approx \left[1.40 \times 10^5 \frac{I_c}{\int I_L d\nu} \frac{f_{nm}}{n} \nu^{2.1} F\right]^{0.87}, \quad 90 \text{GHz} < \nu < 150 \text{GHz} \quad (3)$$

where I_c and I_L are the continuum and line intensities, respectively, f_{nm} is the absorption oscillator strength between principal quantum numbers n and m, ν is the line frequency in GHz, and F is the correction for the fraction of the free-free emission I_c due to He$^+$,

$$F \equiv 1 - \frac{N_{He}}{N_H}, \quad (4)$$

where N is the column density of the species. Oscillator strengths for mm-wave RRLs have been calaculated by Goldwire (1969) and by Menzel (1970); frequencies; by Lilley and Palmer (1968); and departure coefficients, by Walmsley (1990).

The restriction of Eq. 3 to the 2–3 mm wavelength range stems from the approximation for the free-free absorption coefficient used (see Gordon 1989). Substitution of Eq. 1 shows that at mm-waves, the actual electron temperature relates to the LTE value by

$$T_e = T_e^* b_m^{0.87} \quad \tau_c, \tau_L \ll 1. \quad (5)$$

The oscillator strengths, f_{nm}, normally used for RRLs (see Eq. 3) assume that the orbital quantum states ℓ are equally populated for each value of principal quantum number n (see the discussion by Seaton 1980). While this is most likely the situation in HII regions for cm-wave lines where collisional processes are important, it will not be the case in the optical regime where radiative processes dominate. Fig. 3 shows a plot of the minimum value of

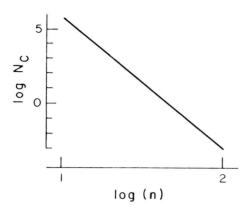

Fig.3–Values of the critical electron density, N_C, as a function of principal quantum number n required to equally populate the ℓ quantum states. From Seaton (1980)

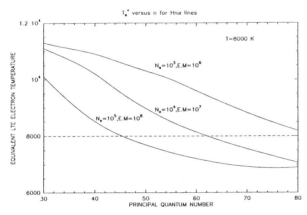

Fig.4–Calculated values of T_e^* for α- type RRLs as a function of principal quantum number for a range of electron densities and emission measures. *Upper curve:* conditions appropriate for a "normal" HII region, *middle curve:* conditions for a compact HII region, *lower curve:* conditions for an ultra-compact HII region. From Walmsley (1989).

N_e as a function of n required to keep the ℓ-levels equally populated. One can see that in the cm-wave range, $N_e > 10^{-3}$ cm^{-3} is adequate. For mm-wave lines where $n \approx 30$, this critical density increases to 10. And, for submm-wave RRLs where $n \approx 20$, the critical density can be as large as 10^3 cm^{-3}—approaching detectable astrophysical situations. At present, however, it appears that changes in oscillator strengths due to variations in level degeneracies are of little concern for observers of mm- wave RRLs.

Fig. 4 shows the variation of T_e^* as a function of principal quantum number for α-type RRLs for conditions appropriate for normal, compact, and ultra-compact HII regions. The LTE electron temperature, obtained with Eq. 3, overestimates the actual temperature because of the underpopulation of the principal quantum levels, *i.e.*, because $b_n < 1$ in Eq. 5.

2. Early Observations

Soon after the detection of RRLs and the discovery of the unexpectedly low electron temperatures derived from them, many astronomers recognized the importance of observations

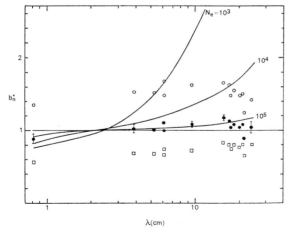

Fig.5–*Lines:* Theoretical values of departure coefficients for a range of volume densities N_e as function of wavelength; *open circles:* values derived from observed RRLs assuming a T_e of 10,000K; *filled circles:* values derived assuming a T_e of 7,000K; *open squares:* values derived assuming a T_e of 5,000K. From Sorochenko and Berulis (1969)

in the mm-wave region to test whether or not the line intensities could be explained by departures from LTE. The idea was that radiation transfer in the mm-wave region was mathematically simple, involving only the departure coefficient as shown by Eq. 5. Comparison of mm-wave and cm-wave observations of RRLs, along with tables of precalculated departure coefficients, would provide a test of the non-LTE theory proposed by Goldberg(1960) and would determine a characteristic T_e for an HII region.

With this objective in mind, Sorochenko *et al.* (1969) and Sorochenko and Berulis (1969) detected H56α lines from the Omega and Orion nebulas at 36.5 GHz. As Fig. 5 shows, comparing departure coefficients derived from these observations and those from cm-wave ones with theoretically calculated ones, they deduced a value of T_e for the Orion Nebula of $(8,800 \pm 700)(1 - 0.1r)$K, where r is the distance from the center of the nebula in arc min. Integrating $T_e(r)$ over the nebula, they obtained a mean temperature for $< T_e >$ of 6,900K. The authors also deduced values of the departure coefficients, b_{57}, and of the RMS turbulent velocities at the 5 positions in the nebula which were observed. Within experimental errors, these values are comparable to those derived 11 years later by Shaver (1980) from copious observations over a large wavelength range, much improved excitation parameters which lowered the temperatures derived from otical emission lines, and a sophisticated transfer theory.

In 1969, however, these low temperatures were not generally accepted by the astronomical community. Observations of the forbidden optical lines from the Orion Nebula indicated characteristic temperatures of 10^4K, in conflict with the lower temperature RRL results. Many astronomers (like the author) interpreted the soviet RRL results in terms of observational difficulties at mm-wavelengths. Some astronomers *did* accept the low electron temperatures, thereby stimulating a lively re-examination of the observational and theoretical basis for the canonical 10,000K electron temperature of HII regions—a controversy which persisted for many years.

But, if history shows that these early results should have been accepted, the measurement uncertainties of contemporary observations illustrates the basis for the skepticism. Waltman *et al.* (1973) observed the H42α at 85.7GHz from the Orion Nebula and used them for a similar analysis. Fig. 6 shows the comparison of their observation with cm-wave

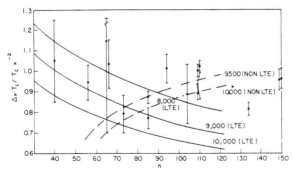

Fig.6–Values of $(T_L/T_c)\Delta\nu\nu^{-2}$ observed for $40 < n > 150$ from the Orion Nebula. *Solid lines:* LTE solutions, *broken lines:* non-LTE solutions for various electron temperatures. From Waltman *et al.* (1973)

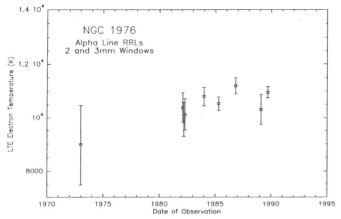

Fig.7–Observed values of T_e^* derived from 2- and 3mm Hα lines from the Orion Nebula.

data available then. In appraising this figure, the authors noted that, while the data appeared to be consistent with a T_e of 7500K, (1) the high-order cm-wave RRLs were inconsistent with this temperature, and (2) the signal-to-noise ratio of the observations were sufficiently large to accommodate a large range of temperatures. This interpretation, although cautious, turned out to be faulty. Simpson (1973)—who also was not believed at the time—and, finally, Shaver (1980) pointed out that (1) Stark broadening weakened the intensities of high-order cm-wave RRLs by merging their wings into the underlying free-free continuum, such that they overestimated $<T_e>$ and (2) that the cm-wave lines give accurate electron temperatures because the partial maser effect was offset by Stark broadening. In retrospect, Waltman *et al.* should not have discounted the 7,500K.

In view of the large system temperatures of 1968-1973 mm-wave radiometers, one might ask how good these measurements really were. Fig. 7 is a plot of the LTE electron temperatures for the Orion Nebula derived from Hα-lines from 1968 through 1990 in the 2 and 3mm bands. The early observations agree well with the more recent measurements, especially considering that the earlier measurements are at lower frequencies which have larger values of b_n.

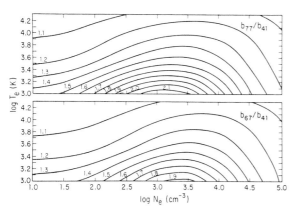

Fig.8–Contours of the ratios of b_{77}/b_{41} and b_{67}/b_{41} as functions of $\log T_e$ and $\log N_e$. From Gordon(1989)

3. Modern Observations

3.1. Finding $<T_e>$ and $<N_e>$ From Pairs of High Frequency Lines

The prospect that mm-wave RRLs from HII regions are free of maser amplification suggested another way to extract characteristic $<T_e>$ and $<N_e>$. From Eq. 5,

$$\frac{b_{n1}}{b_{n2}} = \frac{T^*_{e\,2}}{T^*_{e\,1}}. \qquad (6)$$

With this relationship observations of high frequency RRLs could be compared with tables of pre-computed departure coefficients to determine beam averaged values of $<T_e>$ and $<N_e>$. Specifically using pairs of H40α (99 GHz) and H66α (22 GHz) lines, pairs of H40α (99 GHz) and H76α (15 GHz) lines, and the observed free-free emission, we (Gordon 1989) attempted to make these determinations. Using the departure coefficients calculated by Salem and Brocklehurst (1979), we prepared the isograms shown in Fig. 8. The observed ratio determined from Eq. 6 selected a contour range. The observed free-free continuum determines the quantity $<N_e^2/T_e^{0.35}>$ which, because of the weak temperature dependence, gives a nearly diagonal line cutting from lower left to upper right which cuts across the selected contour ratio. In principle the intersection gives the $<T_e>$ and $<N_e>$ for the HII regions.

In practice, the resulting values of T_e were unrealistically small—as much as a factor of 3 below "reasonable" results obtained by analyses of the cm-wave RRLs. The explanation for the low values of T_e is that some partial maser effect is still present in the 15 and 22 GHz RRLs—which, in retrospect, can be seen in Figs. 2 and 3. These RRLs are too low-frequency to be used for this analysis technique. A calculation indicates that, although the line intensities are diminished with respect to LTE, about 10% of their intensities are due to the term $(1 - \tau_c \gamma/2)$ in Eq. 1.

This analysis technique should still be applicable to pairs of much higher frequency lines, such as a pair where both lines occur in the mm-wave region.

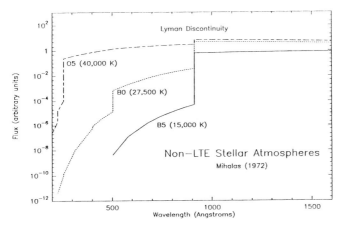

Fig.9–Ultraviolet flux calculated as a function of wavelength for stars of 3 different effective temperatures. No line-blanketing is included. Data from Mihalas (1972)

3.2. EXCITATION BY STELLAR LYMAN LINES

Beigman et al. (1980) and Hoang-Binh (1983) suggested that the "standard" departure coefficients might not be appropriate for small principal quantum levels. They suggested that the shape of the Lyman line and continuum radiation from the exciting star might have a substantive influence on the populations of levels say, $n < 70$, which are dominated by radiative processes. Fig. 9 shows the calculated non-LTE UV emission used for the calculations. The large discontinuity at 912Å—the Lyman Discontinuity—decreases as the effective temperature of stars increase. For cooler stars, ionization by photons with $\lambda < 912$Å would be reduced compared with respect to, say, excitation by Lyman line radiation at $\lambda > 912$Å. This reduction in the ionization rate would correspondingly reduce the number of radiative ionizations out of a bound level n as well as reducing the supply of free electrons available for radiative recombination into level n thereby increasing the importance of radiative excitation of bound electrons from the ground state (Lyman Line Excitation). Note that this mechanism only works where radiative rates dominate collisional rates; specifically, for small values of n such as associated with mm-wave RRLs.

How would a large Lyman discontinuity affect the strength of RRLs? In Fig. 10, Hoang-Binh (1983) showed that, based upon the Mihalas atmospheres and typical HII region conditions, the net effect would be to increase the level populations of $n < 70$ relative to LTE, i.e., to increase the departure coefficients. Because, as Eq. 1 shows, line intensities are a direct function of b_n in the mmwave regime, RRLs involving these levels would be enhanced relative to what would be predicted from LTE transfer theory using standard departure coefficients.

The magnitude of the RRL enhancement would be expected to vary from one HII region to another as a function of Lyman discontinuity of the exciting star(s). In particular, the theory predicted that low-n RRLs from HII regions excited by hot stars—like the Orion Nebula—would appear "normal", whereas those from HII regions excited by cooler stars would appear enhanced. In principle, the size of the enhancement could even be supra-LTE, that is, the departure coefficients of the low ns could exceed 1, giving the RRLs greater intensities than the LTE values.

Observations to detect this effect are difficult to make. H$n\alpha$ transitions involving $n < 60$

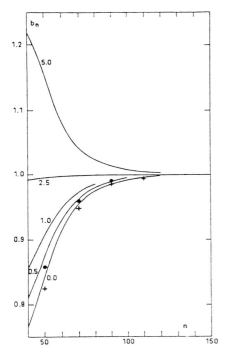

Fig.10–Values of b_n as a function of principal quantum number n. Curves are labeled by a measure of the size of the Lyman discontinuity, 0 indicating no discontinuity. *Filled circles:* departure coefficients from Salem and Brocklehurst (1979); *crosses:* departure coefficients from Burgess and Summers (1976). From Hoang-Binh (1983)

appear in the mm-wave region of the spectrum. In terms of receiver sensitivities, such RRLs are more difficult to observe than at cm-wavelengths. Furthermore, candidate HII regions with strongly enhanced RRLs involve cooler exciting stars, contain less ionization, and therefore are even more difficult to observe. To complicate the observations even more, beam sizes for cm-wave RRLs are often larger than those for mm-wave RRLs, leading to comparison of RRLs from different volumes of gas and, hence, perhaps different $< T_e >$ for the 2 sets of lines. The result would be imossible to interpret.

Nonetheless, several groups have searched for the effect. Hoang-Binh, Encrenaz, and Linke (1985); Sorochenko, Rydbeck, and Smirnov (1988); and Gordon (1989) all looked for this effect by observing mm-wave Hnα lines from the radio bright—and hence easily observable—Orion Nebula. No significant differences from conventional excitation theory were found. But, is the non-detection due to an inapplicable theory, to an insignificant effect in an HII region excited by a very hot star with a small Lyman Discontinuity, or to measurement uncertainties?

Recently, Gordon and Walmsley (1990) used the new IRAM 30-m telescope in Spain to examine the ratios of Hα/Hβ/Hγ/Hδ/Hϵ RRLs in the 2- and 3-mm windows from the same volume of gas. These transitions involve $36 \le n \le 60$. Their technique was to make spectra containing more than one transition. In this way they could compare the relative populations of several n-levels with the same calibration and with the same beam size and hence the same gas volume. In addition to the Orion Nebula, they observed the the HII region NGC 2024, which is believed to be excited by a relatively late type star on the basis of its weak He RRLs. No effect was found.

One explanation (Mihalas 1989) is that the model atmospheres which Hoang-Binh (1983) used to calculate the Lyman discontinuities (Fig. 9) did not include line blanketing. Inclusion of this mechanism would reduce the Lyman discontinuity substantially and, in turn, drive the value of the departure coefficients toward conventional values. For example, the ratio of the Lyman discontinuity changes from 2.4 to 24 as the effective temperature of the exciting stars goes from 40,000 to 30,000K in the unblanketed 1973 Mihalas' models, but only from 2.1 to 12.5 in the more recent, blanketed models of Kurucz (1979)—almost a factor of 2 reduction in the discontinuity.

In addition, the Lyman discontinuity varies as a function of the surface gravity of stars. For cooler stars with large surface gravities, Kurucz's models indicate an additional reduction in the Lyman discontinuities by as much as a factor of 2. If the exciting star of NGC 2024—unknown at this time—has a large surface gravity, and if we consider only line-blanketed model atmospheres, the Lyman discontinuity for NGC 2024 could be a factor of 4 smaller than those considered by Hoang-Binh (1983). For these reasons, enhancement of RRLs by Lyman line excitation may not be a significant effect in practice.

We conclude that, to date, observations indicate that conventional departure coefficients appear to be adequate to describe the intensities of mm-wave RRLs.

3.3. He/H Abundance Ratios

To understand chemical evolution in galaxies, one would like to know the ratio of the volume densities $n(\text{He})/n(\text{H})$ in astronomical objects. Optical observations are vulnerable to uncertain reddening corrections, to anomalous excitation conditions, and to the effects due to variations on line opacities from one HII region to another.

But RRLs are also not ideal for abundance studies. RRLs are measures of the column densities of *ionized* rather than neutral material because of their dependence upon recombination processes. The Stromgren spheres for He^+ and H^+ may not be the same size. Finally, the He RRLs are often blended with RRLs from C and other heavy elements.

Because radiation transfer is simple for mm-wave RRLs, and because of the high angular resolution achievable with a single dish telescope at mm-wavelengths, Peimbert *et al.* (1988) used the 45m Nobeyama radio telescope to observe the H53α and He53α RRLs (43 GHz) from M17 and the Orion Nebula. At these frequencies, the full beam width at half power is 39″.5. Their idea was to observe the RRLs in the vicinity of the exciting stars of the HII regions, and to combine both optical and radio data to find a mutually consistent abundance ratio.

Their results indicate excellent agreement between optical and radio observations for both HII regions. After a small correction for neutral helium based upon optical data, they find values for He/H of 0.099 ± 0.008 and 0.108 ± 0.004 for the Orion Nebula and M17, respectively. For the Orion Nebula, they also found a decrease in the ratio He^+/H^+ outwards from the exciting star θ^1 Ori—as would be expected. This gradient implies that the gradient of the neutral components, He^0/H^0, should vary in the opposite sense. Combining these data with optical data from HII regions in the SMC and in NGC 2363, Peimbert *et al.* conclude that the primordial helium mass fraction is 0.230 ± 0.005 (1σ). Comparing this value to those observed for M17 and the Orion Nebula, they also conclude that the helium to heavy elements enrichment by mass, $\Delta Y/\Delta Z$, is approximately 3.6 ± 0.6.

Presumably, this kind of analysis will be extended to other galaxies as observational sensitivities increase.

4. Conclusion

Almost from the detection of RRLs, mm-wave observations have contributed to this new area of astronomical research. Free of the partial maser effect and of Stark broadening, the radiation transfer is much simpler than in the centimeter or decimeter regimes. Perhaps the first use of these features was the 1969 observations of RRLs at 35 GHz by Sorochenko and colleagues to establish that typical electron temperatures in HII regions were approximately 8,000K, rather than the canonical 10,000K given by the [OIII] optical emission lines.

Because of their weighting by τ_L, mm-wave RRLs tend to reflect the characteristics of the densest parts of the HII regions. Values of $<T_e>$ and $<N_e>$ derived from them will not refer to the lower density areas of the HII region.

As with all RRLs, transitions associated with different principal quantum numbers, n, allow some unique investigations. Millimeter and submillimeter $h\alpha$ lines involve $n < 60$. Since the "classical size" of atoms varies as n^4 in terms of collision cross-sections, small quantum levels are more likely to be probes of radiative rather than collisional processes. Such reasoning forms the basis for studies as to the effects of the Lyman Discontinuity upon the excitation of atoms in HII regions, as has been considered by Beigman *et al.* (1980) and Hoang-Binh (1983).

Mm-wave and submm-wave RRLs, in principal, will permit the study of $n\ell$ quantum degeneracies when receiver sensitivities increase sufficiently to explore low density gas within HII regions, as discussed by Seaton (1980). However, for mm-wave investigations into normal HII regions, the standard oscillator strengths should suffice.

Finally, the small beamwidths associated with mm-wave and submm-wave radio telescope allow us to investigate HII regions with an angular resolution better than a few 10s of arc seconds. Such observations complement those achieved by cm-wave synthesis instruments, permitting comparison of the populations of very different quantum levels from essentially the same volume of gas and the corresponding excitation conditions within the gas. This kind of work will probably continue to be the main contribution of mm-wave RRLs for the near future.

It's a pleasure to thank E. B. Churchwell and C. M. Walmsley for discussions regarding mm-wave RRLs.

References

Beigman, I. L., Gaisinskiy, I. M., Smitnov, G. T., Sorochenko, R. L. 1980, Preprint No. 141, Lebedev Physical Institute, Moscow.
Burgess, A., and Summers, H. P. 1976, *M. N. R. A. S.*, **174**, 345.
Goldberg, L. 1969, *Ap. J.*, **144**, 1225.
Gordon, M. A. 1989, *Ap. J.*, **337**, 782.
Gordon, M. A., and Walmsley, C. M. 1990, submitted to *Ap. J.*.
Goldwire, H. C., Jr. 1969, *Ap. J. Suppl.*, **17**, 445.
Griem, H. 1967, *Ap. J.*, **148**, 547.
Hoang-Binh, D. 1983, *Astr. Ap.*, **121**, L19.
Hoang-Binh, D., Encrenaz, P., and Linke, R. A. 1985, *Astr. Ap.*, **146**, L19.

Kurucz, R. L. 1979, *Ap. J. Suppl.*, **40**, 1.
Lilley, A. E., and Palmer, P. 1968, *Ap. J. Suppl.*, **16**, 144, and a subsequent privately circulated extension.
Lockman, F. J., and Brown, R. L. 1975, *Ap. J.*, **201**, 134.
Menzel, D. H. 1970, *Ap. J. Suppl.*, **18**, 221.
Mihalas, D. 1972, *Non-LTE model Atmospheres for B and O stars*, NCAR- TN/STR-76 (Boulder: NCAR).
_____ 1989, Private communication to the author.
Peimbert, M., Ukita, N., Hasegawa, T., and Jugaku, J. 1988, *Publ. Astron. Soc. Japan*, **40**, 581.
Salem, M., and Brocklehurst, M. 1979, *Ap. J. Suppl.*, **39**, 633.
Seaton, M. J.1980, in *Radio Recombination Lines*, P. A. Shaver, Ed., (Dordrecht: Reidel), pp3-22.
Shaver, P. A. 1980, *Astr. Ap.*, **90**, 34.
Simpson, J. A. 1973, *P. A. S. P*, **85**, 479.
Sorochenko, R. L., Puzanov, V. A., Salomonvich, A. E., and Shteinshleger, V. B. 1969, *Ap. Lett.*, **3**, 7.
Sorochenko, R. L., and Berulis, J. J. 1969, *Ap. Lett*, **4**, 173.
Sorochenko, R. L., Rydbeck, G., and Smirnov, G. T. 1988, *Astr. Ap.*, **198**, 233.
Walmsley, C. M. 1989, Private communication to the author.
_____ 1990, *Astr. and Ap. Suppl.*, in press.
Waltman, W. B., Waltman, E. B., Schwartz, P. R., Johnston, K. J., and Wilson, W. J. 1973, *Ap. J. (Letters)*, **185**, L135.

MAPS OF THE 64α RADIO RECOMBINATION LINES IN ORION A

T.L. Wilson, L. Filges
Max-Planck-Institut für Radioastronomie
Auf dem Hügel 69, D-5300 Bonn 1, FRG

ABSTRACT. Following a brief review of radio continuum and recombination line studies of Orion A, new 40" resolution maps in the 64α lines of hydrogen and helium results are presented. The velocity distribution shows that simple models of "Champagne Flow" do not agree with the data. A search for cold ionized gas toward the KL nebula has revealed no measurable differences in the LTE electron temperatures toward this region and other parts of Orion A. To within the uncertainties, there is no difference between the radial velocities of the H64α and He64α lines.

1. INTRODUCTION

1.1 Background Information

The H II region Orion A is a high surface brightness source of thermal emission located about 500 pc from the sun (see e.g. Glassgold et al. 1983). The largest contributor to the ionization is the star, θ^1 Orionis which has a surface temperature of >40,000 K. Because of its nearness and high surface brightness, Orion A is the most studied galactic source of thermally ionized gas, and one of the few regions for which detailed analyses of the gas motions have been made (von Hoerner 1951, Münch 1958, Osterbrock 1974). Because there is a wealth of data, Orion A can be used to test the applicability of theories of gas dynamical processes.

1.2 Radio Continuum Mapping of Orion A

In the radio range, high resolution continuum maps of the extended thermal emission have been made with resolutions ranging from 40" to 6" (Webster and Altenhoff 1970, Martin and Gull 1975, Johnston et al. 1983, Wilson and Pauls 1984, Akabane et al. 1985). A practical limit for the angular resolution of single dish maps is ~10". Since Orion A is close to the equator, aperture synthesis data taken with east-west arrays will not produce a circular beam. Martin and Gull (1976) had combined data from north-south spacings of the Owens Valley Array with east-west spacing from the Cambridge 1 Mile Telescope. This heroic effort is no longer required since the Very Large Array of the NRAO has north-south antenna spacings. Because the total extent of the H II region is more than 10', the source is larger than the VLA primary beam at wavelengths shorter than 6 cm, and multi-field maps are needed. The 6 cm continuum map presented by Johnston et al. (1983) was made for a single field center, as a by-product of spectral line observations in 1980. At that time, only 13 antennas of the VLA could be used and the FWHP of the VLA antenna was 10'. Because of these restrictions, the source distribution was incompletely sampled, and only about 30% of the total continuum flux density was recorded. With the present VLA operation one would expect that most, but perhaps not all of the single dish flux density will be recorded. To obtain all of the source flux density, it may be necessary to include short antenna spacings (corresponding to extended spatial structures) from single dish measurements. Such hybrid maps should lead to complete radio maps with angular resolutions comparable to or better than those obtained in the optical range. The angular resolutions of aperture synthesis radio maps are limited by the signal-to-noise ratio, not by seeing. With the most sensitive VLA receiver system, at 8 GHz, in 8 hours, with a 1" resolution, one can detect regions with emission measures of $5000 cm^{-6}pc$. This limit is 10-100 higher than that obtainable with the Palomar Observatory Sky Survey.

1.3 Geometry and Large Scale Dynamics of the Ionized Gas

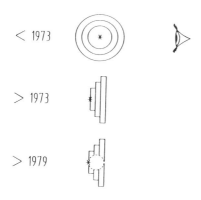

Fig.1 Sketches expressing opinions about the shape of Orion A as a function of time. The observer is at the right. The star represents the source of excitation. In the bottom-most picture, the region interior to the dashed line has a low abundance of gas and dust (see e.g. Zuckermann 1973, Balick et al. 1974 and Pankonin et al. 1979). Since 1973, the estimates of T_e have also changed; see Section 1.5.

In Fig. 1 we show sketches of our impression of the structure of Orion A. Before 1973, Orion A was thought to have a spherical shape. From measurements of the radio recombination lines of ionized carbon, C II, it was concluded that the ionized and molecular gas clouds were in contact (Zuckerman 1973, Balick et al. 1974). This idea gave rise to the alternate picture of Orion A marked "after 1973". The radial velocities were interpreted as the flow of hot gas off the cooler surface of a more massive cloud. Models of the ionized gas have been made by Brown (1975), Walmsley (1980), Shaver (1980) Wilson and Jäger (1985), and Tsivilev et al. (1986). An additional refinement was added by Pankonin et al. (1979) when it was noted that the radial velocities toward the center of Orion A were more positive than around the edges. Such observations may be consistent with a "stellar bubble", that is, a region free of dust and gas close to the exciting stars.

1.4 The Excitation and Transfer of Radio Recombination Line Radiation

Radio recombination lines were predicted by Kardashev (1959), and first measured by Dravskikh and Dravskikh (1964), Dravskikh et al. (1964), Höglund and Mezger (1965), and Sorochenko and Borodzich (1966). The first results showed that the radio recombination lines were present, but the low signal-to-noise ratios and larger-than-expected influence of systematic errors allowed only qualitative conclusions. The results were interpreted in terms of transitions emitted by an LTE population of electrons. The LTE electron temperature, T_e^* depends only on directly measured quantities (see e.g. Eq.(1) of Pauls and Wilson ,1977). This has a great advantage, since with one set of measurements one could determine the (LTE) electron temperature of an H II region. In 1965, it was assumed that T_e^* equalled T_e, the actual, beam averaged electron temperature. From the first set of H109α line measurements, made at 6 cm, the T_e^* values were 40% lower than the T_e values obtained from ratios of the forbidden lines of oxygen (see e.g. Osterbrock 1974). In order to reconcile the radio and optical values of T_e, Goldberg (1966) proposed the first realistic treatment of the populations of high principal quantum number states. The equation relating the LTE electron temperature, T_e^*, to T_e, is given by (Brocklehurst and Seaton 1972, Wilson et al., 1979)

$$T_e = b \, T_e^* (1 + 1/2 (1-\beta)\tau) \frac{g(T_e^*)}{g(T_e)} \qquad (1)$$

where b is the departure coefficient, ß depends on the difference of b values for the upper and lower levels, τ is the continuum optical depth and g is the Gaunt factor. Both b and ß are functions of electron density and temperature. b is less than 1, and ß is positive for hydrogen. Walmsley (1989) has calculated b and ß for transitions in the millimeter wavelength range; for the centimeter wavelength transitions these were made by Brocklehurst (1970). From the dependence on b, at short cm and millimeter wavelengths, the radio recombination line emission arises from the dense cores of the H II regions. The term containing ß gives rise to line masering, which lowers T_e^* below T_e. The b factor raises T_e^* above the value of T_e. Collisional broadening will also influence the line emission. This effect is not explicitly present in Eq (1). Line broadening was examined by Griem (1967); this will tend to redistribute the line intensity from center to wings, where the line emission may be underestimated by incorrect baseline fits, or instrumental baseline effects caused by resonant reflections in the telescope structures. In the centimeter wavelength range, all three factors, line masering, the b factors and collisional broadening affected the relation of T_e^* and T_e. Since these factors would have different amounts of influence on α, ß , γ and higher order transitions, studies were made of such transitions which occured at nearly the same wavelength (Hjellming and Churchwell 1969, Hjellming and Davies 1970, Davies 1971, Smirnov et al. 1984). From these data it was clear that in the centimeter range, T_e^* from alpha lines gave a lower bound to T_e. Estimates based on ß lines were a better approximation to T_e, but collisional broadening could cause the line intensity to be underestimated. A simple and very useful stategy was to measure α lines in the short centimeter and millimeter range. In that wavelength range, τ is very small and line masering is unimportant, at least in Orion A. Also collisional broadening will have no effect at even the highest densities known to be present in this source. These efforts were begun by Sorochenko, Berulis and collaborators and later Waltman et al. (1975). A summary is given in Berulis et al. 1975. The Berulis et al. (1975) data gave T_e^* values of 7000K for Orion A. From Eq. (1) the value of T_e must be lower than that of T_e^*. Such results, if correct, would bring the entire interpretation of radio recombination line formation into question. More accurate values of T_e^* were measured for a wavelength of 1.3 cm by Chaisson and Dopita (1977) and Pauls and Wilson (1977). The latter authors also showed that at 1.3 cm, the b and ß terms balanced and by chance T_e equalled T_e^*. The T_e^* values obtained by Pauls and Wilson have been since confirmed by a number of other groups and their value of T_e = 8200 to 8500 K for the core of Orion A is now the accepted one.

The treatment in Eq. (1) is based on a single region of uniform density and temperature. Brocklehurst and Seaton (1972, hereafter BS) extended this treatment to a model incorporating a range of electron densities. BS applied to their formalism to Orion A using a spherical model consisting of 9 layers. Because of line masering, the relative placement of regions with different has a significant influence on the total line intensity. The agreement of data and prediction were good, if a large portion of the line intensity emitted at wavelengths longward of 10 cm was underestimated because of incorrect baseline fits. The specific model proposed by BS for Orion A was improved by Lockman and Brown (1975, hereafter LB) in two ways. First, LB incorporated a slab geometry, and second, they separated the line-to-continuum ratios and linewidths in comparing model and data. Such a separation reduces the effect of baseline subtraction. The specific model used by LB predicted a rise in T_e with distance from the exciting stars. This could be justified on the basis of photon hardening (Hjellming 1966), but is inconsistent with recombination line mapping results (Wilson and Jäger 1985). A model incorporating a variable density and constant T_e can be made to agree with the data (Walmsley 1980, Shaver 1980) if the local electron density is much larger than the RMS electron density. This may imply a great deal of clumping. However, in early interferometer maps of the radio continuum in Orion A, most of the structure seems to be smooth on a scale of 5" (=4 x 10^{16}cm). Another possibilty is that the slabs have a depth which is only 10% of their diameter. If so, the source has a sheet-like geometry.

1.5 Electron Temperatures and Element Abundances

As pointed out by Peimbert (1967) and Seaton (1980), the T_e values obtained from optical and radio data give upper and lower limits to the range of T_e values. Peimbert (1983) favored fluctuations in T_e in order to raise the low oxygen abundance to the solar value. Radio measurements of T_e values by Pauls and Wilson (1977) and Pankonin et al. (1979) show good agreement with the optical values (Peimbert and Peimbert 1977). Optical estimates for Orion (O.C. Wilson et al. (1959), Spitzer (1968, p. 32; 1978, p. 80)) have declined from 10,000K (1959), to 9000K (1968) to 8000K (1978). From this it would appear that oxygen is underabundant in the gas phase in Orion A. This conclusion is supported, in a more direct way, by the far infrared measurements of oxygen fine structure line by Simpson et al. (1985). Since 1973, T_e estimates for the outer region have declined from 12000K (LB) to ~7000K (Wilson and Jäger 1985).

1.6 Gas Motions in Orion A

The systematic motions in Orion A have been studied using optical spectral lines (O.C. Wilson et al. 1959), but the ionization structure, turbulence and extinction will have a strong influence. The radio recombination lines are not affected by extinction, but are subject to line masering (see Eq.(1)) and lower angular resolution. The first effect has been used by Brocklehurst and Leeman (1971) and Tsivilev et al. (1986) to show that Orion A is expanding at a speed of a few kms^{-1}. The second effect can be used to average over the turbulence and obtain a better impression of the large-scale motions.

2. NEW HIGH RESOLUTION MAPS OF ORION A

2.1 Objectives

Single-dish maps of limited regions in radio recombination lines have been made with resolutions of 40" (Pauls and Wilson 1977, Hasegawa and Akabane 1984). The goals of our new, extensive mapping program are to: (1) measure the distribution of radial velocities over the core of Orion A, in order to compare the data with models of dynamics, (2) determine the electron temperatures over the source, in particular to compare T_e^* (assumed to equal T_e at 1.3 cm; see Pauls and Wilson 1977, Seaton 1980) toward the KL nebula with T_e elsewhere in Orion A, and (3) compare the radial velocities of helium with that of hydrogen.

2.2 Data Taking and Analysis

The H64α line and continuum data were taken between October 1988 and March 1989 with the 100-m telescope of the MPIfR, together with a K-band maser and a 1024 channel autocorrelation spectrometer. The data were taken in excellent weather. The system noise temperature on cold sky was less than 100K. The observing procedure was as follows. First, a series of continuum scans were made by rapid scanning back and forth in Declination. Then spectra were taken for about 2-3 hours at various positions where continuum data were taken. Following this, continuum scans were repeated. The spectra were analyzed using gaussian fits. In all cases the lineshapes were well represented by gaussians.

2.3 The Results

In Fig. 2(a) we show the radio continuum map of Orion A made at a wavelength of 1.3 cm (Wilson and Pauls 1984). The dots show the locations of the Trapezium stars. The hatched region northwest of the Trapezium marks the location of the vibrationally excited molecular hydrogen, H_2^*; the KL infrared nebula (diameter <10") is directly north of this region The hatched area about 100" south of this region is Source 6 (hereafter S6) of Batrla et al. (1983). From the 6 cm H_2CO line map of Johnston et al. (1983), it was concluded that this region is in front of some of the ionized gas, and thus must be inside the H II region. In Fig. 2(b) we show a contour map of radial velocities for the H64α line. The radial velocities were obtained from gaussian fits. In Fig. 2(c) we show a series of rectangles superimposed on the lowest continuum contour. The average values and RMS scatter for various measured quantities are given in Table 1.

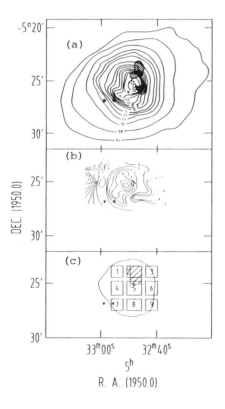

Fig.2 (a) The positions of the Trapezium stars are shown by 4 dots. The θ^2 Orionis stars are shown as *. The shaded area to the NE of the Trapezium is prominent region of vibrationally excited H_2 emission. This source is immediately south of Orion-KL. About 100" south, also shown shaded, is the prominent molecular source S6 (Batrla et al. 1983). (b) The contours of radial velocity, obtained from gaussian fits to the H64α line profiles, are superimposed on the symbols shown in (a). The influence of the SW bar is clearly seen in the -2 and $-3kms^{-1}$ contours. The $-4kms^{-1}$ contour to the W of the Trapezium may be related to KL and S6. (c) The numbered boxes show regions where the averages in Table 1 have been formed. The hatched region refers to the source of vibrationally excited H_2. This is referred to as "KL/H_2^*" in Table 1.

2.4 Discussion

2.4.1 Electron temperatures

TABLE 1: H64α and He64α LINE RESULTS

Position[a]	T_e^* (10^3K)	V(H)-V(He) (kms^{-1})	Turbulent Velocity (kms^{-1})
KL/H$_2$*	8.4(0.5)	-0.6(2.0)	18.2(0.8)
1	7.7(0.6)	0.6(1.8)	19.1(0.8)
2	8.2(0.5)	-0.7(2.1)	18.6(0.7)
3	8.4(0.9)	-0.8(3.0)	19.4(1.0)
4	8.0(0.9)	-0.9(1.7)	20.0(1.4)
5	8.4(0.4)	+0.4(1.9)	18.9(1.6)
6	8.3(1.4)	-2.2(3.1)	20.4(0.8)
7	8.3(1.2)	0.7(2.6)	19.6(0.5)
8	8.4(1.0)	1.5(1.8)	20.6(2.0)
9	9.2(1.0)	1.3(4.6)	21.2(1.7)

[a] values in parantheses are RMS scatter

From Table 1, toward the KL nebula, the value of T_e^* is not significantly different from that toward other regions. This finding does not agree with the result of Hasegawa and Akabane (1984), who reported that, toward the KL nebula the value of T_e^* (measured at 7mm) was lower than that found for the fully ionized gas. Since we have repeated our measurements on different days, and interspersed line and continuum measurements we feel confident about our results.

2.4.2 Radial velocity Distribution

There is general agreement with the 1' resolution data of Pankonin et al. (1979), although our data are more detailed and extensive. The most positive radial velocities are located about 1' and 4.5' east of the Trapezium stars. As in "Champagne Flow" (Tenario-Tagle 1979), the radial velocities of the ionized gas are more negative than those measured for the neutral gas. However, one would not expect the west-east velocity trend to be negative-positive-negative-positive. Also, it is unclear why the continuum maximum and the peak of the radial velocities do not coincide with the Trapezium stars. Pankonin et al. (1979) suggested that we are viewing Orion A from an angle, and this may explain some of the positional differences. The radial velocity of -4 kms^{-1} in the western part of Orion A bends around the location of the KL/H$_2^*$ region and S6. Such a bending is not present in the continuum contours. Although there are irregularities elsewhere in the velocity distribution, this variation would seem to be significant. This radial velocity structure may be caused by a larger neutral gas density in the molecular gas toward the H$_2^*$ and S6 regions. One might expect that the CII line intensity would be largest there; this is not the case (Kuiper and Evans 1978, Jaffe and Pankonin 1978). The -2 and -3kms^{-1} contours outline the SW bar clearly.

Continuum maps made at longer wavelengths emphasize lower density ionized gas. These maps (see e.g. the map of Mills and Shaver 1968) show a slowly declining intensity to the west, and a sharper boundary to the east. This would indicate a free expansion of the ionized gas in the west, but a neutral boundary in the east. The neutral boundary in the east may wrap around the ionized gas. In the optical the "dark bay" may be a part of this structure.

The widths of radio recombination lines exceed thermal widths, even for spectra taken with an angular resolution of 40" (=0.1 pc=3 x 10^{17}cm). The variation of the radial velocities between neighboring positions appears to exceed the noise. This indicates the presence of structure on a scale less than 40".

2.4.3 Velocity Differences Between Hydrogen and Helium

Tsivilev et al. (1986) had reported a significant difference between the radial velocities of ionized hydrogen and helium, in the sense that the radial velocity of the helium is more negative. Our data in Table 1 are based on the gaussian fits to individual profiles. We find no significant difference, but the helium and carbon recombination lines are blended; thus the decomposition is affected by the signal-to-noise ratio.

We thank Mr. A. Przewodnik for help in preparing Table 1.

3. REFERENCES

Akabane, K., Sofue, Y., Hirabayashi, H., Inoue, M., Nakai, N., Handa, T. 1985 Publ. Astron. Soc. Japan 37, 123
Balick, B., Gammon, R.H., and Doherty, L.H. 1974 Ap.J. 188, 45
Batrla, W., Wilson, T.L., Bastien, P., Ruf, K. 1983 Astron.Ap. 128, 279
Berulis, J.J., Smirnov, G.T., Sorochenko, R.L. 1975 'Observations of the Excited Hydrogen Radio Line H56α in Several H II Regions', in T.L. Wilson and D. Downes (eds.) H II Regions and Other Topics, Springer Heidelberg, pp 329-342
Brocklehurst, M. 1970 Mon. Not. Roy. Astron. Soc. 148, 417
Brocklehurst, M. and Leeman, S. 1971 Ap. Letters 9, 35
Brocklehurst, M. and Seaton, M.J. 1972 Mon. Not. Roy. Astron. Soc. 157, 179
Chaisson, E.J. and Dopita, M.A. 1977 Astron. Ap. 56, 385
Davies, R.D. 1971 Ap. J. 163, 479
Dravskikh, Z.V.and Dravskikh, A.F. 1964 Astron. Tsirk. No. 282
Dravskikh, Z.V.and Dravskikh, A.F., Kolbasov, V.A. 1964 Astron. Tsirk. No. 305
Glassgold, A., Huggins, P.J., Schucking, E.L. (eds.) 1982 Symposium on the Orion Nebula to Honor Henry Draper, Annals of the New York Academy of Sciences, 395
Goldberg, L. 1966 Ap.J. 144, 1225
Griem, H. 1967 Ap.J. 148, 547
Hasegawa, T., Akabane, K. 1984 Ap.J. 287, L91
Hjellming, R.M. 1966 Ap.J. 143, 420
Hjellming, R.M. and Churchwell, E. 1969 Ap. Letters 4, 165

Hjellming, R.M. and Davies, R.D. 1970 Astron. Ap. 5, 53
Höglund, B. and Mezger, P.G. 1965 Science 150, 339
Hoerner, S. von 1951 Zs.f. Ap. 30, 17
Jaffe, D.T. and Pankonin, V. 1978 Ap.J. 226, 869
Johnston, K.J., Palmer, P., Wilson, T.L., Bieging, J.H. 1983 Ap.J. 271, L89
Kardashev, N. 1959 Sov. Astron. 3, 813
Kuiper, T.B.H. and Evans, N.J. II 1978 Ap.J. 219, 141
Lockman, F.J., Brown, R.L.: 1975, Ap.J. **201**, 134
Martin, A.H.M.M., Gull, S.F. 1976 Mon. Not. Roy. Astron. Soc. 175, 244
Mills, B.Y. and Shaver, P.A. 1968 Australian J. Phys. 21, 95
Münch, G. 1958 Rev. Mod. Phys. 30, 1035
Osterbrock, D. 1974 The Astrophysics of Gaseous Nebulae, Freeman, San Francisco
Pankonin, V., Walmsley, C.M., Harwit, M. 1979 Astron. Ap. 75, 34
Pauls, T. and Wilson, T.L. 1977 Astron. Ap. 60, L31
Peimbert, M. 1967 Ap.J. 150, 825
Peimbert, M. and Torres-Peimbert, S. 1977 Mon. Not. R. Astron. Soc. 179, 217.
Peimbert, M. 1982 'Physical Conditions of the Orion Nebula Derived from Optical and Ultraviolet Data' in Symposium on the Orion Nebula to Honor Henry Draper, Annals of the New York Academy of Sciences, 395, pp. 24-31
Seaton, M.J. 1980, 'Theory of Recombination Lines', in P.A. Shaver (ed) Radio Recombination Lines, Reidel, Dordrecht, pp3-22
Shaver, P.A. 1980 Astron. Ap. 90, 34
Simpson, J.P., Rubin, R.H., Erickson, E.F., Haas, M.R. 1986 Ap.J. 309, 553
Smirnov, G.T., Sorochenko, R.L., Pankonin, V. 1984 Astron. Ap. 135, 116
Sorochenko, R.L., Borodzich, E.V. 1966 Soviet Physics-Doklady 10, 7
Spitzer, L. 1968 Diffuse Matter in Space, Wiley Interscience, New York
Spitzer, L. 1978 Physical Processes in the Interstellar Medium, Wiley-Interscience, New York
Tenario-Tagle. G. 1979 Astron.Ap. 71,59
Tsivilev, A.P., Ershov, A.A., Smirnov, G.T., Sorochenko, R.L. 1986 Sov. Astron. Letters 12, 355
Walmsley, C.M. 1980, 'Interpretation of H II region Radio Recombination Lines', in P.A. Shaver (ed) Radio Recombination Lines, Reidel, Dordrecht, pp 37-51
Walmsley, C.M. 1989 'Millimeter Recombination Lines', submitted to Astron. Ap.
Waltman, W.B., Waltman, E.B., Schwartz, P.R., Johnston, K.J., Wilson, W.J. 1973 Ap.J. 185, L135
Webster, W.J. and Altenhoff, W.A. 1970 Ap. Letters 5, 233
Wilson, O.C., Münch, G., Flather, E.M., Coffeen, M.F. 1959 Ap.J. Suppl.4, 199
Wilson, T.L., Bieging, J.H., Wilson, W.J. 1979, Astron.Ap.71, 205
Wilson, T.L. and Pauls, T. 1984 Astron. Ap. 138, 225
Wilson, T.L. and Jäger, B. 1985 Astron. Ap. 184, 291
Zuckerman, B., Palmer, P., Penfield, H., Lilley, A.E. 1967 Ap.J. 149, L61
Zuckerman, B. 1973 Ap.J. 183, 863

ESTIMATE OF ELECTRON DENSITIES IN HII REGIONS FROM OBSERVATIONS OF PAIRS OF α-TYPE RECOMBINATION RADIO LINES

A. F. DRAVSKIKH, Z. V. DRAVSKIKH
Special Astrophysical Observatory
of the USSR Academy of Sciences
Leningrad Branch,
196140 Leningrad, Pulkovo, the USSR

ABSTRACT. A method of determining the electron density in ionized hydrogen regions from observations of pairs of excited hydrogen recombination α-type radio lines is presented.

1. INTRODUCTION

Electron densities in HII regions are defined nowadays from radio measurements in the continuum with the use of geometrical models of the regions and their distances.
   In this paper we are going to show the possibility of a more direct measurement of the electron density in HII regions on the basis of spectral measurements of pairs of α-recombination hydrogen radio lines.
   We can separate Doppler and Stark components and determine the electron density of an HII region by measuring of the profiles of two α-recombination lines. One of these lines must be observed at a frequency high enough that there is no Stark broadening of this line; the other line, at a lower frequency so that its profile contains both a Doppler component connected with thermal and turbulent movements and a Stark component caused by effects of pressure.

2. METHOD

It is known (Lang 1978) that a line containing both Doppler and Stark components has a Voigt profile. It is a superposition of two profiles having Gauss and Lorentz forms, respectively. Using the tables from (Finn and Mugglestone 1965) we can obtain the following relation for a normalized line profile

$$I_0 \Delta \nu = f(\Delta \nu_L / \Delta \nu_D), \qquad (1)$$

where I_0 is the peak intensity of a Voigt profile, $\Delta \nu$ is the FWHM line width, $\Delta\nu_L$ and $\Delta\nu_D$ are Stark and Doppler widths.

The left part of equation (1) is the constant value for purely Doppler profiles

$$(I_0\Delta\nu)_D = 2(\ln 2/\pi)^{1/2}. \qquad (2)$$

The form of the curve (1) is presented in Fig. 1. In the $\Delta\nu_L/\Delta\nu_D$ from 0 to 1 region this curve is well enough approximated by the following function

$$I_0\Delta\nu = 1.0362 - 0.3060(\Delta\nu_L/\Delta\nu_D + 0.1)^{1/2} - 0.0143\sin(\pi\Delta\nu_L/\Delta\nu_D), \qquad (3)$$

the rms error being 0.001.

We can also obtain the function $\Delta\nu/\Delta\nu_D = f(\Delta\nu_L/\Delta\nu_D)$. It is presented in Fig. 2, where the crosses are the values linearly interpolated from table (Finn and Mugglestone 1965). The solid line is the approximating function

$$\Delta\nu/\Delta\nu_D = 1 + 0.639\Delta\nu_L/\Delta\nu_D + 0.144(\Delta\nu_L/\Delta\nu_D)^2, \qquad (4)$$

the rms error being less than 0.0019.

The method of calculation of $\Delta\nu_L$ from the Voigt profile observed is the following.

The observed line profiles are normalized (see Appendix). In the normalized high-frequency profile we have $(I_0\Delta\nu)_D = 2(\ln 2/\pi)^{1/2} = 0.94$. In the normalized low-frequency profile we measure the product of $I_0\Delta\nu$. Then from Fig. 1 or from equation (3) we can define $\Delta\nu_L/\Delta\nu_D = b$, and from the equation (4) we obtain $\Delta\nu_L$

$$\Delta\nu_L = \Delta\nu/(0.639 + b^{-1} + 0.144b). \qquad (5)$$

For calculation of the electron temperature T_e^* for the LTE case, we use the expression (2.121 and 2.122, Lang 1978)

$$\Delta\nu_D \Delta T_L / T_c = 2.176 \cdot 10^{-4} \nu^{2.1} T_e^{-1.15}, \qquad (6)$$

where ν is in GHz, and $\Delta\nu_D$ is in kHz.

T_e is calculated from the high-frequency line because the expression (6) is correct only in the case when the line has a purely Doppler profile.

The Stark line width for α-hydrogen lines in radio range is defi-

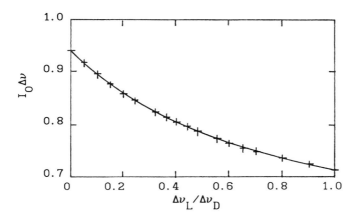

Fig.1. The peak intensity*line width product $I_0 \Delta v$ as a function of the Stark/Doppler width ratio $\Delta v_L / \Delta v_D$ for the Voigt profile.

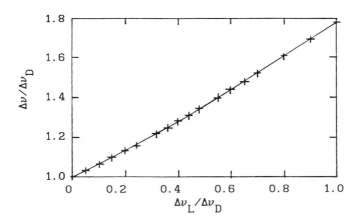

Fig.2 The variation of line width ratio $\Delta v / \Delta v_D$ with Stark/Doppler width ratio $\Delta v_L / \Delta v_D$ for the Voigt profile.

ned by the expression (2.288, Lang 1978)

$$\Delta\nu_L = (2.4 \ 10^{-6} N_e n^4 / T_e^{1/2})[1/2 + \ln(6.64 \ 10^{-6} T_e n)] \ (Hz), \quad (7)$$

from which we obtain the electron density N_e^* because all the other parameters are known.

In the common case of non-LTE the calculated electron temperature T_e^* and the electron density N_e^* must be corrected.

The actual T_e is given by (Brown, Lockman and Knapp 1978)

$$T_e = T_e^* b_n (1 + \gamma \tau_c / 2), \quad (8)$$

where b_n is the ratio of the actual population to the LTE population, γ is the coefficient proportional to the variation of the level population, τ_c is the optical depth of the nebula in the continuum. We obtain b_n and γ from the tables published in (Salem and Brocklehurst 1979). For a small optical depth in continuum (for $\tau_c << 1$) we have

$$\tau_c = T_a / \eta T_e, \quad (9)$$

where T_a is the antenna temperature in the continuum, and η is the beam efficiency.

Taking as the first approach the values of T_e^* and N_e^* (for the LTE) according to (Salem and Brocklehurst 1979) we obtain the coefficients b_n' and γ', then using the formulae (9),(8),(7) we calculate τ_c', T_e', N_e'. To obtain the final values of T_e and N_e it is necessary to carry out the second approach taking T_e' and N_e' as initial values.

It should be pointed out that T_e^* in the equation (8) remains fixed in both iterative loops.

3. CONCLUSION

This method of determining the electron density in HII regions with a pair of α-recombination lines along with the traditional one (Mezger and Henderson 1967) permits one to define the distances to investigated nebulae more exactly.

Naturally, the measurements of N_e can be carried out with (β, γ, δ) recombination lines as has been done in (Smirnov, Sorochenko and Pankonin 1984). But these lines are considerably weaker than α-lines and consequently give less precision in comparison with α-lines.

In connection with the possibility of determining N_e in HII regions from α-lines, the interesting possibility of measuring N_e in compact radio sources associated with galaxies and quasars arises.

In (Dravskikh and Parijskij 1981) it is shown that in compact radio sources associated with galaxies and quasars, recombination lines can be detected in absorption. The continuum spectra of these compact sorces must have the turnover point ν_0 and this turnover must be caused by free-free absorption of the compact source's synchrotron radiation by the ionized gas surrounding the compact source.

The radiation from the ionized region surrounding the compact source can not be detected in the continuum at extralactic distances. But it is this ionized region that is responsible for the lines in absorption in the spectrum of the compact source. So our technique of measuring the electron density N_e in HII regions from observations of a pair of recombination α-lines can also be used for the ionized regions surrounding the extragalactic compact sources. This is very important when there are no another radio methods of estimation of the electron density in these regions.

4. APPENDIX. NORMALIZATION OF RADIO LINE PROFILES

Two α-radio lines one of which has a Doppler (Gauss form) profile and the other a Voigt profile can be normalized (areas under these line profiles reduced to 1) in the following way.

At frequencies where $\tau_c \ll 1$ in HII regions, the antenna temperature in the line ΔT_L is given by

$$\Delta T_L = \eta(\nu) T_e \tau_L, \qquad (1A)$$

where $\eta(\nu)$ is the beam efficiency which depends on frequency in a complicated way, T_e is the electron temperature of an HII region, τ_L is the optical depth and $\sim \nu^{-1}$. Thus

$$\Delta T_L = d\, \eta(\nu)\, \nu^{-1}, \qquad (2A)$$

where d is the coefficient of proportionality.

The width of the Doppler line $\Delta\nu$ is proportional to frequency, and the product is

$$(\Delta T_L \Delta\nu)_D = g\, \eta(\nu), \qquad (3A)$$

where g is the coefficient of proportionality.

It is known that the product $(\Delta T_L \Delta\nu)_{ND}$ does not depend on frequency in normalized Doppler profiles, and it must be described by

$$(\Delta T_L \Delta \nu)_{ND} = 2(\ln 2/\pi)^{1/2}. \tag{4A}$$

To extract the coefficient $\eta(\nu)$, which depends on frequency in an indefined way, from $(\Delta T_L \Delta \nu)_D$ we divide $(\Delta T_L \Delta \nu)_D$ by the continuum temperature T_c

$$T_c = \eta(\nu) \, T_e \, \tau_c, \tag{5A}$$

where τ_c is the optical depth in continuum, and $\tau_c \, \nu^{-2.1}$. Then

$$(\Delta T_L \Delta \nu)_D / T_c = k \, \nu^{2.1}, \tag{6A}$$

where k is coefficient of proportionality.

Dividing both parts of (6A) by $\nu^{2.1}$ and multiplying by some coefficient p, we can reduce the right-hand part to the value

$$k \, p = 2(\ln 2/\pi)^{1/2} = 0,94.$$

As a result we obtain the product of the peak intensity and line width for the normalized Doppler profile

$$(\Delta T_L \Delta \nu)_{ND} = p(\Delta T_L \Delta \nu)_D / T_c \nu^{2.1}. \tag{7A}$$

Thus the factor normalizing the Doppler line profiles is

$$R_D = p / T_c \nu^{2.1} \tag{8A}$$

or

$$R_D = 2(\ln 2/\pi)^{1/2} / (\Delta T_L \Delta \nu)_D. \tag{9A}$$

It is necessary to carry out an analogous normalization for a line with a Voigt profile. The normalizing factor will have the same form (8A) where the coefficient p is the same for both lines, and T_c and ν refer to the measurements of the Voigt line.

From (8A) and (9A)

$$p = 2(\ln 2/\pi)^{1/2} \, T_{cD} \nu^{2.1} / (\Delta T_L \Delta \nu)_D, \tag{10A}$$

$$R_V = 2(\ln 2/\pi)^{1/2} \, T_{cD} \nu_D^{2.1} / (\Delta T_L \Delta \nu)_D T_{cV} \nu_V^{2.1}. \tag{11A}$$

5. REFERENCES

Brown R.L., Lockman F.J., and Knapp G.R. (1978) 'Radio recombination lines', Ann.Rev.Astr.Ap. 16, 445-485.

Dravskikh Z.V., and Parijskij Yu.N. (1981) 'O recombinatsionnykh liniyakh vozbuzhdyonnogo vodoroda v kompaktnykh radio istochnikakh, svyazannykh s galaktikami i kvazarami', Astr.Zh. 58, vyp. 3, 486-489.

Finn C.D. and Mugglestone D. (1965) 'Tables of the line broadening function $H(a,v)$', M.N.R.A.S. 129, 221-235.

Lang K. (1978) Astrofizicheskie formuly, Mir, Moskva.

Mezger P.G. and Henderson A.P. (1967) 'Galactic HII regions. 1. Observations of their continuum radiation at the frequency 5 GHz' Ap.J. 147, 471-489.

Salem M. and Brocklechurst M. (1979) 'A table of departure coefficients from thermodynamic equilibrium (b_n factor) for hydrogenic ions', Ap.J.Suppl. 39, 633-651.

Smirnov G.T., Sorochenko R.L., and Pankonin V. 'Stark broadening in radio recombination lines towards the Orion nebula', Astr.Ap. 135, 116-121.

INTERFEROMETRIC OBSERVATIONS OF HII, CII, AND H⁰ REGIONS IN ORION B

K.R. Anantharamaiah[1], W.M. Goss[2], P.E. Dewdney[3]
[1] *Raman Research Institute, Bangalore 560 080, INDIA*
[2] *NRAO, Very Large Array, Socorro, NM 87801, USA*
[3] *DRAO, Penticton, B.C. CANADA V2A 6K3*

ABSTRACT. The spatial and velocity distribution of the narrow H166α and C166α lines in Orion B, observed with an angular resolution of $\sim 45''$, are found to be similar. This strongly suggests that the CII and H⁰ regions, which are adjacent to the HII region, are spatially coincident. The heavy element line (X166α) has a different distribution and its identification as due to sulfur, based on its velocity, is not certain. A new H166α feature is detected at a velocity of +48 km s^{-1} in the south-western portion of OrionB.

1 Introduction

The galactic HII region Orion B (also known as W12 or NGC2024) is a particularly interesting object as it exhibits a narrow ($\Delta V \sim 4$ km s^{-1}) hydrogen recombination line, in addition to the usual broader ($\Delta V \sim 30$ km s^{-1}) hydrogen line. Strong narrow carbon recombination lines are also observed from this source and another narrow recombination line due to a heavier element, possibly sulfur, has also been detected. The narrow lines are stronger at low frequencies, where they are dominated by stimulated emission (Pankonin *et al* 1977). While the broad hydrogen lines arise in the hot ($T_e \sim$ 8000–10000K) fully ionized HII region, the narrow lines of hydrogen, carbon and the heavy element (X for unknown) are believed to originate in adjoining partially ionized regions at much lower temperatures ($T_e \sim$ 150K). These regions are usually referred to as CII, H⁰, and XII regions. Zuckerman and Ball (1974) and Hill (1977) have proposed that the narrow hydrogen lines are formed in the outer regions of ionization fronts, where the temperatures are low. On the other hand, Pankonin *et al* (1977) and Krugel and Tenario-Tagle (1978) suggest that the narrow hydrogen line region may be formed as a result of ionization by soft X-rays from the stellar wind of the exciting star. In the last Recombination line meeting Pankonin (1979) had reviewed our knowledge about the partially ionized medium.

So far, recombination lines from OrionB have been observed only using single-dish telescopes (e.g. Ball *et al* 1970, MacLeod, Doherty, and Higgs 1975, Pankonin *et al* 1977, Krugel *et al* 1982). The spatial relationship between the HII, CII, H^0, and XII regions is not clearly known. In particular, since it is thought that the partially ionized regions are adjacent to the fully ionized HII region, it is desirable to find out whether the CII, XII and H^0 regions entirely coincide. The interferometric observations, presented here, were aimed at obtaining this information. Although Roelfsema, Goss, and Wilson (1987) and Roelfsema, Goss, and Geballe (1988, 1989) have observed the narrow lines from CII and H^0 regions using interferometers, so far the two regions have not been observed in the same source.

2 Observations and Data Reduction

Observations were made in the 20 cm band using the compact D-configuration of the VLA[1]. H166α (1424.734 MHz) and C166α (1425.444 MHz) were observed separately in order to obtain adequate velocity resolution. In both the observations, we used 64 spectral channels with a velocity resolution of 2.57 km s^{-1}. The H166α data were centered at a V_{lsr} of 9 km s^{-1} and the C166α data at 20 km s^{-1} and included the heavy element line and any possible helium line. The observations were made in a single 12 hour session and the two data sets were taken alternately (changing every 30 minutes) in order to get similar UV coverage. Amplitude and phase were calibrated using frequent observations of 3C138. The frequency response was corrected using observations of 3C147 made in the begining and end of the session.

For both the transitions, 64 line images were formed by Fourier transforming the visibility data corresponding to each channel. Continuum emission was removed by subtracting, from each image, an average of the outer channel-images where no line was expected. Where necessary, the line images were deconvolved using the standard 'clean' method. The 64 line images were put into a data cube, thus forming two data cubes, one for the hydrogen lines and the other for the carbon and the heavy element lines. The angular resolutions were 47" × 46" for the hydrogen lines and 52" × 47" for the C and X lines. A separate continuum image was formed using the average of all the visibilities of the central three-quarters of the two observed bands.

3 Data Analysis and Results

The continuum image is shown in Figure 1. The sharp rise in the intensity in the southern edge of the source is believed to be due the ionization front interacting with

---

[1] VLA is a part of the National Radio Astronomy Observatory, which is operated by the Associated universities Inc., under a co-operative agreement with the NSF.

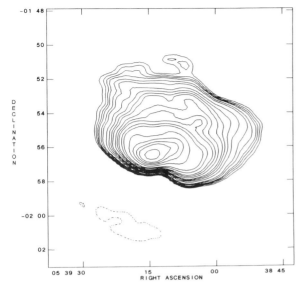

Figure 1: Continuum emission of OrionB at 1424.5 MHz. Contour levels (-3, 3, 4, 5, 6, 8, 10, 15, 20, 25, 30, 40, 50, 60, 80, 100, 150, 200, 250, 300, 400) × 10 mJy/beam. Angular resolution is $40.8'' \times 38.9''$.

the dense outer molecular cloud. Higher angluar resolution continuum observations of this source near 1667 MHz have been recently discussed by Barnes *et al* (1989).

The continuum image was used as a template to look for line emission in the two data cubes. The broad H166α line from the HII region was found over more than half of the continuum region in Figure 1. The peak of the line is shifted north-west from the continuum peak. The narrow H166α and C166α lines were found over a much smaller region centered on the continuum peak. The X166α line was also found, but with a different distribution compared to the other narrow lines. No helium line was detected.

In order to obtain the correct spatial distribution of all the components, further analysis of the data cubes were carried out using the Groningen Image Processing System (GIPSY). The data were analysed pixel by pixel by fitting an appropriate number (ranging from 1 to 3) of gaussian components.

In Figure 2, we show a profile made by averaging over the reigon where the narrow hydrogen line was found. The broad H166α line seen in Fig 2, shows an assymetry towards positive velocities. This assymetry becomes even more pronounced in profiles made over south-western regions of the source, and develops into a separate component with a velocity of ~ 48 km s^{-1}. In these regions, we fitted three gaussian components to the observed profiles. In Figure 3, we show the average profile of the third component, made after subtracting the narrow and broad H166α lines. We believe that this component is real, as it persisted over several contiguous pixels. In the central regions, two components (narrow and broad H166α) were fitted and in the northern and eastern regions only one component (broad H166α) was fitted.

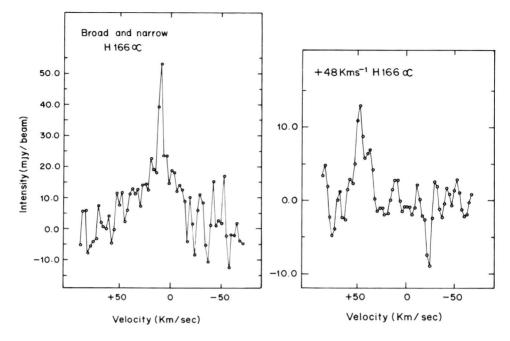

Figure 2: H166α recombination line averaged over the central 2′ × 1′ region. The profile shows the narrow and the braod lines from the H⁰ and HII regions respectively.

Figure 3: The +48 km s^{-1} feature found in the south-west region. The profile was obtained after subtracting the narrow and broad H166α lines over the same region.

The spatial distribution of each of the three components were obtained by subtracting from the original data cube, the sum of the other two gaussin components from each pixel.

In Figure 4, we show a profile made over the central region of the C166α data. The strong narrow line is the C166α line and the weaker narrow line at lower velocity is the heavy element line, which is designated as X166α . In many pixels, the X166α line was found to be much stronger than shown in Figure 4. One or two gaussian components were fitted, as warranted by the data, to all the pixels where line emission was found. The width of both the components were assumed and fixed at 4 km s^{-1} during the fit. The spatial distribution of the two components were obtained by subtracting one or the other fitted component from the original data cube. Repraesentative parameters of the gaussian components from the two data cubes are given in Table 1.

In all, we obtained five components of emission from the gaussian fits. These are the broad H166α line from the HII region, the narrow H166α , C166α , and

X166α lines from the partially ionized regions, and a new H166α component near $V_{lsr} = +48$ km s^{-1}. The spatial distribution of these components, together with that of the continuum, are shown in Figure 5.

Table 1: Represantative line parameters for OrionB

Line	Position+ ra dec	ΔS_L (mJy)	$\frac{\Delta S_L}{S_C}$ (10^{-3})	V_{lsr} (kms^{-1})	ΔV (kms^{-1})
H166α(HII)	+30, −15	37.0±3.0	5.7±0.5	7.8±1.5	47.0±3.8
H166α(H^0)	+30, −30	73.8±12.3	11.4±1.9	9.7±0.3	3.4±0.8
C166α	+30, −30	82.3±8.0	12.8±1.2	9.8±0.3	4.0†
	+45, 00	56.7±7.2	10.3±1.3	9.0±0.3	4.0†
X166α	+30, −30	31.0±8.0	4.8±1.2	−0.4±0.7††	4.0†
	+45, 00	36.1±7.3	6.55±1.3	0.7±0.4††	4.0†
H166α (+48 km s^{-1})	−75, 00	12.7±4.0	4.3±1.4	44.7±2.7	16.0±6.0

+ Offset in arcsec from postion $05^h39^m13^s, -01°56'04''$. † Held fixed during gaussian fit. †† With respect to carbon line.

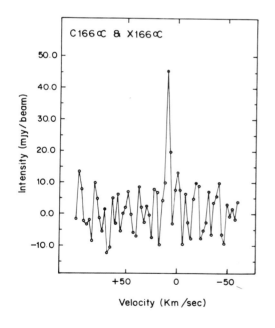

Figure 4: C166α and X166α lines obtained by averaging over the central $2' \times 1'$ region.

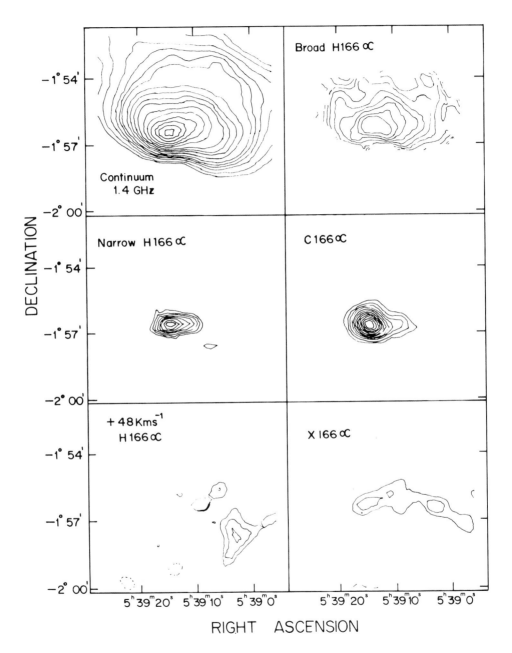

Figure 5: Spatial distribution of continuum and different line components in Orion B. *Top left:* Continuum. Contour levels: 0.07, 0.14, 0.28, 0.42,...*contd.*

.... 0.56, 0.7, 1.05, 1.4, 1.75, 2.1, 2.8, 3.5, 4.2, 4.9, 5.6, 6.3 Jy/beam. *Top right:* Velocity integrated intensity. Contours: 17, 34, 68, 102, 136, 170, 255, 340, 425, 510, 680, 850, 1020, 1190, 1360, 1700 mJy/beam km s^{-1}. *Middle left:* Intensity at V_{lsr} = 9 km s^{-1}. Contours: 20 to 90 in steps of 5 mJy/beam. *Middle right:* Intensity at V_{lsr} = 10.2 km s^{-1}. Contours: 20 to 100 in steps of 5 mJy/beam. *Bottom left:* Intensity at V_{lsr} = 47.5 km s^{-1}. Contours: 15, 20, 25 mJy/beam. *Bottom right:* Intensity at V_{lsr} = -0.6 km s^{-1} with respect to the C166α. Contours: 15, 20 mJy/beam.

4 Discussion

The distribution of the broad H166α line is shown in the top right panel of Figure 5. The peak of the line emission is displaced from the continuum peak, which may be attributed to either variation in the electron temperature, or the line becoming weaker near the peak due to increased continuum optical depth. The line to continuum ratio varies across the source, which is consistent with a gradient in electron temperature as obtained by MacLeod *et al* (1975) and Krugel *et al* (1982) using higher frequency lines. The temperatures derived from the present data are not reliable due to increased continuum opacity at 20 cm. The central velocity of the line varies from about 2–5 km s^{-1} in the north-west portion of the source to about 12–15 km s^{-1} in the south-east portion, consistent with the observations of MacLeod *et al* (1975) and Krugel *et al* (1982). The width (FWHM) of the lines are 25–35 km s^{-1} over most of the line emitting region, except near the sharp southern edge (Fig 1), where they are 40–55 km s^{-1}. Pressure broadening can account for the increased line widths if the electron density in this region is $2 - 3 \times 10^3$ cm^{-3}.

The narrow H166α and C166α lines are both centered on the continuum peak, confirming the importance of stimulated emission at this frequency (Pankonin *et al* 1977). These observations have essentially detected the partially ionized medium which is in front of the HII reigon. The H^0 region is slightly smaller than the CII region, but they overlap in the centre. The difference in their size is partly due to the difficulty in separating the broad and narrow H166α lines when the narrow line becomes weaker. The central velocity of the two lines (see Table 1) are nearly identical in the overlapping region and within the errors the width of the two lines (3.5–4 km s^{-1}) are also identical. The CII region shows a variation in velocity from about 8 km s^{-1} in the west to about 11 km s^{-1} in the east. The same trend also appears in the narrow H166α lines. These data, therefore, strongly suggest that the CII and H^0 regions are spatially coincident. The possible difference in their extent is perhaps due to the difference in their ionization potential.

The distribution of the narrow X166α line, shown in the bottom right panel of Figure 5, is distinctly different from that of the CII and H^0 regions. The velocities are also different. From Table 1, the velocity difference between the C166α and X166α lines are 10.2 ± 0.8 km s^{-1} at one pixel and 8.3 ± 0.8 in another pixel.

This recombination line has been identified as due to possibly silicon (MacLeod et al 1975), or sulfur (Pankonin et al 1977), or a blend of several elements heavier than carbon. Although Pankonin et al (1977) favour sulfur, based on the velocity difference of the X166α line with respect to the C166α line, the different spatial distribution observed here coupled with the different velocities of the two regions show that the identification is not certain. Other arguments based on the strength and distribution of the line region should be used for a proper identification.

The H166α line near +48 km s^{-1} (Fig 3) has not been detected in any earlier single-dish observations. Since the line is weak (peak $\sim$ 13 mJy) and the region is small ($\sim$ 2′), this line would not be seen in the 166α single-dish spectrum obtained by Pankonin et al (1977) using a beam of 8′.7. The velocity of this feature is unusual and suggests that ionized material is streaming out on the far side of the HII region at a velocity of $\sim$40 km s^{-1}. The spatial distribution of this component (bottom left panel of Fig 5) is unlike all the other components discussed above. In fact, in the south-western region, there is no spatial overlap with any other components, except the continuum. An examination of the profiles at different pixels shows that the line becomes narrow ($\sim$5 km s^{-1}) towards the south-western edge of the source where there is no other line emission. It is therefore tempting to associate this line with a partially ionized medium which is seen on one side of the source, but then it is difficult to explain the large velocity of this feature.

Acknowledgements. We are grateful to P.R. Roelfsema for advice on the use of GIPSY.

REFERENCES

Ball, J.A., Cesarsky, D., Dupree, A.K., Goldberg, L., Lilley, A.E., 1970, *Astrophys. J. Lett*, **162**, L25
Barnes, P.J., Crutcher, M., Bieging, J.H., Storey, J.W.V., Willner, S.P., 1989, *Astrophys. J.*, **342**, 883.
Hill, J.K., 1977, *Astrophys. J.*, **212**, 692.
Krugel, E., Thum, C., Martin-Pintado, J., Pankonin, V., 1982 *Astron. Astrophys. Suppl. Ser*, **48**, 345.
Krugel, E. and Tenorio-Tagle, G., 1978, *Astron. Astrophys.*, **70**, 51.
MacLeod, J.M., Doherty, L.H., Higgs, L.A., 1975, *Astron. Astrophys.*, **42**, 195.
Pankonin, V., Walmsley, C.M., Wilson, T.L., Thomosson, P., 1977, *Astron. Astrophys.*, **57**, 341.
Pankonin, V., 1979 in *Radio Recombinatin Lines*, ed. P.A. Shaver, D Reidel (Dodrecht).
Roelfsema, P.R., Goss, W.M., Wilson, T.L., 1987, *Astron. Astrophys.*, **174**, 232
Roelfsema, P.R., Goss, W.M., Geballe, T.R., 1988, *Astron. Astrophys.*, **207**, 132
——————, 1989, *Astron. Astrophys.*, **222**, 247.
Zuckerman, B. and Ball, J.A., 1974, *Astrophys. J.*, **190**, 35.

THE HELIUM ABUNDANCE IN THE HII REGION DR21

A. P. TSIVILEV
P. N. Lebedev Physical Institute
Academy of Sciences of the USSR
Leninsky Prospect 53 Moscow 117924 USSR

ABSTRACT. The 36.5-GHz (H,He)56α and 22.4-GHz (H.HE)66α radio recombination lines have been observed in the DR21 with RT22(FIAN). These data combined with the other show that the measured relative abundance y^+ of ionized helium appears to increase with the angular beamwidth, φ_A, peaking at $\varphi_A = 2'-2.5'$, and then to decrease as φ_A further increases what is more complicated than in previous papers. This behaviour is interpreted as a blister-type model nebula which implies the high actual helium abundance $y_0 = (26\pm11)\%$ in the DR21 and its expansion rate from 5km s^{-1} to 20km s^{-1} by approaching the observed increasing of the hydrogen radial velocity with the frequency decreasing. Discussion has concerned slightly the explanation possibilities of the high helium abundance too.

1. UNTRODUCTION

The determination of the relative helium abundance $y_0 = N(He)/N(H)$ in cosmic nebulae allows us to obtain information about Big Bang element synthesis and chemical evolution of Galaxy. One of the reliable way is measuring by radio recombination lines, RRL, the ratio of the integral line intensities, $y^+ = \int T_L^{HE} d\nu / \int T_L^H d\nu$. At present time it is clear that the transformation from the apparent y^+ to the actual y_0 is not simple and it requires an individual approach to each source using most of the available data often combined with a model calculation.

DR21 is one of the most studied galactic nebulae. But its apparent abundance y^+ varies from < 3% to 11.6% by number in the single-dish observations (Churchwell et al. 1978, Lockman and Brown 1982), and from 3.5% to 9.5% in VLA high angular resolution observations (Roelfsema 1987). To explain this previous discussions considered simple geometric effect (e.g., Pankonin 1977, Pitault 1980). To obtain the additional information on the large scale ionization structure in DR21 the RRL observations of (H,HE)56α and (H,He)66α pairs are made which are thought not to be strongly affected by non-LTE conditions and by Stark broadening. These new data combined with the other and the model computation allow to employ a new interpretation way and to assume the

131

high helium abundance ~26% in DR21.

2. OBSERVATIONS AND RESULTS

DR21 was observed with the RT22(FIAN) radio telescope at the Pushchino station of the Lebedev Physical Institute on several different occasions from December 1986 to April 1987 (H,HE56α pair with 1.9' angular resolution) and from Junaury to April 1989 (H,He66α with 2.6' resolution). The 64-channel filtertype spectrum analyzer with the 500 kHz resolution was used at 36.5 GHz frequency and the 128-channel one with the 125 kHz resolution was used at 22.4 GHz. The source position was $\alpha = 20^h 37^m 14^s$, $\delta = 42°9'5"$. The receiver parameters and the data processing method were the same as in the previous paper (Tsivilev et al.1986) but taking into account that now the hydrogen and helium lines were observed simultaneously. Table 1 contains the obtained lines parameters where the ±1σ errors are parenthesized. The line profiles are shown in Fig.1. It is strange that the helium RRL was not detected at ~5 GHz by single-dish

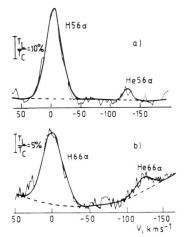

Figure 1. Profiles of the (H,HE)56α lines (a) and the (H,He) 66α lines (b) observed in the DR21. Smooth curves are the best-fit Gaussian profiles, the baselines are dashed.

observations because it has been detected at frequncies somewhat lower and somewhat higher (Vallee 1987). The analysis of the H110α spectra obtained with the Max Planck 100m telescope (Smirnov 1985) shows the peculiarity at the radial velocity, V_{LSR}, which corresponds to the helium position. Two profiles are fitted to this spectrum instead of the single in previous case that allows possibly to detect the He110α line. This result is listed in Table 1 also. The obtained unlikely large width of He110α line is probably due to that the reached sensitivity is not enough to separate the C110α profile from the He110α. The analysis of He,C100α and He,C125α spectra (Vallee 1987) shows the integral carbon intensity is about ~50-70% of the helium one, whence the obtained y^+ ~15% at 5 GHz is reduced to ~9%. Then the available DR21 data are selected in such a way that if y^+ measurement at the same frequecy with the same radio telescope was not single then the newest data which are also listed in Table 1 were taken.

TABLE 1. Hydrogen and helium RRLs in DR21.

Lines	Beam-width(')	T_L/T_C %	ΔV km s^{-1}	V_{LSR} km s^{-1}	y^+ %	Ref.
H56α	1.9	46.0(0.4)	33.6(0.8)	-1.9(0.2)*		
He56α		5.6(0.7)	21.0(3.0)	-1.1(1.0)	7.6(1.5)	1
H66α	2.6	19.2(0.4)	31.4(0.8)	1.4(0.3)		
He66α		3.5(0.4)	25.8(4.1)	-1.9(1.5)	13.8(2.8)	1
H66α	0.7	18.0(2.0)	33.5(0.2)	-0.7(0.1)		
He66α		1.0(0.1)	26.3(3.5)	-0.3(1.3)	4.5(1.0)	2
H76α	0.9	12.0(0.5)	33.7(0.2)	-0.7(0.1)		
He76α		0.68(.05)	25.1(2.5)	0.7(0.8)	4.2(1.5)	2
H76α	2.2	10.4(1.0)	34.6(0.2)	-1.0(.07)		
He76α		1.1(0.2)	38.3(1.5)	-1.5(0.6)	11.6(1.4)	3
H85α	3.2	8.4(0.8)	34.4(0.2)	-0.3(0.1)		
He85α		0.8(0.1)	24.4(2.7)	0.9(1.1)	7.1(0.5)	3
H85α	3.0	6.5(0.7)	32.0(2.0)	-4.6(1.0)		
He85α		0.4(0.2)	28.0(4.0)	-2.6(2.0)	4.9(2.8)	4
H86α	3.5	6.9(0.5)	37.9(0.6)	0.4(0.2)		
He86α		0.6(0.1)	20.0(5.0)	-3.8(1.9)	4.6(1.1)	5
H76α ***	0.3	11.9(0.6)	32.2(0.3)	-1.4(0.4)		
He76α		0.6(.09)	22.1(4.0)	2.4(1.7)	3.6(0.5)	6
H110α	0.3	1.9(.08)	42.7(1.1)	-0.3(0.6)		
He110α		0.14(0.05)	36.0(12.)	-2.0(5.0)	5.1(1.4)	6

		T_L, ^{0}K				
H90α	1.5	1.19(0.11)	38.6(0.6)	3.0(0.8)		
He90α		0.11(0.01)	27.7(2.4)	-0.2(0.9)	6.6(1.0)	7
H110α	2.6	0.99(0.02)	44.3(1.3)**	- - -		
He110α+ +C110α		0.09(0.02)	73.4(15.3)	-6.0(6.0)	9.0(5.0)	1

*-The $V_{LSR} \approx -0.5$ km s^{-1} if the blended line H80γ will be accounted for. **-This line has been fitted by Voigt, here ΔV_L is total half-power width and Lorenz width is $\Delta V=14.1(3.5)$. ***-Used data which are synthesised by Roelfsema over core region of 20". References: 1-this paper, 2-Thum et al.(1980), 3-Lockman and Brown (1982), 4- Churchwell et al. (1974), 5-Lichten et al. (1979), 6 -Roelfsema (1987), 7-Smirnov(1985).

The main result is that the y^+ appears to increase as the half-power angular beamwidth, φ_A, increases, peaking at φ_A =2'-2.5', and then to decrease as the φ_A further increases (Fig.2), namely: y^+=(4±0.4)% for φ_A<1', y^+=(9.7±1)% for φ_A=1.5' - 2.6' and y^+=(6.8±0.5)% for φ_A 3-3.5'. All this is turned out to be more complicated than in the previous papers (e.g., Pankonin et al.1977, Pitault 1980). It is intresting also that if one takes only the single-dish data where the main HII region RRL profiles (at n ~90-110) are separated from the blended HI region RRL ($V_{LSR} \approx -8$ km s^{-1}), then the hydrogen V_{LSR} are seen clearly to increase as the principal quantum number, n, increases (Fig.4). Possibly the heluim V_{LSR} have the similar behaviour (Vallee 1987).

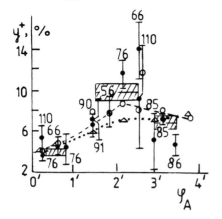

Figure 2. Relative helium abundance y^+ observed in DR21 as a function of angular beamwidth, φ_A. The 91α data has been taken from Pankonin et al. (1977). Vertical bars mark $\pm 1\sigma$ errors. Figures indicate the principal quantum number, n, of RRLs transions. Open circles are the model computed values for the corresponding angular resolutions and transitions, for n $\leq$ 76 which are connected by dashed curve. Three versions of computations agree within 10%. Shaded regions are the average y^+ values over the corresponding interval of φ_A. For comparison the simple case ($R_h = 1$, standard $y_0 = 12\%$, version D) is shown as triangles connected by dotted curve.

3. DISCUSSION

3.1. MODEL AND DR21 STRUCTURE. Pankonin et al.(1977), Smirnov(1985) showed that DR21 was described by the electron density, N_e, distribution of a core-halo type. Roelfsema (1987) pointed out that DR21 had the blister-type structure and this HII region was located on the far side of the molecular cloud. In such a scheem the proposition can be made in a natural way that into the cloud (region core) all quanta which ionize helium and hydrogen atoms (N_c^{HE} and N_c^H) are absorbed, i.e. here the HII region is limited by ionization, then as for the outer cloud's region (halo) not all such quanta are absorbed, i.e. the one is limited by density here. In this case the ratio between the HII and HeII region sizes in the core can strongly differ in a natural way from that in the halo so that $y_h^+ > y_c^+$ if they are observed separately. Really the core and the halo are observed by a single-dish simultaneously. So the apparent y^+ depends on the sum of contributions of the core and the halo where each is proportional to the part of its flux density filled the beam, hence y^+ depends on the observation angular resolution. For large beamwidths the equation can be written as the first approximation:

$$y^+ \approx S_c/S \times y_c^+ + R_h \times y_0 \times S_h/S \qquad (1)$$

or

$$y_0 \approx 1/R_h \times \{(y^+ - S_c/S \times y_c^+)/(1 - S_c/S)\} \; ; \; R_h = S_h^{He}/S_h$$

where S is the full flux density, S_c - the core flux density, S_h - that of the halo, S_h^{He} - that of the halo HeII region. Assuming $R_h = 1$ it can be shown by the equation (1) that the y_0 uncertainty generally is as:

$$\sigma(y_0)/y_0 \approx \{(y^+ - y_c^+)/(y^+ - y_c^+ \times S_c/S)\} \times S/S_h \times \sigma(S_c/S) \quad (2)$$

Both the y_0 and $\sigma(y_0)$ values clearly depend on the S_c/S ratio, hence it is necessary to evaluate it. Practically all high angular resolution observations agree in that the DR21 core consists of some clumps placed over 30"-40" of its central region. At 5 GHz Harris (1973) estimated $S_c \approx$ 13.1 Jy and Dickel et al.(1983) obtained S_c <15.5 Jy for the structure of 30" size. The full S at 5 GHz is about ~19-22 Jy (Goudis 1976, Churchwell et al .1978). At~5 GHz Dickel et al.(1986) obtained $S_c \approx$13.8 Jy for the structure of ~20" size, Roelfsema (1987) found the sum flux density of the clumps S_c = 16.2 Jy, whereas according to Dent (1972) the full S is expected to be of ~20 Jy at this frequency. So at ~5 GHz the S_c/S ratio is expected to be in ~0.59-0.82 interval, at ~15 GHz - in ~0.68-0. Guilloteau et al.(1983) has obtained the S_c/S =0.61 at~24 GHz. Therefore the DR21 S_c/S value is probably in ~0.6-0.8 interval. Fig.2 shows the y'_+ observed with φ_A >1' group to about ~7-10% interval. The decreasing y^+ for φ_A > 2.5' (Fig.2) indicates that for our scheem in the halo the HeII region is also smaller than the HII region, i.e., $R_h \leq 1$. Really the S_c/S and R_h values depend on the beamwidth of observations and it is necessary to account for the effects of optical depth, non-LTE and so on. To account for it and as the next approximation step the model of a blister-type has been computed. One has the core-halo N_e distribution based on Smirnov(1985) model which well describes the RRL widths and the S frequency chart. Fig.3 shows the model scheem and Table 2 lists three versions (B,C and D) of satisfactory model parameters. The calculations with no maser effect but with the departure coefficients

TABLE 2. The model parameters of DR21. D= 2 kpc Vt = 23 km s^{-1}

No layer	T_e, K	N_e, cm^{-3}	R, pc	f	Vex, km s^{-1}	N(He$^+$)/N(H$^+$), %	
-3	6000	270	**	1	18	0	
-2	7000	700	0.64	1	5	22	B
-1	8200	4.2×10^4	0.077	0.095	0	22	
1	8200	4.2×10^4	0.077	0.365	0	0	
-3	7000	300	**	1	20	0	
-2	7000	540	0.64	1	6	34	C
-1	8200	4.2×10^4	0.077	0.067	0	34	
1	8200	4.2×10^4	0.077	0.393	0	0	
-3	6000	235	**	1	18	0	
-2	7000	735	0.64	1	5	18	D
-1	8200	4.2×10^4	0.077	0.094	0	18	
1	8200	4.2×10^4	0.077	0.285	0	0	

f is the filling factor=$(N_{e,RMS}/N_{e,loc})^2$. Vt-turbulent velocity. Vex-expansion rate. **-The last layer has the form of cylinder with the base radius of

1.1 pc and height of 0.64 pc. The 22,34,18% are the Version's y_0. B,C,D indicate corresponding versions.

for actual atomic level population are found to agree much better with the observational integral contrast of RRLs and only they are presented here. The D,B,C versions have the following S_c/S and R_h values: 0.6 and 0.77; 0.65 and 0.71; 0.71 and 0.55, respectively. One can see that the

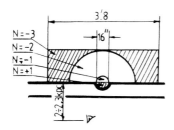

Figure 3. Model structure of the DR21 HII region. ■ - boundary of the molecular cloud; ▨ - region of ionized hydrogen and neutral helium; ☐ - region of both ionized hydrogen and helium; ▽ - observer position; No - number of hemisphere layer according to Table 2.

model y_0 increases strongly with the S_c/S increase whereas the R_h reduces slightly with that. The angular size of 2.2' can be accepted as the HeII region size in the halo. Then using the low frequency (~0.4 GHz) map of DR21 (Ryle and Downs 1967) which might be the most representative of the halo structure one can show that the $R_h \approx 0.55$ is the low limit on the R_h. Consequently, the corresponding $y_0 = 34\%$ from C version is the upper limit for y_0. The $y_0 = 18\%$ is the low value according to the D version whose $S_c/S = 0.6$ is the low value of S_c/S interval. So the DR21 model heluim abundance will be $y_0 = (26\pm11)\%$ as a middle value with uncertainty obtained on the basis of (1,2) equations and using the found S_c/S and R_h intervals. Today such helium abundance is not unusually high since for example Roelfsema (1987) has measured $y^+ \geq 20\%$ in the HII region W3.

As for the DR21 core Harris (1973), for example, has found that it consists of at least four clumps with sizes of ~4"- 7". It has been assumed that the ionization of the clumps can come from outer region and hence HeII regions are placed on the clumps periphery. It can explain why the He line emission maximum do not coincide with that of the H line and why y^+ varies over the DR21 core in VLA observations (Roelfsema 1987). The reason why Rolfsema does not observe a high y^+ in the DR21 core can be following: their angular resolution (~4"×4") is not enough to resolve the HeII clumps which are expected to be of ~(4"-7"), $xy^+_c/y_0 \approx 0.6"-1"$ sizes. Otherwise it is probably enough for the W3 observation (Roelfsema 1987). To prove the presented model it is necessary to map the DR21 core with $\varphi < 1"$ and to measure the y^+ in some positions shifted at ~50" from the DR21 center with $\varphi_A \sim 40"$ (for example with Max Planck 100m telescope) where the y^+ is expected to be of ~20% (B version), ~29% (C) and ~17% (D).

The important evidence of a blister-type structure is a large-scale mass motions within the nebula. The obtained dependence of the hydrogen lines V_{LSR} versus n (Fig.4) is probably due to such motions (DR21 is expanding in the direction from the observer). The model based on this dependence expects the HII region expansion rate from ≈ 5 km s^{-1} to ≈ 20

km s^{-1} correspondingly to the radial distance from the center. Whence one can evaluate the minimum age of this HII region ~1.5×10^5 years. By assuming that the ionizing quanta, N_c^{He} and N_c^H, are absorbed completely into the molecular cloud (core) one can evaluate the effective temperature of ionizing source $T_{eff} \approx 35800^0$K using Copetti and Bica (1983) calculation.

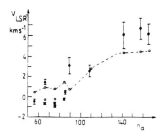

Figure 4. The V_{LSR} dependence of hydrogen RRLs on the principal quantum number, n. The H109α data are taken from Smirnov (1985); the H158α and H166α data, from Pankonin et al. (1977); the H142α data, from Barcia et al. (1985). Open circles are the model values for the corresponding resolution and frequency, which for φ_A >1' are connected by dashed curve. It is accepted that the center of the expanding has V_{LSR} = -0.5 km s^{-1}.

Since the blister-type structure probably can be common among the HII regions the presented y^+ interpretation way can be used more than once. For example in Orion A it was found that the apparent y^+ increases with the angular distance from the nebula center (Tsivilev et al.1986, see their Fig.2), authors considered that the blister structure was more likely for the y^+ data and had implicitly used only the assumption that the ratio of HeII and HII region sizes at the Orion center differed from that in the Orion outer part. It is intresting that Roeifsema et al.(1987) observe with the VLA the similar y^+ behaviour in the Sgr B2 component 3 (see their Fig.4) which can be explained in the similar way too.

3.2. EXPLANATION POSSIBILITIES OF HIGH HELIUM ABUDANCE. The ways where high helium abundance is a local peculiarity of the galactic objects is discussed in detail by Roelfsema (1987). They are: at first, the stellar mass loss or ejected shell, which both are enriched with the helium mass during the stars evolution, at second, the special radiation field leads to an underionization of hydrogen in comparison to helium which itself has really the standard abundance. If the local high abundance is real, then one can add that the galactic HII regions can be of two types: with high $y_0 \geq$ 20% as W3 (Roelfsema 1987) and DR21 (this work) or with standard y_0 ~10-12 % as Orion A (Tsivilev et al.1986) and Sgr B2 (Roelfsema et al.1987).

Nevertheless there is some sense to discuss the helium enrichment as a possible global cosmic process. As for a Big Bang nucleosynthesis Zel'dovich and Novikov (1975) has outlined the standard model primordial nucleosynthesis can not give the helium abundance by mass, Y_p, larger than 33%. The helium abundance obtained here will be Y= (51±11)% by mass. If the galactic evolution enrichment Y_g is assumed to be of ~3% then remaining value of ~ 48% is still larger than the maximum of standard model Y_p. Consequently, differences will be probable from the standard primordial nucleosynthesis. According to cosmology theories (e g., Zel'dovich and Novikov 1975) the primordial helium producing is determined mostly by producing the neutron to proton ratio, n/p. There is a large number of papers and reviews concerning various types of such differences which can lead to increasing of n/p ratio and hence of Y_p (e.g., Zel'dovich and Novikov 1975).

Recently Matsumoto et al. (1988) has discovered that the cosmic microwave background radiation has the increment over the blackbody spectrum at the Wien part. To explain this Yoshioka and Ikeuchi (1987) has examined the inverse Compton hypothesis based upon the explosion sceanario for a giant pregalactic objects which are assumed to be Population III. They propose the total necessitated energies should be of the order of $\varepsilon_p(z) \approx 10^{-14} \times (1+z)^4$ ergs cm^{-3}, where z is the cosmological redshift. Let us assume hypothetically that the origin of such energies is the thermonuclear reactions inside the Population III objects leading to the helium mass production, whose abundance can be evaluated, using present paper, as: $\Delta Y = Y - Y_p - Y_g \approx 18\%$. The binding energy of helium nucleus is about $\sim 3 \times 10^{-5}$ ergs, so the total energy produced in this way will be of $\varepsilon_y \approx 0.4 \times 10^{-10} \times \Delta Y \times \Omega_b \times (1+z)^3$ ergs cm^{-3}, where Ω_b is the ratio of the barion density to critical density which is taken to be of 10^{-29} g cm^{-3}. The $\varepsilon_y / \varepsilon_p \approx 4 \times 10^{-3} \times \Delta Y \times \Omega_b / (1+z)$ ratio will become of about 1 at z ~ 45 and $\Omega_b \approx 0.06$ what are not strongly in disagreement with Yoshioka and Ikeuchi (1987) conclusions. Is it a casual nature that the energy of the possible thermonuclear ΔY burning at some z is comparetive with the observed microwave background increment?

To select these explanation possibilities further investigations are required as, for example, the search for a heavier elements abundance distribution within the nebulae which should be correlated with the observed helium distribution, or a theoretical search for the possibility of such radiation field where N_c^{He} are overabundant in comparison with N_c^H, and so on. In principle, the presented explanation scheem does not contradict all the possibilities described here because the equation (1) can include only the apparent values (y^+, y_c^+, R_h, y_o), not actual. But if the helium is intermixed evenly within a nebula then this scheem allows to evaluate the actual heluim abundance.

The author is very grateful to R.L.Sorochenko and G.T.Smirnov for the fruitful discussions and to Z.P. Abaeva for the manuscript handling.

REFERENCES

Barcia A., Gomez-Gonzalez J., Lockman F.L., and Planesas P. (1985) 'Radio recombination lines from the galactic plane in Cygnus' Astr.Ap.147,237-240.

Churchwell E. and Mezger P.G., Huchtmeier W.(1974) 'Helium abundance in galactic HII regions', Astr.Ap.32,283-308.

Churchwell E., Smith L.F., Mathis J., Mezger P.G., and Huchtmeier W. (1978) 'Gradient of HII region electron temperatures and helium abundance in the Galaxy', Astr.Ap. 70, 719-732.

Copetti M.V.F. and Bica L.D.(1983)'The volumes of He$^+$ and H$^+$ in HII regions: the Orion Nebula', Ap.Space Sci.91,381-394.

Dent W.A.(1972) 'A flux-density scale for microwave frequecies'

Ap.J.177,93-99.
Dickel H.R., Goss W.M.,Rots A.H., and Blount H.M. (1986) 'VLA observations of the 6 cm and 2 cm lines of H_2CO in the direction of DR21', Astr.Ap. 162,221-231.
Dickel H.R.,Lubenow A.F.,Goss W.M.,Forster J.R.,and Rots A.H.(1983) 'VLA observations of H_2CO in DR21', Astr.Ap.120,74-84.
Goudis C.(1976) 'The radio and infrared spectrum of DR21',Ap.Space Sci.39,L1-L6.
Guilloteau S.,Wilson T.L.,Martin R.N.,Batrla W.,and Pauls T.A.(1983) 'Ammonia toward DR21:a weak maser in ortho-NH_3?', Astr.Ap. 124, 322-325.
Harris S.(1973)'5 GHz observations of small-scale structure in DR21', M.N.R.A.S.162,5p-10p.
Lichten S.M.,Rodrigues L.F.,and Chaisson E.J.(1979) 'A hydrogen and helium radio recombination-line survey of galactic HII regions at 10 GHz', Ap.J.229,524-532.
Lockman F.J. and Brown R.L.(1982) 'A survey of ionized helium in galactic HII regions', Ap.J.259,595-606.
Matsumoto T.,Hayakawa S.,Matsuo H.,Murakami H.,and Sato S.(1988) 'The submillimeter spectrum of the cosmic background radiation', Ap.J.329,567-571.
Pankonin V.,Thomasson P.,and Barsuhn J.(1977) 'A survey of radio recombination lines from HI regions and associated HII regions', Astr.Ap.54,335-344.
Pitault A.(1980) 'The "helium problem" in the source DR21', Astr.Ap.91,374-375.
Roelfsema P.R.(1987) 'Radio recombination line observations of HII regions and planetary nebulae', Ph.D. thesis. University of Groningen.
Roelfsema P.R.,Goss W.M.,Whiteoak J.B.,Gardner F.F.,and Pankonin V. (1987) 'VLA hydrogen and helium 76α radio recombination line observations of Sgr B2', Astr.Ap.175,219-230.
Ryle M. and Downes D.(1967) 'High-resolution radio observations of an intense HII region in Cygnus X', Ap.J.148,L17-L21(345-349).
Smirnov G.T.(1985) 'Electron density in the HII regions DR21 and W3 from Stark broadened radio recombination lines ', Sov.Astron.Lett.11(1), 7-11.
Thum C.,Mezger P.G.,and Pankonin V.(1980) 'The helium abundance of galactic HII regions', Astr.Ap.87,269-275.
Tsivilev A.P.,Ershov A.A.,Smirnov G.T.,and Sorochenko R.L.(1986)'The Orion A helium abundance', Sov.Astron.Lett.12,355-359.
Vallee J.P.(1987) 'The neutral interface adjoining the HII region DR21', Ap.Space Sci.129,339-351.
Yoshioka S. and Ikeuchi S.(1987) 'On the origin of possible deviation from the blackbody spectrum of the cosmic microwave background radiation. Inverse Compton hypothesis based upon the explosion scenario' , Ap.J. 323, L7-L11.
Zel'dovich J.B.,Novikov I.D.(1975) 'The structure and evolution of Universe', "Science" Publisher,Moscow.

Garay (left) and Cersosimo (right) discussing the importance of southern hemisphere astronomy.

RADIO RECOMBINATION LINES FROM STELLAR ENVELOPES:

PLANETARY NEBULAE

Yervant Terzian
Cornell University
NAIC
Ithaca, New York
USA

1. INTRODUCTION

We think we understand that planetary nebulae are formed from the mass ejected by red giant stars when the central remnant star becomes hot enough to ionize the expanding circumstellar material. In recent years we have been trying to understand the evolutionary details and the various physical processes which take place during the formation of planetary nebulae. This has been an area of research rich in the interplay of observation and theory, and although we have made significant progress, we do not yet have a satisfactory detailed understanding of the entire evolutionary progression of the formation of planetary nebulae (see for example the most recent IAU Symposium Proceedings; Planetary Nebulae, 1989, ed. S. Torres-Peimbert).

Studies in this area have been enhanced due to the identification of proto-planetary nebulae and very young planetary nebulae, such objects like CRL 618, NGC 6302 and NGC 7027. These nebulae have an ionized expanding envelope of high density near the remnant star and are surrounded by cold molecular material like CO and H_2, and by a significant amount of dust that radiates strongly in the infrared. The continuous spectra of the three objects mentioned above are shown in Figure 1. The spectrum for CRL 618 is from the compilation by Martin-Pintado et. al (1989a), that for NGC 7027 is from the compilation by Terzian (1989), and the data for NGC 6302 is from Payne et. al (1988) and from Gómes et. al (1989). These spectra show the characteristic strong infrared emission by the circumstellar dust with a range in dust temperatures from ~70 K to 280 K. The radio spectra show the emission due to free-free transitions with cutoff frequencies which are very high, indicating high electron densities of the order of 10^4 cm^{-3} for NGC 7027 and NGC 6302 and as high as 10^7 cm^{-3} for CRL 618. When the central star of a planetary nebula reaches surface temperatures $> 2 \times 10^4$ it begins to ionize the surrounding gas. This results in a compact HII region with significant optical depth at high radio frequencies.

As the central star evolves and reaches surface temperatures of $\geq$ 50,000 K, the ionizing photons increase and the HII region moves

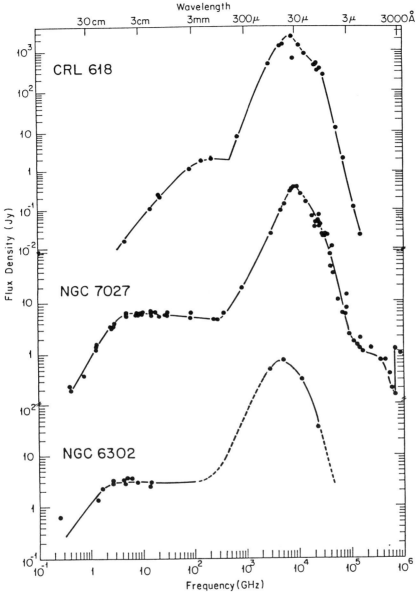

Figure 1. Continuum spectra of the compact nebulae CRL 618, NGC 7027, and NGC 6302 (see text for references). The strong infrared emission is due to warm radiating dust, and the radio emission is due to free-free transitions. At low radio frequencies these objects radiate like black-bodies.

outwards ionizing the neutral gas and the radio emission increases. After several thousand years, the star stabilizes and the HII region slowly expands outwards resulting in a lower electron density of the plasma, and the optical depth of the HII region becomes small at high radio frequencies. Eventually due to the nebular expansion, the radio emission diminishes and eventually becomes almost undetectable. The lifetime of a detectable HII region around such a central star is estimated to be of the order of ~30,000 years.

2. RADIO RECOMBINATION LINES

Radio recombination emission lines from the ionized envelopes of planetary nebulae are normally superpositions on the flat thermal spectra of these objects at frequencies where the nebulae are optically thin. A survey of such line emissions from planetary nebulae has been given by Terzian (1980). Although the number of identified planetary nebulae in the galaxy is more than one thousand, only about ten have shown detectable radio recombination lines. The main reason for this low yield is the weak thermal radio fluxes due to the low ionized masses of these nebulae. Table 1 gives a summary of the observational results. The highest frequencies (lowest principal quantum levels) for such lines are the H30α, H35α, and H41α lines at 231.9, 147.0, and 87.6 GHz respectively, from the high density proto-planetary nebula CRL 618 (Martin-Pintado et. al., 1989a). Figure 2 shows the spectra of these lines observed with the 30-m telescope of IRAM at Pico de Veleta, Spain. The emission is probably due not only to the ionized gas ejected in the star's red giant phase, but also to an ionized stellar wind generated by the central star. The Table gives the line to continuum temperature ratios, or the continuum and line fluxes (S_c, S_l) as given by the various authors. In addition the half power line widths, and the radial velocities with respect to the local standard of rest indicated by the recombination lines, are tabulated.

Most of the detected lines correspond to hydrogen $\Delta n = 1$ transitions. A few $\Delta n = 2$ lines have been detected from the nebulae IC 418 and NGC 7027 as shown in Figure 3 (Walmsley, et. al. 1981). Helium lines have been detected from three sources including doubly ionized helium from NGC 6302 and NGC 7027. No carbon recombination lines have been reported so far from these objects.

Some of the most recent observations of radio recombination lines from planetary nebulae were made by Garay et. al. (1989) using the Very Large Array. These authors detected the H76α lines from the compact nebulae NGC 6369, NGC 6543, NGC 7009, and IC 418. The recombination line emission VLA maps of NGC 6543 and NGC 7009 are significantly different from the continuum maps, but those of NGC 6369 and IC 418 closely resemble the continuum maps. The observed differences may be due to line amplification of stimulated emission due to an inhomogeneous density distribution for NGC 6543, and for NGC 7009 the differences may be explained as electron temperature variations within the nebula (Garay, et. al. 1989). These authors have also observed the H110α transition from IC 418 and the collisionally broadened lines

Table 1
PLANETARY NEBULAE RADIO RECOMBINATION LINES

Planetary Nebula	Line	Freq. (GHz)	T_l/T_c	ΔV (km/s)	V(LSR) (km/s)	Ref.
IC 418	H76α	14.7	0.150	26±2	43±1	WCT
			0.075	26±8	41±4	B
			0.124	27±1	43±1	GGR
	H85α	10.5	0.096	39±3	46±3	TB
	H110α	4.9	0.025	36±2	42±1	GGR
	H95β	14.9	0.041	26±4	42±1	WCT
	He76α	14.7	(0.019)	-	-	WCT
CRL 618	H30α	231.9	$T_l = 0.22$ K $S_c = 2.0$ Jy	30±9	-23±3	MBBGP
	H35α	147.0	$T_l = 0.20$ K $S_c = 1.8$ Jy	32±1	-22±1	MBBGP
	H41α	87.6	$T_l = 0.06$ K $S_c = 1.1$ Jy	30±9	-23±3	MBBGP
NGC 1514	H140α	2.4	$T_l = 0.011$ K $S_c = 0.3$ Jy	26±3	43±1	PT
NGC 6210	H140α	2.4	≤ 0.0104	-	-	WCT
NGC 6302	H76α	14.7	$S_l = 78$ mJy $S_c = 2.4$ Jy	40±3	-31±2	GMRG
	H110α	4.9	$S_l = 13$ mJy $S_c = 2.4$ Jy	56±2	-32±1	GMRG
	He$^+$121α	14.7	$S_l = 26$ mJy	27±6	-29±3	GRG
NGC 6369	H76α	14.7	0.058	43±6	-91±2	GGR
NGC 6543	H76α	14.7	0.122	41±12	-45±5	B
			0.120	41±2	-50±1	GGR
	H85α	10.5	0.057	57±12	-55±3	THMD
	H94α	7.8	0.078	34±10	-50±4	GC
	H109α	5.0	0.036	46±9	-57±2	CTW
NGC 6572	H76α	14.7	0.069	33±3	8±2	WCT
	H140α	2.4	≤ 0.0030	-	-	WCT
NGC 6853	H140α	2.4	≤ 0.0044	-	-	WCT
BD+30°3639	H76α	14.7	0.124	50±5	-14±2	WCT

PLANETARY NEBULAE RADIO RECOMBINATION LINES
Table 1 (continued)

Planetary Nebula	Line	Freq. (GHz)	T_l/T_c	ΔV (km/s)	V(LSR) (km/s)	Ref.
M1-78	H76α	14.7	0.048	37±5	-72±2	WCT
	H85α	10.5	0.030	54±12	-77±3	THMD
	H109α	5.0	0.018	65±12	-76±4	CTW
NGC 7009	H76α	14.7	0.069	45±6	-36±2	GGR
NGC 7662	H76α	14.7	(0.05)	27±5	(5)	WCT
	H85α	10.5	0.039	49±24	-4±4	THMD
NGC 7027	H56α	36.5	0.155	38±3	26±1	EB
	H66α	22.4	0.087	37±1	26±1	PTM
	H76α	14.7	0.051	48±2	23±1	WCT
			0.039	48±3	24±1	CM
			0.030	41±7	24±3	B
			0.039	44±1	24±1	R
	H85α	10.5	0.023	43±1	23±3	TB
	H90α	8.9	0.017	49±5	25±2	CTW
	H94α	7.8	0.010	44±12	17±4	GC
	H109α	5.0	0.0035	65±12	24±2	CTW
	H110α	4.9	0.0097	71±3	20±1	CM
			0.004	55±15	29±3	R
	H95β	14.9	0.007	29±5	24±1	WCT
	H113β	8.9	0.0035	77±14	29±6	CTW
	He76α	14.7	0.0083	27±4	17±2	WCT
	He⁺105α	22.4	0.0160	36±4	25±1	PTM
	He⁺121α	14.7	0.0098	31±4	17±2	WCT
			0.0061	48±12	20±4	C M

References:

WCT: Walmsley, Churchwell and Terzian, 1981;
B: Bignell, 1974;
GGR: Garay, Gathier and Rodriquez, 1989;
TB: Terzian and Balick, 1972;
MBBGP: Martin-Pintado, Bujarrabal, Bachiller, Gómez-Gonzales, and Planesas, 1989a;
PT: Terzian and Phillips, 1989;
GMRG: Gómes, Moran, Rodríguez and Garay, 1989;
GRG: Gómez, Rodríguez and Garcia-Barreto, 1987;
THMD: Terzian, Higgs, MacLeod and Doherty, 1974;
CTW: Churchwell, Terzian and Walmsley, 1976;
PTM: Pankonin, Thum and Mezger, 1979;
CM: Chaisson and Malkan, 1976;

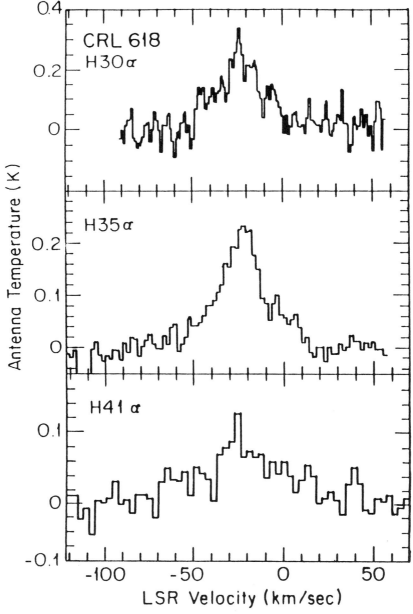

Figure 2. The high frequency radio recombination lines from the proto-planetary nebula CRL 618 observed by Martin-Pintado, et. al. (1989a).

PLANETARY NEBULAE RADIO RECOMBINATION LINES
Table 1 (continued)

References (continued):

GC: Goad and Chaisson, 1973;
EB: Ershov and Berulis, 1989;
R: Roelfsema, 1987.

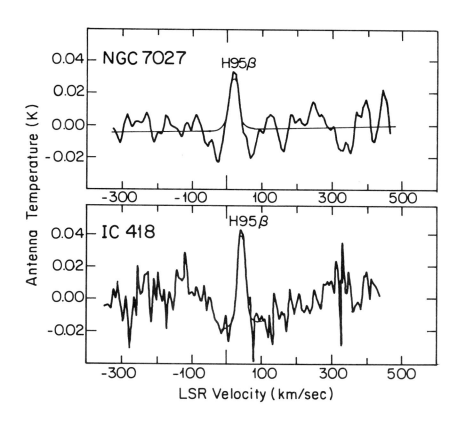

Figure 3. The H95β lines from NGC 7027 and IC 418 observed with the Effelsberg 100-m telescope.

indicate an electron density of 1.8×10^4 cm^{-3}. Figure 4 shows the H76α and H110α lines from IC 418 observed by Garay et. al (1989) using the VLA.

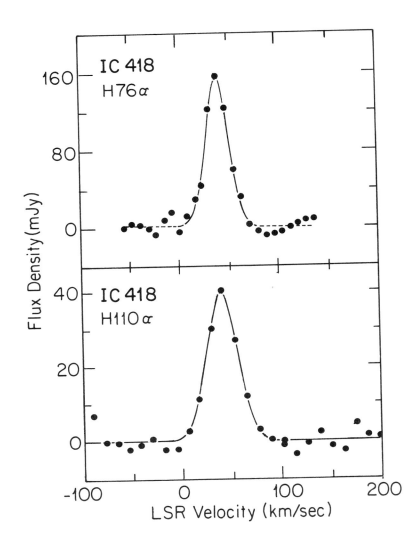

Figure 4. Very Large Array observations of the H76α and H110α lines from IC 418 made by Garay et. al. (1989) showing a broader H110α line due to the contribution of electron collisions.

In an attempt to estimate the distance to the nebula NGC 6302 Gómez et. al (1989) observed the H76α and H110α lines with the VLA, and from the broader H110α line, due to electron pressure broadening (Figure 5), they derived an electron density of 2.5×10^4 cm^{-3}. With this density and a toroidal geometrical model they then derive a distance of 2.2 ± 1.1 pc for NGC 6302, which results into an IRAS luminosity of 1.3×10^4 $L_\odot$.

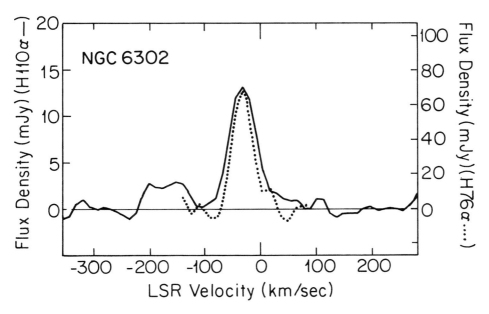

Figure 5. The same observations as in Figure 4 for NGC 6302 made by Gomez et. al. (1989).

The lowest frequency used to detect Hnα lines from planetary nebulae has been 2.4 GHz corresponding to the H140α transition. Such observations provided upper limits and were made by using the Arecibo 305-m antenna as reported by Walmsely et. al (1981). More recently Terzian and Phillips (1989) have observed the H140α line from NGC 1514

with the Arecibo telescope and Figure 6 shows a marginal detection.
   The radial velocities for the Hnα lines of NGC 7027 given in Table 1 may indicate a tendency to decrease with increasing principal quantum number n, as suggested by Ershov and Berulis (1989), but given the errors in the velocity measurements such an effect has not been clearly established.

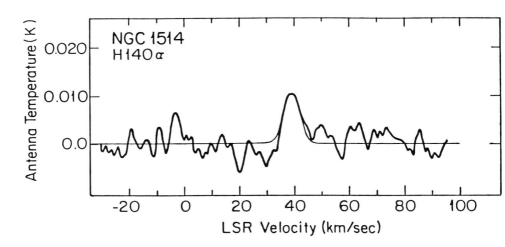

Figure 6.   The H140α line from NGC 1514 observed with the Arecibo 305-m radio telecope.

3. ELECTRON TEMPERATURES

One of the useful results of detecting radio recombination lines from thermal sources has been the derivation of the electron temperatures of the emitting regions. Significant work has already been done in this area where LTE and non-LTE derivations are widely used. In general the electron temperatures of planetary nebulae derived from radio recombination line observations assuming LTE conditions are in very good agreement with the temperatures derived from optical forbidden lines. This suggests that non-LTE effects play a minor role for planetary nebulae. Table 2 presents a summary of electron temperatures derived from radio recombination lines. It can be seen that on the average the non-LTE values are ~12% higher than the LTE ones.

Table 2
PLANETARY NEBULAE
ELECTRON TEMPERATURES FROM RADIO RECOMBINATION LINES

Nebula	Line	T_e(LTE)	T_e(non-LTE)	Ref.[1]
IC 418	H76α	8400 K	9800 K	WCT
	H76α	9600	-	GGR
CRL 618	H30α	13500	-	MBBGP
	H35α	12400	-	MBBGP
	H41α	(17800)	-	MBBGP
NGC 1514	H140α	(14000)	-	PT
NGC 6302	H76α	18000	-	GMRC
	H110α	21000	-	GMRC
NGC 6369	H76α	12100	-	GGR
NGC 6543	H76α	6400	7800	WCT
	H76α	6600	-	GGR
NGC 6572	H76α	12900	15000	WCT
BD30+3639	H76α	5900	7600	WCT
NGC 7009	H76α	9700	-	GGR
NGC 7027	H76α	11700	14000	WCT
M1-78	H76α	16200	17500	WCT
NGC 7662	H76α	(19000)	(19000)	WCT

[1]See Table 1

4. LINE WIDTHS

The first analysis of the line widths of radio recombination lines from planetary nebulae was made by Terzian and Balick (1972), where they determined that the relatively broad lines were primarily due to thermal broadening and nebular expansion. Electron collisional broadening and turbulence was found to be negligible. Walmsely et. al. (1981) were able to show that for the high density nebulae NGC 7027 and M1-78 the linewidths for $n \gtrsim 100$ were affected by electron collisional broadening as described by Brocklehurst and Seaton (1972). Figure 7 shows the available Hnα data for the line widths of NGC 7027, indicating increased widths with principal quantum level n. More recently Garay et. al (1989) have also measured this effect for IC 418. Such measurements allow derivations of the electron densities of these objects independent of distances. The derived densities from this method are as follows:

M1-78	7×10^4 cm^{-3}
NGC 7027	3×10^4
NGC 6302	2.5×10^4
IC 418	2×10^4

These densities are in good agreement with other optical and radio determinations using different methods.

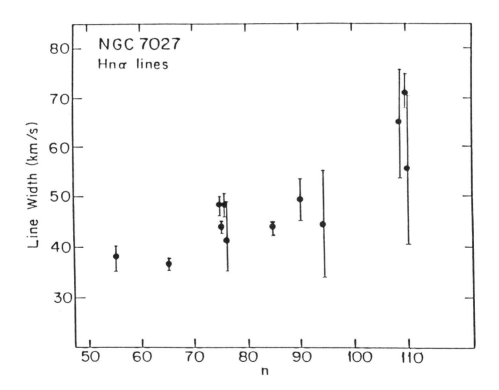

Figure 7. Observed line widths as a function of principal quantum number from NGC 7027

5. THE ANOMALOUS SOURCE MWC 349

The source MWC 349 remains of unknown nature. It has strong stellar winds and a small bipolar radio continuum configuration and some of its features are not unlike some of the known proto-planetary nebulae. Recently Martin-Pintado et. al. (1989b) have reported the detection of the recombination lines H29α, H30α, H31α, at ~λ1mm, and the H41α at ~3.3 mm from MWC 349. The line shapes and intensities of the λ1mm lines are significantly different from the λ3.3mm line. The λ1mm lines show double peaked profiles, and could be explained by significant non-LTE effects, while the λ3.3 mm line has a single gaussian shape and can be accounted for by LTE emission of an isothermal stellar wind at an electron temperature of 6000 K.

More recently MWC 349 has shown new surprises. Martin-Pintado et. al. (1989c) from observations made a few months apart, report time variability in the λ1mm radio recombination lines emitted by this source. These results confirm that the λ1mm line emission is enhanced by significant maser action. The nature of this source is still not well defined, but it could represent the birth of a proto-planetary nebula or it may be a region of star formation.

6. CONCLUSIONS

Significant radio recombination line emission observations have been made from planetary nebulae confirming the thermal character and the recombination theory for these objects. However, due to their weak emissions only a small number of objects have been detected in the radio recombination lines. It should be possible to detect many more nebulae with the high sensitivity radio telescopes at cm wavelengths. Such measurements with n $\geq$ 100 could detect the electron collisional broadening of the recombination lines and yield accurate electron densities of the emitting regions.

The derived electron temperatures form the radio recombination lines are in good agreement with those found from optical methods, and it appears that the departures from LTE are small.

Although no carbon radio recombination lines have been observed from planetary nebulae, these should be detectable near the boundaries of the HII regions and such emissions should be searched for.

Emphasis should be put in observing the compact, young nebulae, that are associated with strong infrared emission. This emission indicates the presence of dust and most often these sources cannot be well studied at optical wavelengths due to the strong extinction by the dust.

Thanks are due to Anthony J. Phillips for reading the manuscript and making some useful suggestions. This work was supported in part by the National Astronomy and Ionosphere Center, which is operated by Cornell University under a management agreement with the National Science Foundation.

REFERENCES

Bignell, R. C., 1974, Ap.J., 193, 687.
Brocklehurst, M., and Seaton, M. J., 1972, Monthly Notices Roy. Astron. Soc., 157, 179.
Chaisson, E., and Malkan, M., 1976, Ap.J., 210, 108.
Churchwell, E., Terzian, Y., and Walmsley, M., 1976, Astron. and Astrophys., 48, 331.
Ershov, A. A., and Berulis, I. I., 1989, Astronomical Journal Letters, Academy of Sciences, USSR, 15, No. 5, 413.
Garay, G., Gathier, R., and Rodríguez, L. F., 1989, Astron. and Astrophys., 215, 101.
Goad, L., and Chaisson, E., 1973, Mem. Soc. Royale Des Sciences de Liège, V, 115.
Gómez, Y., Rodríguez, L. F. and Garcia-Barreto, A. J., 1987, Rev. Mexicana Astron. Astrof., 14, 560.
Martin-Pintado, J., Bachiller, R., Thum, C., and Walmsley, M., 1989a, Astron. and Astrophys., 215, L13.
Martin-Pintado, J., Bujarrabal, V., Bachiller, R., Gómez-Gonzalez, J., and Planesas, P., 1989b, Astron. and Astrophys., in press.
Martin-Pintado, J., Thum, C., and Bachiller, R., 1989c, Astron. and Astrophys., 222, L9.
Pankonin, V., Thum, C., and Mezger, P., 1979, (private communication).
Payne, H. E., Phillips, J. A., and Terzian, Y., 1988, Ap. J., 326, 368.
Roelfsema, P. R., 1987, Thesis, University of Groningen, The Netherlands, 150.
Terzian, Y., and Balick, B., 1972, Astrophys. Letters, 10, 41.
Terzian, Y., Higgs, L., MacLeod, J., and Doherty, L., 1974, Astron. J., 79, 1018.
Terzian, Y., 1980, Radio Recombination Lines, ed. P. A. Shaver, (D. Reidel Publ. Co.), 75.
Terzian, Y., 1989, Planetary Nebulae, ed. S. Torres-Peimbert, (Kluwer Academic Publ.), 17.
Terzian, Y., and Phillips, J. A., 1989, (private communication).
Torres-Peimbert, S., (ed)., 1989, Planetary Nebulae, (Kluwer Academic Publ.).
Walmsley, M., Churchwell, E., and Terzian, Y., 1981, Astron. and Astrophys, 96, 278.

RADIO RECOMBINATION LINES FROM COMPACT PLANETARY NEBULAE

GUIDO GARAY
Departamento de Astronomía, Universidad de Chile
Casilla 36-D, Santiago, Chile

ABSTRACT. We summarize recent results of H76α and H110α recombination line observations, made with high angular resolution, toward five compact planetary nebula. The H76α line widths averaged over the source are greater than 40 km s^{-1} for all, except one, nebulae. We attribute the large widths to expansion motions of the ionized gas. The H110α line widths are significantly wider than the H76α widths, suggesting that broadening by electron impacts is important at the higher quantum number. Electron densities derived from the collisionally broadened line widths are in good agreement with those determined from optical and radio continuum data. The peak H76α line to continuum ratios among the five nebulae are spread by more than a factor of four. We conclude that the observed large range is mainly due to the considerable extent of their electron temperatures. For three nebulae the H76α line maps are noticeable different from the continuum maps. We suggest that this result is primarily due to line enhancement by stimulated emission in regions with emission measure larger than average.

1. INTRODUCTION

Young planetary nebulae (PN) are regions of ionized gas created shortly after their central star has become hot enough to ionize the neutral material surrounding it. Such nebulae are characterized by having small angular sizes, high electron densities ($N_e \geq 10^3$ cm^{-3}) and high emission measures (EM $\geq 10^6$ pc cm^{-6}) (Gathier 1987). Compact planetary nebulae are often surrounded by dust and neutral material ejected during their precursor red giant stage, which severely limits their study at optical wavelengths.

Radio emission, being unaffected by dust, offers a unique way to determine the physical conditions of the ionized gas during the early phases of PN evolution. In particular, radio recombination lines should constitute an excellent probe to investigate such conditions. From the line to continuum ratio it should be possible to derive the electron temperature of the gas. From the linewidth, which should show the effect of pressure broadening, it should be possible to derive the electron density. Very few observations of radio recombination lines from PN are, however, available, primarily because of their weak radio emission. A review on radio recombination line observations of planetary nebulae is given by Terzian (1990).

In this paper we present recent results of radio recombination line observations of compact PN made with the Very Large Array. A detailed discussion is given elsewhere (Garay, Gathier and Rodríguez 1989). The sources and their parameters are given in Table 1.

2. RADIO CONTINUUM

The essential information obtained from radio continuum observations of PN are their radio spectrum and projected geometry. Our observations with high sensitivity and angular resolution enable us to make a precise determination of the geometry of the PN. In general, our continuum maps made with the VLA show a decrease in emission at the centre of the PN. In particular, the radio maps of IC 418 and NGC 6369 exhibit fairly high symmetric morphologies suggesting that their ionized gas is distributed in spherical shell structures. These shells are most likely ionization bounded regions of gas, surrounding the recently turned-on central star (Kwok 1985). The radio continuum map of NGC 6302 exhibits a ring-like elongated shape, which Rodríguez et al. (1985) interpreted as due to radio emission arising from a circular toroid tilted at an angle of ~45° with respect to the line of sight.

The radio spectra of the PN in Table 1 can be well modelled in terms of free-free emission from an homogeneous source of ionized gas (see Fig. 5 of Garay, Gathier and Rodríguez). Using the source geometry, and assuming that the distance to the PN is known, the radio spectrum can thus be employed to derive mean physical parameters of the nebula (e.g., Churchwell, Terzian and Walmsley 1976). The electron temperatures and densities of the compact PN derived in this way, given in columns 5 and 7 of Table 1, are in quite good agreement with the values determined from optical observations.

TABLE 1. PARAMETERS OF COMPACT PLANETARY NEBULAE

Source	Observed			Derived				
	Line	Δv (km s^{-1})	T_L/T_C	T_e (K)	$EM/10^6$ (pc cm^{-6})	$N_e/10^3$ (cm^{-3})	V_{exp} (km s^{-1})	T_e^* (K)
IC 418	H76α	27.0	0.124	9500.	6.0	20.	8.5	9600.
	H110α	35.7	0.025					
NGC 6302	H76α	40.	0.033	21000.	15.	20.	13.	18000.
	H110α	56.	0.0055					
NGC 6369	H76α	42.8	0.058	13000.	1.6	3.8	18.	12100.
NGC 6543	H76α	41.1	0.120	7500.	1.7	6.3	19.	6600.
NGC 7009	H76α	44.5	0.069	12000.	1.6	7.0	20.	9700.

3. RADIO RECOMBINATION LINES

3.1 Global Profiles

The essential data obtained from radio recombination line observations are the line center velocity, the line width, and the line to continuum ratio. Line parameters of the compact PN obtained by fitting the global profiles with a single Gaussian are given in columns 3 and 4 of Table 1.

3.1.1 **Linewidths.**
The widths of the H76α line are considerable broader than those expected from the thermal motions and turbulence within the ionized gas, of ~25 km s^{-1}. Possible mechanisms leading to an increase of the line widths are pressure broadening and systematic motions of the gas. At the densities of the compact PN pressure broadening is, however, negligible for principal quantum numbers smaller than ~80. Thus, we attribute the broadening of the H76α lines to the expansion motions of the nebula. The expansion velocities derived from the H76α line widths, given in column 8 of Table 1, are in good agreement with those determined from optical measurements.

The widths of the H110α line from IC 418 and NGC 6302 are, on the other hand, significantly broader than those of the H76α line, suggesting that electron impacts becomes the dominant mechanism of line broadening at the higher quantum number. From the difference in the observed line width of two recombination lines it is possible to derive an average value of the electron density, independent of distance estimates. The densities derived using this method for IC 418 and NGC 6302, of 1.8×10^4 and 2.5×10^4 cm^{-3}, respectively, are in good agreement with the rms electron densities derived from the radio continuum observations.

3.1.2 **Line to continuum ratio.**
As indicated in Table 1, a large range of values in the peak H76α line to continuum ratio is observed among the five compact PN. This result could be attributed to either enhancement of the line by stimulated emission; optical depth effects; or due to a wide range of electron temperatures among the PN. To investigate the actual reason, we plot in Figure 1 the integrated intensities under the H76α line versus the electron temperatures determined from both radio continuum and optical data. Clearly, there is a strong correlation between those parameters; a least square linear fit for log $T_L/T_C \times \Delta v$ = C log T_e + D, gave C = -1.3 ± 0.1 and D = 5.7 ± 0.2. Two conclusions can be drawn from Figure 1. First, the large range span by the observed line to continuum ratios is primarily due to the wide range of electron temperatures among the five PN. Second, the slope C agrees well with that expected from an LTE analysis of recombination lines, of -1.15, suggesting that departures from LTE at the frequency of 15 GHz and at the average physical conditions of the compact PN are not important. The mean electron temperatures derived from the H76α line to continuum ratio, on the assumption that the gas is in LTE, are given in column 9 of Table 1.

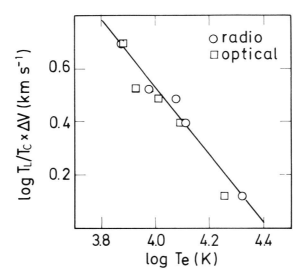

Figure 1. Plot of integrated H76α line intensity versus electron temperature for compact planetary nebula. Squares and circles indicate, respectively, electron temperatures derived from optical and radio continuum data. The line is a least squares fit to the data.

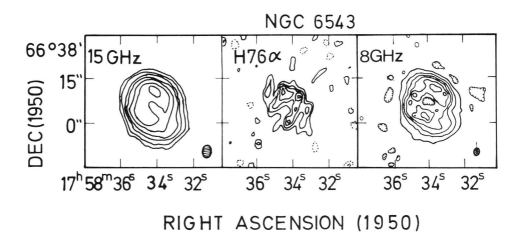

Figure 2. Comparison of continuum and line emission from NGC 6543. Shaded ellipses indicate half power beam areas. **a** 15 GHz continuum map. The HPBW is 3".4x4".3. **b** H76α line map. The HPBW is 3".4x4".3. **c** 8.1 GHz continuum map. The HPBW is 2".5x2".2.

3.2 Line Maps

The most striking result of our H76α line observations with high angular resolution is the noticeable difference between the line and continuum maps of NGC 6302, NGC 6543 and NGC 7009. This result is illustrated in Figure 2, which shows maps of the radio continuum and H76α line emission from NGC 6543, made with an angular resolution of 4.3"x3.4". Possible effects which could account for the differences are (van Gorkom 1980; Rodríguez et al. 1985) :(i) pressure broadening; ii) non-LTE effects; iii) gradients of the electron temperature; iv) large velocity gradients; and v) large abundance of doubly ionized helium.

The regions of strongest line emission in NGC 6543 and NGC 6302 are also those of largest emission measure. To demonstrate this point we show in Figure 2c a map of the radio continuum emission at 8.1 GHz from NGC 6543, made with an angular resolution of ~2" (Terzian et al. 1974), which exhibits substantially more fine structure than our lower angular resolution continuum map. A comparison of Figures 2b and 2c shows that there is a significant correlation between the peaks in the H76α line emission map and the fine radio continuum structure, which, of course, correspond to the regions of ionized gas with the largest emission measure.

Since non-LTE line enhancement is expected to be more important in regions with high emission measure, we suggest that the large line to continuum ratios observed toward the locations of high emission measure are due to stimulated line emission. Not all the differences between the line and continuum maps are due to this effect, however. For instance, part of the differences seen in NGC 6302 might be due to the presence of large amounts of doubly ionized helium (Gómez, Rodríguez and García-Barreto 1987); while for NGC 7009 variations in the electron temperature across the nebula are thought to be the cause of the dissimilarities. In summary, we believe that non-LTE effects, the contribution to the continuum emission from twice ionized helium and gradients in electron temperature across PN can account for the difference between the line and continuum maps.

4. DISTANCES

Knowledge of the distance to a planetary nebula is essential to determine its physical properties and Galactic location. However, determining distances to PN is a difficult and controversial problem (e.g., Lutz 1989). Recently, Gómez et al. (1989) suggested a promising method of finding distances to PN that makes use of radio recombination lines observations. Combining the electron density, derived from the pressure broadening of recombination lines, with the geometry of the nebula and its flux density, derived from the radio continuum observations, they determined distances to IC 418 and NGC 6302 of 0.2 and 2.2 Kpc, respectively. This technique has the advantage that no a priori assumptions have to be made about the nebular mass or the luminosity of the central star.

5. CONCLUSIONS

We review recent observations, made with the VLA, of radio recombination line emission from compact planetary nebulae. Five nebulae, IC 418, NGC 6302, NGC 6369, NGC 6543 and NGC 7009, were observed in the H76α line; the first two were also observed in the H110α line. The main results and conclusions are summarized below.

The observed radio continuum morphologies of the PN exhibit fairly high symmetric shapes with clear central depressions, suggesting that the ionized gas is distributed in shell or toroidal structures.

The widths of the H76α line averaged over the source are $\geq$ 40 km s^{-1} for all PN, except IC 418. The large widths are adequately explained by broadening due to the thermal and expansion motions of the PN. The widths of the H110α line are significantly broader than the H76α line widths. The electron densities derived from the pressure broadened line widths are in good agreement with optical and radio continuum values.

The peak H76α line to continuum ratios among the PN ranges from 0.033 to 0.124. This large extent is mainly due to the wide range of electron temperatures exhibited by the PN. The average electron temperatures derived from an LTE analysis of the H76α recombination line are also in good agreement with those determined from optical and radio continuum data.

The H76α line maps of NGC 6302, NGC 6543 and NGC 7009 are significantly different from the radio continuum maps. We suggest that the differences are due to stimulared emission arising from regions of large emission measure.

The author is the grateful recipient of an Henri Chrétien Award.

REFERENCES

Churchwell, E., Terzian, Y., and Walmsley, M. 1976, Astr. Ap., 48, 331.
Garay, G., Gathier, R., and Rodríguez, L.F. 1989, Astr. Ap., 215, 101.
Gathier, R. 1987, The Late Stages of Stellar Evolution, eds. S. Kwok, S.P. Pottasch (Dordrecht:Reidel), p. 371.
Gómez, Y., Rodríguez, L.F., and García-Barreto, J.A. 1987, Rev. Mexicana Astron. Astrofís., 14, 560.
Gómez, Y., Moran, J.M., Rodríguez, L.F., and Garay, G. 1989, Ap. J., 345, 862.
Gorkom, J.H. van 1980, Ph.D. Thesis, University of Groningen, Holland.
Kwok, S. 1985, A. J., 90, 49.
Lutz, J.H. 1989, Planetary Nebulae, IAU Symp. 131, ed. S. Torres-Peimbert (Dordrecht:Reidel), p. 129.
Rodríguez, L.F., García-Barreto, J.A., Cantó, J., Moreno, M.A., Torres-Peimbert, S., Costero, R., Serrano, A., Moran, J.M., and Garay, G. 1985, M.N.R.A.S., 215, 353.
Terzian, Y., Balick, B., and Bignell, C. 1974, Ap. J., 118, 257.
Terzian, Y. 1990, these proceedings.

RADIO RECOMBINATION LINE MASER EMISSION IN MWC349

J. MARTIN-PINTADO[1], R. BACHILLER[1] AND C. THUM[2]
1) Centro Astronómico de Yebes, Apartado 148, 19080 Guadalajara (Spain)
2) IRAM, Avd. Divina Pastora 7-9 Nc, 18012 Granada (Spain)

ABSTRACT. We present the discovery of maser emission in radio recombination lines at wavelengths between 1 and 2 mm towards MWC349. The maser emission takes off for radio recombination lines with quantum numbers smaller than 36. For larger quantum numbers, the line intensities are consistent with thermal emission from an isothermal ionized stellar wind expanding at constant velocity. Like most of the molecular line masers, the radio recombination line maser shows strong time variability. From the variability of the radio recombination lines near 1 mm we estimate a brightness temperature larger than 10^6 K.

1. Introduction

Radio recombination lines (RRLs) were discovered in interstellar space by Dravskikh and Dravskikh (1964), Sorochenko and Borodzich (1965) and Höglund and Mezger (1966). Soon after, Goldberg (1965) showed that recombination lines at centimeter wavelengths were expected to be formed in non-LTE (Local Thermodynamic Equilibrium) and stimulated emission must occur. The extent to which non-LTE effects influence the line intensities of the RRLs has been subject of debate in the last two decades (see e. g. Shaver 1980 and reference therein). The most accepted idea is that RRLs at wavelengths longer than 1 cm arising in conventional galactic HII regions are emitted under near-LTE conditions (Shaver 1980) and stimulated emission is not very important. At millimeter wavelengths, most galactic HII regions are optically thin and stimulated emission is negligible (Walmsley 1979; and Gordon in this volume). However, some HII regions produced by ionized stellar winds can be optically thick even at millimeter wavelengths (Panagia and Felli 1975, Wright and Barlow 1975) and therefore RRLs maser emission might occur, provided that the stimulated emission coefficient, β_n, is negative.

MWC349 is the prototype of this kind of objects, it has the radicontinuum spectrum, $S_n \propto \nu^{0.6}$, typical of an ionized isothermal stellar wind with constant expansion velocity (Olnon 1975). Radio recombination line observations at centimeter wavelengths indicated that these lines are emitted under LTE conditions (Altenhoff et al. 1981 and Escalante et al. 1989). This is in contrast with the recombination lines at millimeter wavelengths, which are enhanced by maser emission with optical depths smaller than -1 (Martín-Pintado et al. 1989a and 1989b). In this paper we present a summary of the observational properties of the thermal and maser emission toward MWC349.

2. Observations and results

All observations presented here were carried out with the IRAM 30-m telescope at Pico Veleta (Spain). Fig.1 shows a compendium of RRLs observed at millimeter wavelengths toward MWC349. This figure clearly show that the shape and the intensity of the RRLs at millimeter wavelengths are strongly dependent on quantum numbers. For quantum numbers smaller than 36, the millimeter recombination lines have gaussian line shapes, similar to those found for larger quantum numbers (H76α and H66α) at centimeter wavelengths (Altenhoff et al. 1981; and Escalante et al. 1989). RRLs with quantum numbers larger than 36 show completely different line shapes with double peaked profiles.There are also large difference between the line intensity of H41α and those of the 1 mm RRLs. The peak flux density of the H30α (~26 Jy) is a factor of 70 larger than that of H41 (0.37 Jy). The line intensities of the 2 and 1 mm RRLs also strongly depend on quantum number. The lines are systematically more intense for smaller quantum numbers. For instance, assuming the same aperture efficiency for all RRLs at 1 mm, the peak line intensity of the H29α line is a factor of ~2 larger than that of the H31α line.

3. Discussion

3.1. CHARACTERISTICS OF THE HII REGION SURROUNDING MWC349

At millimeter wavelengths, nearly all ultracompact HII regions are optically thin and the intensities and the profiles of the recombination lines are expected to change smoothly with wavelength (see Wilson et al. 1987 and Martín-Pintado et al. 1988). The interpretation of the recombination lines arising in MWC349 is, however, more complicated than in an optically thin HII region because of the density structure and the velocity field of the ionized gas in this source (Altenhoff et al. 1981; and Felli et al. 1985). MWC349 has the typical radio continuum spectrum of an ionized stellar wind with constant expansion velocity. This corresponds to a density that decreases with the square of the distance from the star. Due to this density structure, the HII region is partially optically thick in the continuum even at millimeter wavelengths (Panagia and Felli 1973, Wright and Barlow 1973). The size of the optically thick core depends upon frequency and hence recombination lines at different wavelengths sample different regions. As the quantum number of the RRLs decreases, one sees the ionized gas closer to the star. Thus, both the line shape and the line intensities of the millimeter recombination lines can perhaps be explained by changes of the electron temperature and the kinematics of the ionized gas with radius.

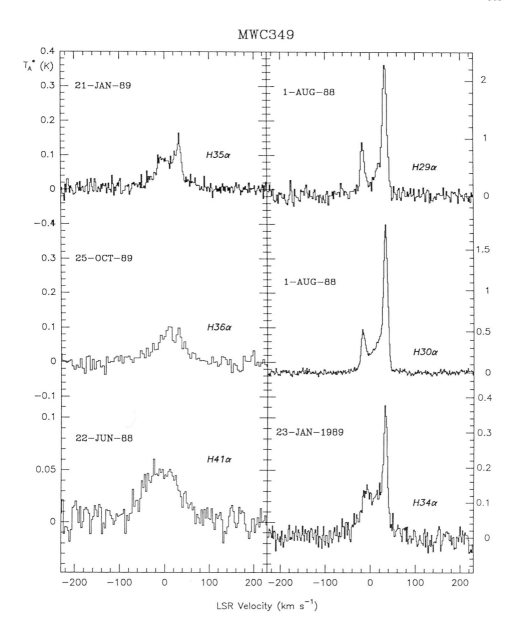

Figure 1. Radio recombination lines at millimeter wavelengths observed with the IRAM 30-m telescope towards MWC349.

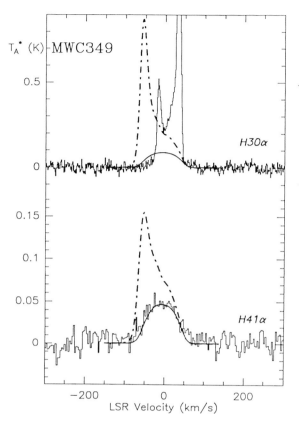

Figure 2. Comparison between the observed radio recombination lines (RRLs) towards MWC349 (thin lines) and those obtained from model calculations. For all calculations we considered that RRLs arise in an isothermal (6000 K) envelope expanding with a constant velocity of 50 km s^{-1}. The electron density varies with radius as $n_e = A\ r^{-2}$. In the formation of the RRLs we have considered the effects of collisional broadening and turbulent velocity of 10 km s^{-1}. The thick lines show the result of the model when the RRLs are formed under LTE conditions and the dashed-dotted lines when non-LTE effects are taking into account. For the non-LTE calculation we have used the departure coefficients derived by Walmsley (1989) for millimeter recombination lines and the physical conditions in MWC349 following Brocklehurst (1970).

3.2. LTE EMISSION FROM AN ISOTHERMAL EXPANDING ENVELOPE

Martín-Pintado et al. 1989a have made model calculations for the transfer of the continuum and recombination lines arising in an spherical expanding envelope with an electron density law given by $n_e = Ar^{-2}$. For the formation of the recombination lines, the model includes the effects of pressure broadening (Brocklehurst and Seaton 1972).

Fig. 2 shows the results of the model calculations superimposed on the measured recombination lines (thin lines). The thick lines show the profiles obtained with the model assuming that the lines are formed in LTE in an isothermal (Te=6000 K) expanding envelope with a constant expansion velocity of 50 km s^{-1}. The very good agreement between the measured and the predicted profiles for the H41α line, indicates that this line is consistent with LTE emission from an isothermal expanding envelope with constant velocity. The electron temperature, 6000 K, required to explain the line intensity is also in good agreement with those determined from radio continuum observations at 15 and 22 GHz (White and Becker 1985).

The results of the LTE model for the 1 mm RRLs are, however, clearly inconsistent with the observations (see Fig. 2). The observed line intensities are much stronger than predicted and the line shapes and radial velocities are also very different. Thus, it is clear that LTE emission from an isothermal expanding envelope can explain the line intensities of the H41α line but not the RRLs around 1 mm.

3.3. STIMULATED EMISSION FROM AN ISOTHERMAL ENVELOPE

For the typical densities of HII regions, negative absorption or stimulated emission is expected in RRLs at centimeter wavelengths. For recombination lines at millimeter wavelengths, stimulated emission is negligible because the continuum opacities are very small. However, in MWC349 the electron densities where the millimeter RRLs arise ($n_e \sim 10^7$ cm^{-3}) are much larger than in galactic HII regions giving rise to larger line optical depths. For the radius at which the continuum optical depth is 0.5, the line optical depths of the H41α and H30α are ~ 0.3 and ~ 1 respectively. Walmsley (1989) has made calculations of the departure coefficients, b_n and β_n, for the quantum numbers corresponding to the millimeter RRLs following the code of Brocklehurst (1970). For the physical conditions in MWC349, he found negative values (< -1) of β_n for both the H41α H30α lines. Thus, substantial stimulated emission or even maser emission with optical depths smaller than -1 should be observed. Furthermore, the smaller opacities for H41α and the larger collisional broadening for a given density make stimulated emission less important for this line than for H30α, in good agreement with the observations.

That stimulated emission in MWC349 is of importance for the RRLs around 1 mm is illustrated in Fig. 2 which shows, as dashed-dotted lines, the recombination lines obtained from the isothermal expanding envelope model taking into account non-LTE effects with the departure coefficients obtained by Walmsley (1989). Though the free parameters and the geometry used in these calculations do not fit to the data, the model illustrates that maser emission occurs in MWC349. For the physical condition in MWC349, the model predicts stimulated emission for both H30α ($\tau \sim -2.5$) and H41α ($\tau \sim -0.5$) . As observed, the model shows that stimulated emission for H30α is more important than for H41α . The model also gives the observed trend in the line intensities for the 1 mm RRLs, with stimulated emission being more important for H29α than for H31α.

4. Time variability of recombination lines near 1 mm.

Radio recombination lines arising from typical HII regions do not show any time variability. On the order hand, time variability is usually associated to most of the molecular line transitions which show maser emission. Fig. 3 shows the first detection of time variability in recombination lines. The line shape and the integrated line intensities of the recombination lines near 1 mm, H29α and H30α, have undergone changes in a time scale of six months. During this period of time, the intensity of the redshifted

MWC349

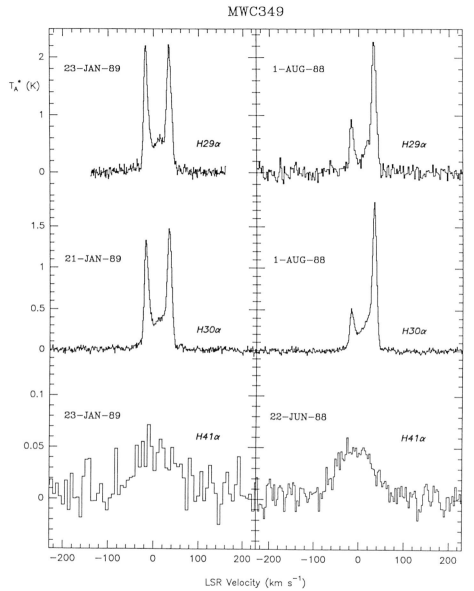

Figure 3. Radio recombination lines at millimeter wavelengths observed in two different epoches with the IRAM 30-m telescope toward MWC349. In the right panel we show the recombination lines measured in August 1988 and in the left we present the new observations made in January 1989. While the recombination lines near 1 mm, H29α and H30α, show large variation in the blueshifted component, the 3.3 mm recombination line, H41α, does not show any variability.

component has remained constant, but the blueshifted one has increased by a factor of 2.5. This behavior is in contrast with the 3.3 mm recombination line which does not show any significant change.

From the time scale of the intensity variations, Martín-Pintado et al. (1989b) have estimated an upper limit for the size of the maser region of $<4\ 10^{14}$ cm ($<0.02"$ at a distance of 1.2 Kpc, Cohen et al., 1985). For this size, the source brightness temperature is $>10^6$ K. This temperature is a factor 28 larger the effective temperature of the hottest star, ~35000 K, in the binary system (see Hartmann et al., 1980) and confirmed that maser emission enhance the RRLs at wavelengths between 1 and 2 mm. Long term monitoring of MWC 349 in line and continuum emissions at different wavelengths will help in elucidating the dynamics, the physical conditions and the evolution of this remarkable ionized wind. Using the 30-m telescope, we have began a long term monitoring of the radio recombination line emission at millimeter wavelengths

Acknowledgments: This work has been partially supported by the Spanish CICYT under grant number PB88-453.

References

Altenhoff, W. J., Strittmatter, P. A., Wendker, H. J. 1981, *Astr. Ap.*, **93**, 48.
Brocklehurst, M. 1970, *M.N.R.A.S.*, **148**, 417.
Brocklehurst, M. and Seaton, M.J 1972, *M.N.R.A.S.*, **157**, 179.
Cohen, M., Bieging, J.H., Dheher, J.W., Welch, W.J. 1985, *Ap. J.* **292**, 249.
Dravskikh, Z. V. and Dravskikh, A. F. 1964, *Astron. Tsirk.*, **158**, 2.
Escalante, V., Rodriguez, L.F., Moran, J.M., Cantó, J. 1989, *Rev. Mexicana Astron. Astrof.*, **17**, 11
Felli, N., Panagia, N., Stanga, R. 1985, *Proceedings of the ESO-IRAM- ONSALA workshop on "(Sub)Millimeter Astronomy*, p. 501.
Goldberg, L 1966, *Ap. J.*, **144**, 1225.
Hartmann, L., Jaffe, D., Huchra, J.P. 1980, *Ap. J.*, **239**, 905..
Höglund, B., Mezger, P. G. 1965, *Science*, **150**, 339.
Martín-Pintado, J.,Bujarrabal, V., Bachiller, R., Gomez-Gonzalez, J. Planesas, P. 1988, *Astr. Ap.*, **197**, L15.
Martín-Pintado, J., Bachiller, R., Thum, C. 1989a, *Astr. Ap.*, **215**, L13.
Martín-Pintado, Thum, C., Bachiller, R. 1989b, *Astr. Ap.*, **229**, L9.
Olnon, F.M. 1975, *Astr. Ap.*, **39**, 217.
Panagia, N., Felli, N. 1975, *Astr. Ap.*, **39**, 1.
Shaver, P. A. 1980, *Astr. Ap.*, **90**, 34.
Sorochenko, R.L., Borodzich, E.V. 1965, *Reports of the Sov. Academy of Science*, **163**, 3, 603.
Walmley C. M. 1980, in *Radio Recombination Lines*, ed. P. Shaver, Reidel Company, p. 37
Walmsley C. M. 1989, submitted to *Astr. Ap*.
White, R.L. and Becker, R.H. 1985, *Ap. J.*, **297**, 677.
Wilson, T. L., Mauersberger, R., Brand, J., Gardner, F.F. 1987, *Astr. Ap.*, **186**, L5.
Wright, A.E., Barlow, M.J. 1975, *M.N.R.A.S.*, **170**, 41.

Recombination emission from CII-region around Be-Stars

S.P. Tarafdar and K.M.V. Apparao
Tata Institute of Fundamental Research
Homi Bhabha Road, Bombay 400 005, India

ABSTRACT. Various physical processes proposed for the explanation of the flaring emissions of H_α and at 6578 Å in Be stars have been critically examined. The idea that the flaring occurs when X-ray emitting compact companion of Be-star produces the excess ionization of the shell material appears to give best agreement with observation.

1. Introduction

The line emission from Be-stars is well known for its variability. Recently, flaring of H_α-emission has been observed in two Be-star, μ Cen and HR4123 by Baade et al. (1988) and by Ghosh et al. (1989) respectively. The duration of flare is about a month in both stars and the rise and decay times are a few days. Together with the flaring of H_α, an emission at 6578Å appears and dissappears along with the appearance and dissappearance of H_α. Baade et al. (1988) have attributed this emission to CII- doublet at 6578 and 6583 Å. Ghosh et al. (1989) have examined different processes and attributed this emission to CII emission arising from dielectronic recombination of CIII produced by hard X-ray photons from compact companion of Be-star in the CII-region surrounding the HII-region of Be-star. We shall critically review here the different possible processes which could produce this emission at 6578 Å.

Table 1 gives some of the relevant data for the two stars μ Cen and HR4123. The table also contains the observed equivalent width of H_α and 6578 Å emission at the peak emission time. The equivalent width has been converted to line flux using continuum flux from Kurucz's models. We now discuss in the following different hypothesis of flare emission in the two stars.

2. The flare emission is due to ejection of matter from the Be-star:

This idea has been proposed by Peters (1984) and Baade et al (1988). Note that the cooling time of ejected hot matter is very small compared to the duration of the flare, because of high density in envelopes of Be stars. Therefore, according to this idea the enhancement of ejected matter should increase the circumstellar matter so that the dimension of the HII-region increases and thus enhancing the H_α - and 6578 emission. An implicit assumption in this scenario is that the HII-region was density bound at least before the flare. When the HII-region becomes ionization bound, it can emitt an H_α-emission of 7.9×10^{32} ergs s^{-1} and a 6578 Å CII-emission of 2.3×10^{28} ergs s^{-1} for μ Cen (Table 1). The H_α-flux is more than the observed value and hence can be made consistent with observation by reducing the matter in HII-region. The expected 6578 Å emission, however, is smaller than that observed, if the emission is from CII 3p $^2P^o - 3s\ ^2S$ transition as suggested by Baade et al (1988). For HR 4123 the expected emission of both H_α and 6578 Å from an enhancement of the gas envelope are small compared to those observed due to paucity of Lyman continuum photons from the B8 star. Thus this process can only explain H_α-flare in μ Cen but not the 6578 Å emission in it nor H_α and 6578 Å emissions in HR4123.

3. The flare emission is due to non-radial oscillation or surface inhomogenuity of Be-star

The detail how non-radial oscillation of the star could give rise to observed flare emission is not clear, as the detail of the interaction of pulsation of the star with the surrounding nebula has not yet been worked out (cf Baade 1987). However, the rotation of an inhomogeneous surface brightness can increase the extent of HII-region by supplying ionizing photons from hot spots in the star (cf. Balona and Engelbrecht 1986 and Harmanec 1984). The non-radial pulsation may acts very similarly. In both cases then one can think that the envelope is supplied by additional ionizing photons during flare time. This, of course, can explain the H_α-emission from μ Cen. However, this will not be adequate to explain 6578 Å emission, if this is attributed to CII-line, because the expected ratio of the H_α and 6578 Å emissions in HII-region is proportional to $n_C/n_H \approx 3 \times 10^{-4}$, whereas the observed ratio is about 0.1. Moreover, to explain the flare emission of HR4123, an increase of temperature of 10000°k is necessary to have enough ionizing photons to give the observed H_α-emission strength. Note that even in presence of such increase, the 6578Å emission cannot be explained as argued above. Whether such a large amplitude inhomogenuity can exist on the surface or whether such large amplitudes wave can be produced for sufficient duration on the surface of the star are another question. However, it is clear that both the processes will fail to explain the emission strengths of H_α and 6578 Å emission simultaneously, if the latter emission is attributed to $3p^2P^o - 3s\ ^2S$ transition of CII-ion.

4. The flare emission is from transient HII-region produced by X-rays from a compact companion of Be Star:

This explanation is proposed by Ghosh et al (1989). According to this idea a compact object (neutron star, white dwarf or black hole) will emit X-rays due to acretion of matter from circumstellar matter of Be star when the circumstellar gas reaches the compact object. The luminosity of the emitted X-ray could vary from 10^{33} ergs s^{-1} to 10^{38} ergs s^{-1} depending the nature of compact object and the density of the circumstellar gas. The X-rays from compact object will give the extra ionization needed for flare emission. The emission from increased ionized material will give the H_α-flare emission. Note that the extra ionization occurs when compact object is X-ray bright which happens only when circumstellar gas reaches the orbit of the compact object. If the X-ray luminosity is $\sim 2 \times 10^{35}$ ergs s^{-1}, the HII-region around compact object extends about 10^{11} cm (model 3 of Kallman and McCray 1982). This HII-region gives a H_α-emission of 5.0×10^{32} ergs s^{-1}. Note that this flux is in good agreement with the observed H_α-flux in HR4123. The CII-emission from the HII-region is, however, small compared to its observed value. There is, however, a novel way to get CII- emission in the present case. Note that together with HII-region, a CII-region of extension of $8 \times 10^{12} - 2 \times 10^{13}$ cm exists around Be-star. Further, the HII-region around compact objects has a column density of 10^{23} hydrogen nucleii. Thus the photons with energy greater than 3 KeV can penetrate into the CII-region leaving the HII-region around the compact object. Note that this high energy photons will not ionize hydrogen, as ionization crosssection of hydrogen in small. However, the ionization crosssection of CII and higher atomic species will be significant, because of k-shell ionization. Thus the CII-emission at 6578 Å can occur through following sequence of transitions.

$$CII \stackrel{h\nu}{\to} CIII^* \stackrel{\gamma}{\to} CIII \stackrel{e}{\to} CII^* \to CII.$$

the 6578Å emission being produced by last transition which may be dielectronic (Nussbaumer and Storey 1984) or direct recombination. Note that this process, as it occurs in CII-region outside HII-region, can give 6578Å emission flux much higher than that expected from HII-region. The detail of the method has not been worked out. But it is encouraging to note that if 3/10th of 3 KeV and higher photons give rise to a 6578Å photons the production of 8.8×10^{31} ergs s^{-1} of 6578 Å photons will need 10^{44} photons s^{-1} i.e. an energy of 1.4×10^{35} ergs s^{-1}, which is the same order as that necessary for the production of H_α-flare emission. Thus the present model can produce both H_α- and CII, 6578 Å flare emission of HR4123, if the X-ray spectrum of compact companion of Be-star has about equal energy on bothside of 3KeV. The X-ray luminosities needed on either side of 3 KeV for the explanation of the observed emission in μ Cen are respectively 5×10^{34} ergs s^{-1} and 1.4×10^{34} ergs s^{-1}.

5. 6578Å flare emission is CI-recombination emission:

There is a possible alternative explanation of 6578 Å emission that the emission is not due to CII- $3p\ ^2P - 3s\ ^2S$ transition but due $6d\ ^1F - 3p\ ^1D$ transition of CI- at 6578.77Å. This emision will be produced in CII-region as a result of recombination of CII ions. A recombination coefficient of 8.2×10^{-12} cm$^3 s^{-1}$, gives an emission luminosity of 10^{33} ergs s^{-1} from a spherical CII-region of radius 8×10^{12} cm. This should be compared with the observed luminosity of 4.2×10^{31} ergs s^{-1}. The derived large luminosity is due to the assumption of large recombination rate. The assumed recombination rate is the total CII-recombination rate. The proposed transition has an upper level with significantly high quantum number. Hence recombination to this level is expected to be much smaller than the total recombination rate. A perusal of variation of hydrogenic recombination rate with upper quantum number suggests that the recombination rate to 6d ^{1} F may be a factor 50 less than the total recombination rate. This reduced recombination rate will reduce the emission luminosity to 2×10^{31} ergs s^{-1} which is just half of that observed. The increase of CII-region volume due to emission from compact object will, then, give the flare emission at 6578Å. Thus the enhancement of H_α and 6578Å emission could be traced to the appearance of X-rays in the compact object. Also, as H_α and 6578Å emissions are now arising from HII and CII-regions respectively, their relative strength could be different from that when both arises from HII-region.

References

Baade, D. (1987), In Physics of Be Star, I.A.U. Coll. 92, eds. A. Slettebak and T.P. Snow. p. 361 (Cambridge, University Press).
Baade, D., Dachs, J., Waygaert, V.D.R. and Steeman, F. (1988) Astr. Ap. 198, 211.
Balona, L.A. and Engelbrecht, C.A. (1986), Mon. Not. R. Astr. Soc. 219, 131.
Ghosh, K.K., Apparao, K.M.V. and Tarafdar, S.P. (1989), Ap. J. 344, 437.
Harmanec, P. (1984), Bull. Astr. Inst. Czechosl. 35, 193.
Kallman, T.R. and McCray, R., (1982), Ap. J. Suppl. 50, 263.
Nussbaumer, H. and Storey, P.J. (1984), Astron. Ap. Suppl. 56, 293.
Peters, G.J. (1984), Pub. Astr. Soc. Pacific 96, 960.

Table 1

	μ Cen	HR 4123
Spectral Type	B2Vpne	B9
B-V	-0.21	-0.04
E(B-V)	0.04	-0.01
T_{eff}	22500°k	13000°k
R_*(cm)	3.8×10^{11}	2.0×10^{11}
Luminosity in Lyman Continuum (ergs s^{-1})	3.6×10^{34}	1.1×10^{30}
Luminosity in CI-ionizing continuum (ergs s^{-1})	3.7×10^{36}	2.7×10^{34}
Extent of HII-region (n=10^{12}cm^{-3}) (cm) ($\alpha = 2.6 \times 10^{-13}$ cm^{-3} s^{-1}) (R=10^{12} cm)	5.1×10^{8}	1.5×10^{4}
Dimension of CII-region (C/O=3×10^{-4})(cm) $\alpha_c = 8.2 \times 10^{-12}$cm^{+3} s^{-1}	1.9×10^{13}	7.8×10^{12}
H_α emission from HII-region ergs s^{-1} (1.24×10^{-25} ergs cm$^{+3}s^{-1}$)	7.9×10^{32}	2.3×10^{28}
6578 Å CII-emission from HII-region (ergs s^{-1})	2.4×10^{29}	7.0×10^{24}
Equivalent width of H_α	0.9 Å	31.6 Å
Equivalent width of 6578 Å(0.04Å×3)	0.12Å	4.94Å
Observed H_α emission (ergs s^{-1})	6.6×10^{31}	2.7×10^{32}
Observed 6578Å emission (ergs s^{-1})	2.9×10^{30}	4.2×10^{31}

REVIEW OF DECAMETER WAVE RECOMBINATION LINES:
PROBLEMS AND METHODS

A.A. KONOVALENKO
Institute of Radioastronomy Academy of
Sciences of the Ukrainian SSR
4 Krasnoznamennaya str.
Kharkov 310002 USSR

ABSTRACT. There we present the main investigation results of the radio recombination lines in the decameter wave range. The detection 10 yeas ago of carbon lines with quantum numbers n>600 at ν<30 MHz gave new possibilities for studying characteristics of low density interstellar plasma. The most complete data have been received for the source of Cas A. Some problems with regard to choosing a model for the medium forming the carbon lines in the decameter range (n= 768 - 603) and for shorter waves (up to n = 273) have not yet been solved definitively. The methods of high sensetive and interference-proof decameter radio spectroscopy using the UTR-2 radiotelescope allow to record recombination lines of the excited carbon for different components of the interstellar medium.

I. INTRODUCTION

I.I. Detection history

The radio recombination lines (RRL's) of interstellar atoms have become one of the most effective means of radioastronomical investigations. It should be emphasized that an investigation of such lines (as any other radiolines) was always a deal of the high frequency radioastronomy. At the same time, there has always been an interest to investigations of longer-wave lines as is discussed in some papers (Shaver (1975), Pedlar and Davies (1980). Such investigations give a possibility of studying the most rarefied regions of interstellar plasma, which, in spite of low density of charged particles, play a significant role in the energetics, dynamics and evolution of galactic substance. The relevant data allow us also to judge processes concerning deviations from equilibrium conditions, interaction of highly excited atoms with charged particles and radiation.

Ten years ago, the longest-wave RRL was H300α , λ = I.25m (Pedlar et al. (I978). Attempts to record lines of longer wavelength were unsuccessful.

Our analysis, based mainly on the theoretical results of Shaver (I975), showed that the above-mentioned problems might be accessible and interesting to study up to very long radiowaves, i.e. decameters ones (ν < 30 MHz). In the middle of I970-s, we organized radiospectroscopic investigations using the largest UTR-2 telescope of the decameter range. At first, some searching observations of hydrogen RRL's were made, which gave a negative result (Konovalenko and Sodin (I979). However spectral lines at decameter waves were found in the direction of Cas A in I978 by Konovalenko and Sodin (I980), being later identified as carbon RRL's corresponding exceptionally high quantum numbers n > 600 (Blake, Crutcher and Watson (I980), Konovalenko and Sodin (I980). Up to now carbon lines in absorption, except for the decameter range (ν = I6.7 - 30 MHz, n = 732 - 603) (Konovalenko (I984), were also found for meter waves: ν = 42-84 MHz (Ershov et al. (I984), ν = 26-68 MHz (Anantaramaiah et al. (I985), ν = 34 - 325 MHz (Payne et al. (I989). The lines were observed in emission at frequencies more than ~ 200 MHz.

The detection of RRL's of so low frequencies results in some interesting phenomena. Further, a carbon and its ions play a peculiar role in the physical and chemical processes in the interstellar medium. Therefore, alongside with other observation methods of ionized carbon (in the UV - and IR - ranges, or high frequency RRL's), independent investigations of the RRL's at extremely low frequencies are important and interesting.

I.2. The theoretical data

To interpret the observations of the decameter carbon RRL's the results of Shaver (I975) might be used. In conditions characteristic of the decameter range the relative intensity for a maximum in the RRL is described by the following relation:

$$\frac{\Delta T_L}{T_c} = -\tau_L = \tau_L^* b_n \beta_n , \qquad (1)$$

where τ_L is the optical depth in the RRL;
τ_L^* is the equilibrium optical depth:

$$\tau_L^* = \frac{1.92 \cdot 10^3 N_e^2 S}{\Delta \nu_D T_e^{5/2}} \left(1 + 1.48 \frac{\Delta \nu_P + \Delta \nu_R}{\Delta \nu_D}\right)^{-1} ; \qquad (2)$$

T_e, N_e are the electron temperature (K) and density (cm^{-3});
S is the path length along the region (pc);
$\Delta\nu_D$, $\Delta\nu_P$, $\Delta\nu_R$ are the doppler, collisional and radiative components of the broadening (kHz);
b_n is the departure coefficient;

$$\beta_n = 1 - \frac{kT_e}{h\nu} \frac{d\ln b_n}{dn}. \qquad (3)$$

Assuming the populations of hydrogen and carbon levels to be identical, the values of b and $d\ln b_n/dn$ can be taken from data of Shaver (1975), but an extrapolation is necessary for n>500. For n=600-700 $b_n \sim 1$ and $\beta_n < 1$ for $T_e \sim$ 20 K, N_e =0.1-1 cm^{-3} (the lines are in absorption) or $\beta_n < 0$ for $T_e > 50$ K, $N_e < 0.1$ cm^{-3} (the lines are in emission).

The collisional component of the broadening for the data basis of Gee et al. (1976) is described by the approximate formula

$$\Delta\nu_P \approx \frac{N_e n^4}{T_e^{3/2}} \frac{7 \cdot 10^{-3} \ln(1+\beta)}{(2z+z^2)(1.4n+\beta n)} \, (kHz), \qquad (4)$$

where $z = 2.1 + 1.5\beta$;
$\beta = 1.58 \times 10^5 / nT_e$.

The radiative broadening (Shaver (1975) is

$$\Delta\nu_R \approx 7.6 \cdot 10^{-17} n^{5.8} \, (kHz). \qquad (5)$$

For this estimate the experimental averaged temperature of the galactic background at 25 MHz, $T_b = 37 \cdot 10^3$K with a spectralindex $\alpha = -2.6$ has been used.

The total line width at the half-intensity level can be estimated by the formula

$$\Delta\nu = \sqrt{(\Delta\nu_P + \Delta\nu_R)^2 + \Delta\nu_D^2}. \qquad (6)$$

Principally different conditions of forming the low frequency carbon RRL's are considered by Watson et al. (1980), Walmsley and Watson (1982). It is found that for carbon ions at T_e= 50-100 K and $N_e < 1$ cm^{-3}, the dielectronic-like recombination mechanism may be effective, resulting in an essential change of the high level populations, a negative derivative $d\ln b_n /dn$ and coefficient increase $b_n\beta_n$ up to 10-100. This may result in a considerable enhancement of the carbon RRL's in absorption.

2. THE EQUIPMENT AND METHODS

A characteristic feature of the decameter RRL's is their extremely low intensity $I \cdot 10^{-3}$ of the background level. Moreover, radioastronomical investigations at such low frequencies are connected to various difficulties. It is evident that in such situation it is important to understand to what degree the corresponding experiments are possible and real.

2.1. The parameters and operations of the UTR-2

The UTR-2 radiotelescope (Braude et al. (1978), to a considerable extent, meets the set requirement due its large effective area, broadband, rather high angular resolution, long observation of the source, minimum level of parasitic spectral effects. To realize the maximum sensetivity, reliability of line identification and definition of spacial characteristics various radiotelescope configurations are used (the parameters being given for 25 MHz and zenital beam orientation):
(a) the N-S antenna: the size of the phased electrically-driven array is 1800 x 52 m, effective area is $A_{ef} \approx$ 100000 m^2, beam size at half power level is $\theta \approx 20' \times 12°$;
(b) the W-E antenna: the size is 900 x 52 m, $A_{ef} \approx$ 50000 m^2, $\theta \approx 40' \times 12°$;
(c) the summation of the antenna signals (N-S+W-E): $A_{ef} \approx$ 150000 m^2, the beam has complex form;
(d) the multiplication of the antenna signals (N-S x W-E): the beam is symmetrical in a cross-section, $\theta \approx 30'$, there are no regions of low spacial frequencies on the UV-plane, i.e. the radiotelescope is low-sensetive to the components of more than 30'.

2.2. The spectrometer

At present we use two specially designed correlation spectrum-analyzers of Weinreb's type. This analyzer type, as we belive, is the most convenient for corresponding investigations. One equipment set contains the radio receiver and 32-channel correlator, the other having the 128-channel correlator divided into 4 groups of 32 channels and 4 of receivers. Thus, we have an opportunity of simultaneous observations of 5 RRL's which increases the sensetivity, noise-stability and reliability of the measurements. Our work showed that for low interference-effects, particularly when seeking lines, the most convenient parameters are as follows: analysis band of 20 kHz, (± 120 km/s), resolution of 1-3 kHz (12-36 km/s).

Note that for very low frequencies, there is an unique possibility of increasing the sensetivity by a substantial increase of a number of the lines observed. The distance between the lines is $\delta\nu \sim 120$ kHz for the decameter range (at 1420 MHz $\delta\nu \sim 25$ MHz). About 90 lines are within ±5 MHz in the decameter range, and only one is near 1420 MHz.

2.3. The observation methods

When using low frequency radiospectroscopy one has to come into collision with various kinds of interference signals: stationary interferences from the ether, combination noises, outher and inner interferences from digital devices and beating oscillators. The struggle against these interferences consists in using the two principles: (a)interference attenuating; (b)distinguishing of both interference features and useful signals.

The first principle is carried out by taking the following measures:(1)the choice of the observation objects (background source, coordinates); (2) the choice of observating time (season, time of the day); (3) the choice of observating frequency (interference statistics); (4)the increase of devices linearity, filtration; (5) reliable screening of the blocks, transmission lines, matching of lines and elements; (6) attenuation of parasitic radiation of the digital and receiving devices; (7) narrowing of the analysis band ; (8) modulation operation, the accurate balance of the reference and measurement channels; (9) the spacial selection (beam multiplicatin, beam switching).

To have a reliable identification of the found lines and for a further attenuating of interferences we use: (1) long-duration observations; (2) selection of interferences of various levels; (3) simultaneous observations at 5 frequencies and their averaging; (4) the telescope additive operation when seeking the absorption lines; (5) frequency-tune shift over the band; (6) the experiment simultation using the calibrating noise-signal of the same duration; (7) the two independent equipment sets; (8) the spacial tests; (9) the orbital-shift observations.

As a result of the measures taken, the useful telescope operation time near 25 MHz is about 60% of the total operation time of the given program. To illustrate possibilities of the method, Fig. I(a) shows the carbon line, with n = 637-641, found in the direction of DR 21 (resolution of 2.8 kHz, integration time of 330 h, telescope operation of (N-S+W-E). The r.m.s. noise-level is 2.5×10^{-5}. The (b) spectrum shows the fluctuations level when using, insted of the antenna, the noise oscillator with the noise temperature equal to that of the antenna and the integration of 580 hours (the r.m.s. noise is 1.5×10^{-5}). The given spectrum of DR 21 was obtained for about 15 seven-hours (not daily) observations during the day-time in the period of February-March in 1987.For this

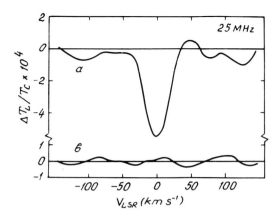

Figure I. The spectra illustrating the measurement sensetivity.

time period the solar activity was very low and the observation conditions were very favourable. From the spring of 1988, the solar activity increased sharply and the observations became possible only at night as a rule.

Thus in spite of the difficulties in the low frequency radiospectroscopy, a high sensetivity of the measurements is quite real.

3. THE INVESTIGATIONS TOWARDS CASSIOPEIA A

3.I. The new data of the decameter range

The observations of the decameter carbon RRL's towards Cas A are given by Konovalenko (1984). Recently we have found the lines of C 764α - C 768α at I4.7 MHz. At present, these are the lowest frequency lines and their quantum number is the highest. Fig.2(b) shows the averaged line (the antenna operation is N-S+W-E, the effective integration time T=50 h, the resolution is 2.8 kHz) in

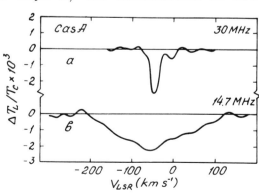

Figure 2. The lowest frequency lines towards Cassiopeia A.

comparing to the line C603α ($\nu \sim$ 30 MHz) from Konovalenko(1984). As seen, the line corresponding to such high n, is considerably broadened. Its parameters agree well with the formula calculations of (I)-(6) over the range of physical conditions N_e =0.05-0.3 cm^{-3}, T_e=I00-20 K.

In paper of Konovalenko (1984) the detection of decameter lines corresponding to the Orion arm (V_r = 0 km/s) and the weaker ones than for the Perseus arm (V_r=-45 km/s) is reported. However the lines near 0 km/s were not observed at frequencies 30 MHz, exept for 325 MHz(Payne et al.(1989). Fig.3 (a) shows the carbon line spectrum measured near 25 MHz(n=637-641) with the analysis band of 20kHz and resolution of 1.4 kHz, T=40 h. Fig.3(b) and 3(c) are with the band of 5kHz, resolution of 0.4 kHz, T=100 h, the line corresponding to the Perseus arm being out of limits of the analysis band. The

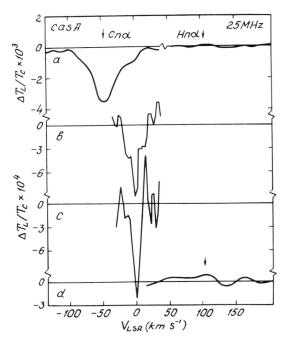

Figure 3. The lines at 25MHz for Cas A.

(b)-spectrum was obtained in summer of 1987(operation of N-S+W-E) and the (c) -spectrum was obtained in the same season of 1989 (operation of N-S). As seen, the lines are observed near zero velocity, their intensity being $1 \cdot 10^{-3}$, the width of 1 kHz. Fig.3(d) shows the spectral region near H 640α line, obtained with resolution of 2.8 kHz and T=100 h, which gives a new upper estimate of the hydrogen RRL's intensity at low frequencies: $\Delta T_L/T_C < 7 \times 10^{-5}$.

3.2. The interpretation problems

Though we have a large amount of data obtained for Cas A, the problem with choosing a model of the medium forming the observed lines are for the present not completely solved. The two alternative models are considered: with the electron temperature, T_e= 35-100 K, the electron density, N_e=0.05-0.15 cm^{-3}, for the agreement of the available experimental data one has to use a mechanizm of the low-temperature dielectronic recombination and a colder model with T_e= 20 K and N_e= 0.3 cm^{-3}(Walmsley and Watson

(1982), Konovalenko (1984), Ershov et al.(1984). Until recent time, the RRL's at >100 MHz were not yet observed, that is why the first model was believed more preferable. However the detection of the carbon emission lines over 200 to 325 MHz by Payne et al. (1989) with the optical depth up to $1.8 \cdot 10^{-3}$ showed a better agreement of the intensity lines with the calculations for the low temperature model, though judging by the totality of the physical conditions and processes known elsewhere, the warm model is more real (Walmsley and Watson (1982), Konovalenko(1987), Payne et al. (1989).

It is evident that one should have additional investigation to identify of models. First of all it is nessecary to understand the accuracy of defining all the line parameters both at the low-and high-frequencies, having decreased the random and, possibly, systematic errors.

An effective method of defining the medium parameters is the measurement of the ratio of intensities α and β -lines, which is expressed by

$$\frac{I_\alpha}{I_\beta} = 3.6 \frac{\mathcal{E}_n \beta_n(\nu)_\alpha}{\mathcal{E}_n \beta_n(\nu)_\beta} = 3.6 \frac{\mathcal{E}_n \beta_n(\nu)_\alpha}{\mathcal{E}_n \beta_n(\nu/2)_\alpha} \qquad (7)$$

and has its peculiarities at low frequencies (Konovalenko (1987). Fig.4 shows the results obtained by using (7) for the warm, $T_e = 100$ K, $N_e = 0.1$ cm^{-3} (a), and cold (b) models. It is seen that optimal frequencies are within 30-60 MHz where the calculation results differ considerably. The experiment using the given method was described by Leht et al. (1989). The observed intensity ratio is better for the warm model but it is desirable to have measurements of higher accuracy

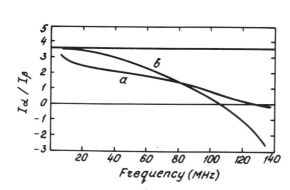

Figure 4. The intensity ratio of and -lines at low frequencies.

If the model with low temperature and increased density consisting of compact ($\sim 1'$)details is valid, the true optical depth will be sufficiently large (up to 10^{-2}), so that it can be recorded with the decameter interferometer URAN with the base of 40 km.Moreover,as at >30 MHz the lines are practically not broadened and are devided into separate components (Payne at al.(1989) corresponding to various regions, high frequency aperture synthesis might locate these regions.

One can apply the method of defining recombination mechanism using the carbon decameter RRL's with respect to their width and polarity (Konovalenko (1984). It is evident that the boundary condition between the existence of the low frequency line in the emission and absorption, as follows from (I), is

$$(\kappa T_e /h\nu) d\ln b_n /dn = 1 . \qquad (8)$$

This condition for the hydrogen approximation of the departure coefficients at 25 MHz (n=640) is shown at Fig. 5 with the shaded zone the boundaries of which correspond to the data on $d\ln b_n/dn$ from Shaver(1975) and Walmsley and Watson (1982). Also, there is given the connection N_e with T_e for different collisional width: (a), (b) and (c) for 0.03, 0.3 and 3 kHz, respectively. It is seen that for the low collisional width the existence of lines in the absorption requires temperatures not higher than 20 K. If the dielectronic-like mechanism is valid, the dependence corresponding to the condition of (8) will be a shaded curve. As seen, practically for all the quantities of N_e and T_e, the carbon line with n=640 will be observed in the absorption. This method may be applied to analyse the absorption lines in the Orion arm (Fig.3). In this case the operation of the dielectronic-like recombination process is quite possible.

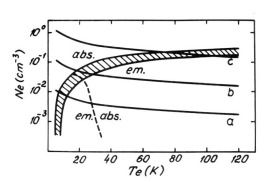

Figure 5. The conditions of observing the decameter lines of Cnα in the absorption and emission.

The methods discussed and also possibilities of the high sensetive spectrum measurements at low frequencies indicate reality of the arising problems to be solved.

4. THE POSSIBILITIES OF INTERSTELLAR MEDIUM BEING INVESTIGATED IN THE DECAMETER WAVE RRL'S

4.1. The investigation objects

The analysis has shown that for a wide range of physical conditions characteristic of cold components of the interstellar medium for $N_e < 1$ cm^{-3}, the carbon RRL's may be more preferable at the decameter waves than at higher frequencies ~1000 MHz (Konovalenko (1987). This makes the low frequency lines a convinient tool for studying various objects. Let us consider the most important of them.

One of the main structure components of the interstellar medium is the diffuse clouds of HI. The electron density caused by the total carbon ionization is to be 0.001-0.1 cm^{-3}, i.e. very low and seeming to be inaccessible for recording. However if the dielectronic-like mechanism enhancing the decameter lines is valid, there appears, when using the latter ones, a possibility of obtaining some information practically inaccessible with other astrophysical observation methods.

Surprisingly out of a large number of the dark clouds investigated at high frequencies, CII-regions were found only in some of them. However it is quite possible that these regions may be found using the decameter radiolines.

The partially-ionized medium in the layer between the HII-region and the ambient neutral gas reflects an interrelation, energetics and evolution of the complex. At high frequencies near a large number of the HII-regions, lines of carbon and heavy elements were detected. The complex and inhomogeneous structure of these regions makes them interesting for decameter observations.

Possibilities of decameter investigations of the RRLs can be also completed by the more detailed observations of the lines of hydrogen and heavy elements, appearing as a result of the high temperature dielectronic recombination.

4.2. Some investigation results for the galactic objects in the decameter Cn lines

It is evident that most of the objects being of interest to the decameter radiospectroscopy may be observed only in the background distributed galactic radiation. The intensity observed in the line is defined as

$$\frac{\Delta T_L}{T_c} = -\tau_L \frac{\Omega_s}{\Omega_a} \frac{T_b}{T_b + T_f} \frac{T_a}{T_a + T_n} \quad , \tag{9}$$

where Ω_s, Ω_a, are the solid angles of the source and antenna beam;
T_b, T_f are the brightness temperatures of the background behind and in front of the object;

T_a, T_n are the antenna temperature at the input of the receiving system and the noise temperature of the latter.

Even for large antennas of the type of UTR-2, the ratio of Ω_s/Ω_a may be much less than 1. If the distance up to the object is less than 1 kpc, the ratio of $T_b/(T_b+T_f)$ is close to 1, but it also may be small for considerable distances. The ratio of $T_a/(T_a+T_n)$ for the UTR-2 is about 1.

Possibly some of the above-mentioned reasons explain the so far few positive results of seeking lines at low frequencies (Konovalenko(1984), Anantharamaiah et al. (1988). However, in spite of the problems pointed out, the elaborated observation methods (see 2.3.) allow to set a task of complex review of the types of objects which are given in the previous section. Favourable conditions might be in the directions of the maximum column densities of HI (l= 0,35,75,180°). We have interesting observation conditions for the carbon RRL's for the regions manifesting as zones of the decreased brightness temperature of the galactic continuum background, which might be caused by presence of the cold partially-ionized gas. For T_e < 100 K and N_e = 0.1-1 cm^{-3}, the optical depth of the continuum may be 0.1 and more at frequencies<30 MHz, which is accessible for recording with the UTR-2.

For last years, we conducted the search of carbon decameter lines in various objects and for all of them where we succeeded in realizing a large integration (up to 100 and more hours) and in having the noise-level of $1 \cdot 10^{-4}$, a positive result was obtained. These are the following regions: G 75.0 + 0.0 ; NGC 2024; S 140; DR 21; L 1407. We also conducted previous investigations of the objects of PerOB 2; W 49; W 3; G 180.0+ 0.0 ; 3C 144; M 16; G 0.0 + 0.0 ; ρOph; G 35.0 + 0.0 and some others; however

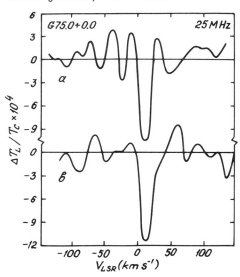

Figure 6. The decameter lines in the region of G 75.0+0.0.

due to the comparatively small integration the 3σ-noise level is only $1 \cdot 10^{-3}$ and the upper intensity estimation can be made only. It is hardly probable that the optical depth in the decameter lines not would considerably exceed the level of $1 \cdot 10^{-3}$. Therefore, for a successful use of the given investigation method, one has to realize the sensetivity of $1 \cdot 10^{-4}$ and even less. To illustrate these suggestions, we give the observation data of the object of G 75.0 + 0.0. The carbon lines of C640α in this object were found by Konovalenko (1984) at V_r = 0 km/s, Fig.6(a). In paper of Anantharamaiach et al. (1988) there are described the investigations made using the 93-m antenna NRAO with relative low integration time, which did not allow to find the line for the given velocity but a datail of it did appear near -77 km/s. In spring of 1989 we conducted repeated measurements with the UTR-2 using of N-S-antenna and T=46h. The spectrum is shown in Fig.6(b) where one can see also the line near 0km/s, but the line at -77 km/s is absent. It is most likely that such discrepancy is connected to insufficiently large integration at the 93-m antenna.

5. CONCLUSIONS

The detection of decameter carbon RRL's gave new possibilities of studying the low-density interstellar plasma. In the direction of Cas A, the Cnα lines are observed both for the decameter range of up to n=768, ν =14.7 MHz and for the shorter waves. The available uncertainies in choosing the medium models may be solved by more detailed investigations. In spite of specific spectroscopy difficulties at the lowest frequencies, a high sensetivity of the measurements is in principle possible. This allows to study various components of the interstellar medium with ionized carbon with an electron density of <1 cm^{-3}. Up to the present, using the UTR-2 radiotelescope, the decameter carbon RRL's were detected successfully in a number of the galactic objects.

ACKNOWLEDGEMENTS

The author greatfully thanks L.G. Sodin for his taking part in many investigations, S.Ya. Braude for his stimulating these investigations, A.A. Golynkin for his taking part in designing new spectral equipment and P. Roelfsema for a reading of the manuscript.

REFERENCES

Anantharamaiah, K.R., Erickson, W.C. and Radhakrishnan, V. (1985)'Observations of highly excited carbon radio recombination lines towards Cassiopeia A', Nature 315, 647-649.

Anantharamaiah, K.R., Payne,H.E. and Erickson, W.S. (1988)'Detection of carbon recombination lines below 100 MHz towards Galactic Centre and M16', Mon. Not. R. Astr. Soc. 235, 151-160.

Blake, D.H., Crutcher, R.M. and Watson, W.D. (1980) 'Identification of the anomalous 26.13 MHz nitrogen line observed towards Cassiopaia A', Nature 287, 707-709.

Braude, S.Ya., Megn, A.V. and Sodin, L.G.(1978) 'Radiotelescope of the decameter range UTR-2', Antennas (Sov.) 26, 3-15.

Ershov, A.A., Iliasov, Yu.P., Leht, E.E., Smirnov, G.T., Solodkov, V.T. and Sorochenko, R.L.(1984) 'The low frequency excited carbon radio lines towards Cassiopeia A. Observations at frequencies 42,57 and 84 MHz', Sov. Astron. J. Lett. 10, 833-845.

Gee, C.S., Percival, I.C. and Richards, D.(1976) 'Theoretical rates for electron excitation of highly excited atoms', Mon. Not. R. Astr. Soc. 175, 209-215.

Konovalenko, A.A. and Sodin,L.G. (1979) 'Negative result of an attempt to detect hydrogen recombination lines at decameter waves', Sov. Astron. J. Lett. 5, 663-664.

Konovalenko, A.A. and Sodin, L.S. (1980) 'Neutral ^{14}N in the interstellar medium', Nature 283, 360-361.

Konovalenko, A.A. and Sodin, L.G.(1981) 'The 26.13MHz absorption line in the direction of Cassiopeia A', Nature 294, 135-136.

Konovalenko, A.A.(1984) 'Observations of carbon recombination lines at decameter wavelengths in the direction of Cassiopeia A', Sov.Astron. J. Lett. 10, 846-852.

Konovalenko, A.A.(1987) 'An investigations of carbon recombination lines in cosmic radioemission of the decameter range', Herald of the Academy of Sciences of the Ukrainian SSR (Sov.) 4, 17-31.

Konovalenko, A.A.(1984) 'Decameter excited carbon lines in certain galactic objects', Sov. Astron. J. yeet. 10, 912-917.

Leht, E.E., Smirnov, G.T. and Sorochenko, R.L.(1989) 'Observations of the atoms in excitation levels close to the limit in the Galaxy', Sov. Astron. J.Lett 15, 396-399.

Payne, H.E., Anantharamaiah, K.R. and Erickson, W.C. (1989) 'Stimulated emission of carbon recombination lines from cold clouds in the direction of Cassiopeia A', The Astrophysical J. 341, 890-900.

Pedlar, A. and Davies, R.D.(1980) 'Low frequency recombination lines', in P.A. Shaver (ed.), Radio Recombination Lines, Reidel, Dordrecht, pp. 171-183.

Pedlar, A., Hart, F., Davies, R.D. and Shaver, P.A. (1978) 'Studies of low frequency recombination lines from the direction of the Galactic centre and other galactic sources', Mon. Not. R. Astr. Soc. 182, 473-488.

Shaver, P.A. (1975) 'Theoretical intensities of low frequency recombination lines', Pramana 5, 1-28.

Walmsley, C.M. and Watson, W.D.(1982)'The influence of dielectronic-like recombination at low temperatures on the interpretation of interstellar, radio recombination lines of carbon', The Astrophysical J. 260, 317-325.

Walmsley, C.M. and Watson, W.D.(1982)'Very high Ridberg states (n=600) of carbon in the interstellar gas', The Astrophysical J.(Letters) 255, L123-L127.

Watson, W.D., Western, L.R., and Christensen, R.B. (1980) 'A new, dielectronic-like recombination process for low temperatures and radio recombination lines of carbon', The Astrophysical J. 240, 956-961.

LOW FREQUENCY RADIO RECOMBINATION LINES TOWARDS CAS A

R.L. SOROCHENKO and G.T. SMIRNOV
P.N. Lebedev Physical Institute, USSR Academy of Sciences
117324 Moscow, Leninsky prospect 53
USSR

ABSTRACT. Radio recombination lines (RRL) of carbon towards Cas A has now been observed over the range of frequencies 16.7 - 225 MHz. The obtained results allow to deduce ISM parameters in Perseus arm where the lines originate. Satisfactory agreement with the entire observation data is achieved by two models: the "hot" one with T_e=50 K and N_e=0.15 cm^{-3} and the "cold" one with T_e=18 K and N_e=0.3 cm^{-3}. The negative result in attempts to detect hydrogen RRLs with presence of carbon lines make it possible to define the upper limit of hydrogen ionization rate.

1. INTRODUCTION.

Konovalenko and Sodin (1980) with Kharkov UTR-2 radiotelescope detected an absorption line in Cas A spectrum at 26.131 MHz. They identified it as the F=5/2-3/2 hyperfine transition of nitrogen, ν=26.127 MHz rest frequency. The authors explained 4 kHz difference as a result of frequency deviation because of Perseus arm radial velocity equal to -40 km s^{-1} at that direction of the Galaxy. The interpretation created great difficulty in explaining the observed line optical depth $\tau_L = 3 \cdot 10^{-3}$. The estimated nitrogen abundance from the measurements $N_N/N_H = 8 \cdot 10^{-3}$ (Konovalenko and Sodin, 1980) exceeded the adopted Galaxy value ($N_N/N_H = 1.2 \cdot 10^{-4}$) by more than an order of magnitude.

Concerning the mentioned difficulties Blake et al. (1980) suggested that the detected spectral line is C631α line rather than nitrogen one. The line frequency, ν=26.126 MHz, corresponds to the nitrogen line frequency within the measurement accuracy. Blake et al.(1980) determined the interstellar medium (ISM) parameters which agreed with the observation data. Assuming local thermodynamic equilibrium (LTE) they obtained T_e=50 K, N_e=0.1 cm^{-3}, EM=0.07 cm^{-6}pc.

Subsequent observations confirmed that the identification of the spectral line at ν=26.131 MHz as a RRL of carbon had been correct. At the frequencies of C630α and C640α RRLs with an amendment for -40 km s^{-1} of radial velocity Konovalenko and Sodin (1981) detected in Cas A spectrum two more absorption lines with $T_L/T_C = -\tau_L = -(2-3) \cdot 10^{-3}$ (T_L and T_C are the temperatures of the line and Cas A in continuum respectively).

For the medium parameters adopted by Blake et al. (1980) for expla-

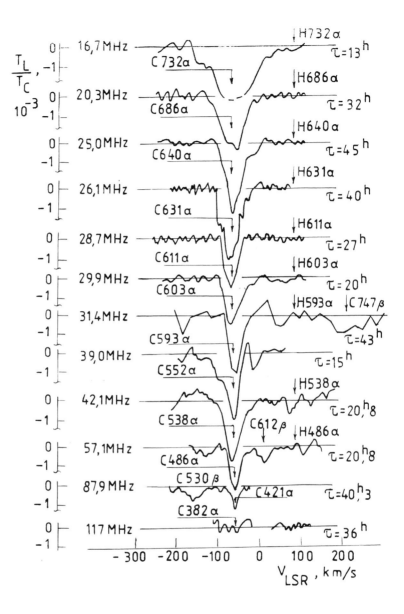

Fig. 1 Carbon recombination lines observed towards Cas A at Kharkov (ν<30 MHz) and at Pushchino (ν>30 MHz). The frequencies are indicated on the left side and the integration time on the right side. Arrows mark the expected position of hydrogen lines, that have not yet been detected

nation of the observed absorption in decameter RRLs, consideration of deviations from LTE (Shaver, 1975) led the to conclussion that at higher frequencies of 80-200 MHz the carbon RRL towards Cas A must be observed in emission with relative contrast $T_L/T_C =10^{-2}$. However the search for C400α line (ν=102.4 MHz) carried out by Ershov et al. (1982) had a negative result with the upper limit of $T_L/T_C =2 \cdot 10^{-3}$.

The RRL measurement results in meter and decameter ranges agree well with each other when the carbon line is assumed to originate in cool ISM with Te=20 K and Ne=0.3 cm^{-3}. The observation data could also be reconciled if Te=50 K and Ne=0.15 cm^{-3}, though in this case the consideration of low temperature dielectronic recombination (DR) is needed. Watson et al. (1980) have shown, the recombination of carbon in ISM can be accompanied by 2P$_{1/2}$-2P$_{3/2}$ transition between fine structure levels of the C$^+$ core and this process must effect the population of highly excited levels.

After calculation of level population in carbon up to n=900 Walmsley and Watson (1982a, 1982b) came to conclussion that it is the DR that is responsible for the observed RRLs of n=630-640. When electron densities Ne=0.01-0.1 cm^{-3} and temperatures Te=50-100 K the RRLs of carbon n=500-700 must be observed in absorption with amplification of intensities of the lines by dozens of times. For hydrogenic population and the above mentioned ISM parameters the C630α-C640α RRLs towards Cas A would be observed in emission rather than in absorption.

Later on a number of low frequency carbon RRLs have been detected towards Cas A. Ershov et al. (1984, 1987), Lekht et al. (1989) at Pushchino radiotelescope DKR-1000 in the range of 31-87.9 MHz detected also in absorption 6 α-lines and 4 β-lines, Anantharamaiah et al.(1985) at 91 meter radiotelescope in Green-Bank observed the lines of carbon in the range of 26 - 68 MHz. The optical depth of α-lines was equal to $=(1.3-4.8) \cdot 10^{-3}$. At the same time attempts to detect C382α RRL at higher frequency of 117.5 MHz were a failure with the upper limit $T_L/T_C <6 \cdot 10^{-4}$ (Ershov et al., 1987). Five more carbon RRLs were detected in decameter range down to the lowest frequency line C732α (ν=16.7 MHz)(Konovalenko, 1984).

Two new important results were obtained by Payne et al. (1989) in recent RRL investigations towards Cas A in Green-Bank. The obsevations carried out in the range of 34-325 MHz with a high spectral resolution yeilded the following: 1)carbon RRLs observed at low frequencies in absorption, at frequencies higher than 200 MHz are observed in emission; 2)splitting in two spectral elements with velocities of about -39 km s^{-1} and -47 km s^{-1} in profiles of the detected spectral lines is observed.

2. THE RESULTS OF RRL OBSERVATIONS TOWARDS CAS A.

Fig.1 shows spectrograms of carbon RRLs detected towards Cas A in the meter and decameter ranges in Kharkov (ν<30 MHz) and in Pushchino (ν>30 MHz); Fig.2 shows carbon RRLs observed in Green-Bank in the range of 34-325 MHz. All available data of carbon RRL measurements are compiled in Table 1. Besides the obsrevations mentioned in section 1 the Table 1 also contains the preliminary results of the recent observations

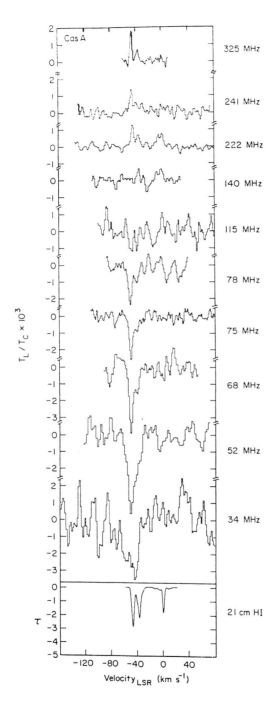

Fig. 2 Carbon recombination lines (Δn=1 transitions) observed towards Cas A at 10 frequencies in the range 34-325 MHz at Green-Bank. The quantum numbers corresponding to these frequencies are n=565, 502, 450, 440, 435, 385, 360, 310, 300 and 272. The HI absorption spectrum is taken from Mebold and Hills (1975). This Figure is taken from Payne et al. (1989).

of C537α-C539α lines carried out in Pushchino (Lekht et al., 1990) and the upper limit obtained by Shaver et al. (1976) in early attempt to detect RRLs towards Cas A. The purpose of the attempt was the search for H352α but the carbon line frequency was within the frequency range of the analysis. For repeated observations carried out by the same group of observers the latest more accurate measurements were taken.

In columns of Table 1 are: frequency range, names of lines, radial velocity, optical depth, half intensity line width, line integral intensity accounting for Voigt profile (see below), references to the original papers.

TABLE 1

Frequency (MHz)	Transition	V_{LSR} (km s^{-1})	τ_{Cn} (10^{-3})	ΔV_L (km s^{-1})	$-\int \tau_{Cn} d\nu$ (Hz)	Reference
325	C272α, C273α	−47.1±0.7	−1.83±0.08	3.2±0.16	10.3±0.7	19
"	"	−37.7±0.8	−0.63±0.05	4.8±0.5		
"	"	− 0.7±0.9	−0.39±0.1	2.2±0.6		
240	C300α–C303α	−48.2±1.8	−1.3 ±0.15	5.0±0.7	8.2±1.4	19
"	"	−40.5±2.2	−0.6 ±0.15	5.0±1.8		
220	C308α, C310α	−45.6±2.0	−1.0 ±0.1	7.6±1.0	8.9±1.3	19
"	"	−36.3±2.2	−0.6 ±0.1	5.7+1.3		
150.1	C352α		<|1.66|		<|6|	21
140	C358α, C359α, C360α, C362α		<|0.7|		<|2|	19
117.5	C382α		<|0.6|		<|2.5|	8
"	C481β	−43.2±4.5	−0.3 ±0.16	15 ± 11		8
115	C384α, C385α	−48.5±2.6	1.2 ±0.5	7.3±2.0	−5.2±2.0	19
"	"	−40.0±2.6	1.2 ±0.5	3.5±1.7		
110	C486β, C487β		<|1.4|			19
102.4	C400α		<|2|		<|15|	6
87.8	C421α	−44.1±0.6	1.34±0.09	15 ± 2	−6.2±0.9	8
"	C530β	−45.4±1.4	0.67±0.08	23.1±4.6		8
84	C427α	−42.4±3.2	1.57±0.42	24.2±8.8	−10.6±4.8	7
78	C435α–C438α	−50 ± 2	2.3 ±0.3	7.5±1	−7.0±1.2	19
"	"	−39.5±2.5	1.0 ±0.25	6.2±2		
75	C441α, C443α	−48.1±1.4	2.5 ±0.1	6.7±0.4	−6.3±0.5	19
"	C445α, C447α	−38.9±1.7	0.9 ±0.1	5.9±1.1		
"	C559β	−51.2±1.5	0.51±0.07	9.7±1.5		19
"	"	−38.0±1.5	0.53±0.07			
70	C450α, C451α	−50.0±2.2	3.2 ± 0.4	7 ± 1	−9.7±1.4	19
"	C456α	−40.0±2.7	1.7 ±0.3	7.5±1.5		
57	C486α	−44.7±1.0	2.35±0.22	17.1±2.5	−8.95±1.5	8
"	C612β	−43.4±3.3	0.79±0.21	21.4±7.9		8
52	C502α, C503α	−51.0±3.3	3.7 ±0.6	7 ±1.5	−12.5±1.9	19
"	"	−42.0±3.3	3.2 ±0.4	11 ±2		
42	C537α–C539α		3.4 ±0.1	19.9±0.7	−11.7±0.5	17
39	C552α	−48.6±2.1	3.88±0.4	29.5±5.2	−21.2±4.5	8
38	C554α, C556α	−49 ± 2	3.5 ±0.4	17.9±2.4	−9.5±1.7	1
34	C564α, C565α	−55.0±4.8	2.26±0.4	27 ±7		19
"	C576α	−42.0±3.3	2.7 ±1.0	8 ±3		
31.5	C593α	−52.4±1.8	4.8 ± 0.7	23.5±7.6	−16.3±5.6	16
"	C747β	−56 ±14	0.89±0.36	72.5±38		16
29.9	C603α	−46.8±3.0	2.7 ±0.3	28.1±3.0	−10.7±1.6	15
28.8	C611α	−46.5±3.1	2.5 ±0.6	31.2±3.1	−10.8±2.8	15
26.1	C631α	−44.3±3.4	3.6 ±0.3	40.2±3.5	−19.7±2.4	15
26	C629α, C634α	−45 ±5	3.8 ±0.5	32.9±5.0	−16.0±3.5	1
25.0	C640α	−46.1±3.6	3.7 ±0.3	38.4±3.6	−18.5±2.0	15
20.3	C686α	−46.7±4.4	2.8 ±0.6	59.1±8.9	−17.6±4.6	15
16.7	C732α	−44.8±18	2.5 ±1.0	126 ±27	−27.5±12	15

Numerous attempts have been also made to detect hydrogen RRLs towards Cas A. All of them were unsuccessful. The estimations of upper limits obtained from these observations are compiled in Table 2. The Table columns are: frequency range, line name, upper limit of optical depth, references.

TABLE 2

Frequency (MHz)	Transition	τ_{Hn} (10^{-3})	References
325	H272α, H273α	<1.5	Payne et al. (1989)
242	H300α	<0.56	Casse and Shaver (1977)
220	H308α	<0.5	Payne et al. (1989)
150	H352α	<1.6	Shaver, Pedlar, Davies (1976)
115	H384α	<1.4	Payne et al. (1989)
75	H443α, H445α, H447α	<0.8	Payne et al. (1989)
57.1	H486α	<0.63	Sorochenko and Smirnov (1987)
42.1	H538α	<1.1	Sorochenko and Smirnov (1987)
26.1	H630α	<0.3	Konovalenko and Sodin (1979)

3. DISCUSSION

3.1 Analysis of Observation Results of Carbon RRLs Towards Cas A.

From the very first observations of carbon RRLs towards Cas A following principal questions arised: 1)Where the detected RRLs originate - in the rarified ISM or in denser regions? 2)What is the source of carbon ionization - is it the diffused UV radiation with 912<λ<1101 Å or some particular B-stars as in the case of high frequency carbon RRL (Brown, 1980)? 3)What are the physical conditions in RRL origination region and first of all what are the temperature and the density?
    Two alternative models have been offered for interpretation of the RRL observations towards Cas A:
1)The "hot" model presuming that the low frequency RRLs in the direction of Cas A originate in comparetively rarified and warm regions with T_e=50-100 K and N_e=0.05-0.15 cm^{-3} (Blake et al., 1980; Ershov et al., 1982,1984,1987; Anantharamaiah et al., 1985; Konovalenko, 1984; Payne et al., 1989).
2) The "cool" model presuming that carbon RRLs originate in cooler and condenser medium with T_e=16-20 K, N_e=0.27-0.4 cm^{-3} (Ershov et al., 1982, 1984, 1987; Walmsley and Watson, 1982; Anantharamaiah et al., 1985; Payne et al., 1989).
    Let us analyse all the results of RRL observations obtained up to now to give an answer to these questions.
    The width of RRL. Fig. 3 shows the observed dependence of the line width of carbon as a function of n. Since the latest observations distinguished two components in RRL profiles (Payne et al., 1989) comparison of the widths was made for more intensive feature with V_{LSR}=-47 km s^{-1}. The width of low frequency unsplitted profiles is dimi-

nished by 8 km s^{-1} which is the distance between the components.

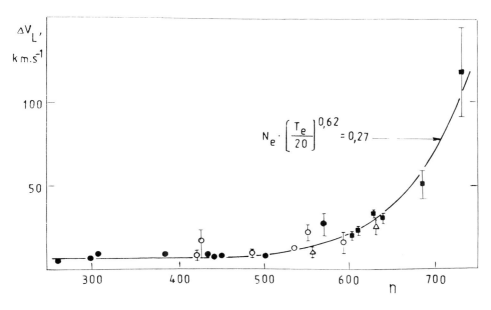

Fig.3 Width of the carbon RRLs ($\Delta n=1$ transitions) towards Cas A as a function of the principal quantum number. The squares are from the Kharkov observations (Konovalenko, 1984). The open circles are from the Pushchino observations (Ershov et al., 1984, 1987; Lekht et al., 1989, 1990). The triangles (Anantharamaiah et al., 1985) and the filled cirles (Payne et al., 1989) are from the Green-Bank observations.

As can be seen from the figure the line width increases with n. Under high excitation levels besides Stark broadening due to collissions with electrons a tangible contribution to the width gives broadening by background radiation. The line has Voigt profile which is a convolution of Gaussian (Doppler) and Lorentz profiles. The full width of a Voigt profile with a presision not worse than 1% can be expressed as (Ershov et al., 1984):

$$\Delta V_L = 0.53 L + \sqrt{0.22 L^2 + D^2} \qquad (1)$$

where L and D are the full widths of the Lorentz and Doppler profiles at half maximum. The width L comprises two terms:

$$L = (\Delta \nu_{col} + \Delta \nu_{rad}) \cdot 3 \cdot 10^5 / \nu, \text{ km s}^{-1} \qquad (2)$$

where $\Delta \nu_{col}$ is the Stark broadening, $\Delta \nu_{rad}$ is the radiation broadening, ν is the frequency. The widths are defined by:

$$\Delta \nu_{col} = 4.23 \cdot 10^{-4} (n/100)^{5.1} (T_e/20)^{0.62} N_e, \text{ kHz} \qquad (3)$$

$$\Delta \nu_{rad} = 3.82 \cdot 10^{-5} (n/100)^{5.65}, \text{ kHz} \qquad (4)$$

Having taken for the feature $V_{LSR}=-47$ km s^{-1}, $D=6.7$ km s^{-1} (Payne et al., 1989) and approximating the experimental results by the equation (1) and accounting for eqs. (2)-(4) one can obtain:

$$N_e(T_e/20)^{0.62}=(0.27\pm0.05), \text{ cm}^{-3}\text{K}^{0.62} \qquad (5)$$

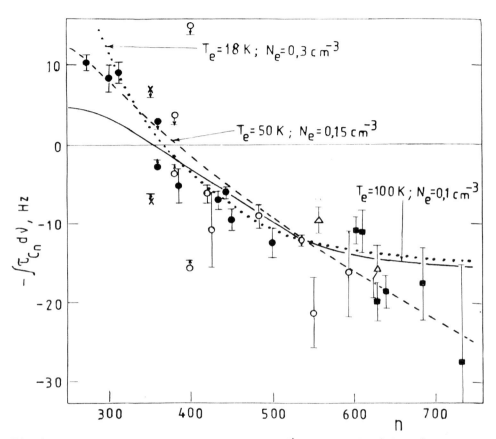

Fig. 4 Observed intensities of the carbon RRLs ($\Delta n=1$ transitions) toward Cas A as a function of n. The crosses are from Shaver (1976); the rest symbols are the same as in Figure 3. The solid line corresponds to the "hot" model, $T_e=100$ K, $N_e=0.1$ cm^{-3}; the dashed line - to the "hot" model, $T_e=50$ K, $N_e=0.15$ cm^{-3}; the dotted line - to the "cold" model, $T_e=18$ K, $N_e=0.3$cm^{-3}. See the text.

So the measurements of the width of carbon RRLs give an important information on physical conditions in the region of RRL origination. The measurements do not allow to define unambiguous values of temperature and density but they set the dependence between each other. This fact makes it possible to impose certain limits. Since by measurements in 21 cm line the ISM temperature towards Cas A does not exceed 100K (Mebold and Hills, 1975) it follows from expression (5) that the electron density in the regions of RRL origination must not be less than 0.1 cm^{-3}. It is nesesary to point out that the expression (5) does not depend on the population mechanism and on the dilution factor which is unknown yet.

RRL intensity. Fig.4 gives the measured intensities of carbon RRLs towards Cas A as a function of n. They are determined as:

$$-\int \tau_{Cn} d\nu = p \cdot \Delta\nu \cdot T_L/T_C \qquad (6)$$

where $T_L/T_C = -\tau_{Cn}$ is the line contrast, $\Delta\nu$ is half intensity width in Hz. The coefficient p accounts for the wings of Voigt profiles, it may be approximated by:

$$p = 1.57 - 0.507 \, e^{-0.85L/D} \qquad (7)$$

which corresponds to the exact data with 1% accuracy. Determination of the intensity by expression (6) and (7) is valid, to be precise, only in case when line parameters are obtained by fitting to a Voigt profile (Ershov et al., 1987). But when the line fits Gaussian profile the use of the expressions is not quite correct. However this error is much smaller than that appearing if line intensity is defined under assumption of pure Doppler broadening $\int \tau_L d\nu = 1.066\Delta\nu T_L/T_C$, especially for higher n. This reason, in particular, can be responsible for exceeding of the calculated values of RRL intensities over the measured values for n>600 in the analysis carried out by Payne et al. (1989).

For line components with $V_{LSR} = -47$ km s^{-1} and $V_{LSR} = -39$ km s^{-1} the Doppler widths D=6.7 km s^{-1} and 5.75 km s^{-1} respectively (Payne et al., 1989). For profiles with unresolved components D was adopted 15 km s^{-1}. The Lorentz width was determined by (2)-(5). The intensity of such profiles was defined as a sum value of both components. In cases when only the upper limits of line optical depth are defined for estimation of the limiting intensity the width of a component $V_{LSR} = -47$ km s^{-1} was taken for the line width or the analyser spectral resolution $\Delta\nu a$ if $\Delta\nu < \Delta\nu a$.

The comparison of experimental data with model structure Fig.4 gives the calculated values of RRL intensity for three cases: 1)"hot"model Te= 100 K, Ne=0.1 cm^{-3}, EM=1.3·10^{-2}cm^{-6}pc, 2)"hot" model Te=50 K, Ne=0.15cm^{-3}, EM=1.34·10^{-2}cm^{-6}pc, 3)"cold" model Te=18 K, Ne=0.3 cm^{-3}, EM=10^{-2} cm^{-6}pc.

The calculated curves are defined by the expression (Shaver, 1975; Ershov et al., 1984):

$$\int \tau_{Cn} d\nu = 2.05 \cdot 10^6 EM \cdot Te^{-5/2} b_n \beta_n \qquad (8)$$

where EM=$\int$NeNc+dl is the emission measure, Ne and Nc+ are the number densities of electrons and ions of carbon respectively, b_n is the coefficient characterising the non-LTE deviations of the population, β_n is the "maser" effect connected with it

$$\beta_n = \left(1 - \frac{20.8 \cdot Te}{\nu} \cdot \frac{d\ln b_n}{dn} \right). \qquad (9)$$

In expression (9) ν is the frequency in GHz.
The curves for the "hot" model were calculated allowing for the DR (Walmsley and Watson, 1982a). Calculations of the hydrogenic population were taken for the "cold" model (Ungerechts and Walmsley, 1978). For intermediate temperatures and densities the interpolation and for n>420

in hydrogenic population the extrapolation has been made. All the curves are refered to the observation results of C537α-C539α lines (Lekht et al., 1990) where the smallest measurement error was reached.

3.2 The Nature of Carbon RRLs Origination Regions.

The available data do not yet allow to unambiguously determine the values of electron temperature and density in the RRL origination region. At the same time all these parameters can be localised in certain limits. As Fig.4 shows there is a great discrepancy with the experimental data in the model with $T_e=100$ K, $N_e=0.1$ cm^{-3}; the measured intensities of high frequency RRL seen in emission considerably exceed the computed values. If for above mentioned model the population calculations ignor DR the discrepancy would be still more drastic: the lines with n<700 according to the calculations must be seen in emission what fully contradicts the observation results (Walmsley and Watson, 1982a).
Satisfactory agreement of calculated and measured RRL intensities for simultaneous compliance with the line widths (expression (5)) is reached in both "hot" and "cold" models with the following ISM parameters:
1) The "hot" model with $T_e=50$ K, $N_e=0.15$ cm^{-3} and with levels population accounting for DR. The model provides a good agreement of calculated and measured RRL intensities for both high and low frequencies. At the same time the turnover from absorption to emission in this model corresponds to somewhat higher n than it follows from the experimental data. The model predicts the turnover at n=385 whereas C384α-C385α lines are observed in absorption though with low signal/noise ratio. Yet the crossover is utterly critical to the density and with some refinements of the model and the results of the measurements better compliance may evidently be achieved.
2) The "cold" model with $T_e=18$ K, $N_e=0.3$ cm^{-3} and hydrogenic level population. The turnover from absorption to emission in this model takes place at n≈370 what better agrees with experimental data. The model however predicts a strong growth of RRL intensities for n<300; the measured intensity of C272α, C273α lines is noticeably lower than the calculated one.
The above given analysis points out that the RRL origination regions are rather densed formations. Since $N_e=0.15-0.3$ cm^{-3} the density of hydrogen in these regions must not be less than $N_H=450-900$ cm^{-3} if one proceeds from the solar abundance of carbon $N_C/N_H=3.3\cdot10^{-4}$. Allowing for the fact that only a part of carbon can be in the gas phase the density of clouds in the RRL origination region may be still higher probable $N_H = (1-3)\cdot10^3$ cm^{-3}.
An assumption has been put forward that the RRL origination regions of carbon are linked with the molecular clouds (Ershov et al.,1984,1987). Such clouds were discovered towards Cas A after observations of series of molecular lines. Goss et al. (1984) with an angular resolution of 10" picked out a number of H2CO clouds with an average size of 0.3 pc in Perseus arm in Cas A solid angle. According to the latest measurements of H2CO, CO, NH3 and OH lines the density of these clouds amounts to

$(1-3)\cdot 10^3 \mathrm{cm}^{-3}$ (Troland et al., 1985; Heiles and Stevens, 1986). The data on 21cm line point out that the distribution of neutral hydrogen towards Cas A in Perseus arm forms envelopes around molecular clouds with the density $N_e \approx 300$ cm^{-3} and average thickness $l \approx 0.19$ pc (Goss et al., 1984). According to Federman (1979) such HI envelopes are necessary for protection of the molecular gas from interstellar UV radiation.

Lines of formaldehyde from each cloud are narrow $\Delta V_L = 1-2$ km s^{-1} and have different radial velocities in the range from -36 km s^{-1} to -48.5 km s^{-1} (Goss et al., 1984). Profiles of H2CO and CO lines which represent an integral over the Cas A solid angle are very much alike. They have a specific two-humped appearance with climaxes at $V_{LSR} = -48$ km s^{-1} and $V_{LSR} = -39$ km s^{-1} and are approximately equal to the width of carbon RRL for n<450 where Stark and radiation broadenings are still negligible.

The obtained data imply that carbon RRL towards Cas A originate in two or more compact formations "HI envelope-molecular cloud" rather than in large diffused clouds. Ionization of carbon in this case is produced by outer radiation. An additional argument for such a conclussion is the restricted results in attempts to detect carbon lines at low frequencies in other directions of the Galaxy. If carbon lines originated in rarified ISM likewise 21cm line they might be observed at low frequencies throughtout the Galactic plane in absorption against the nonthermal background. Nevertheless the measurements did not confirmed the fact. Out of 19 investigated directions in the Galactic plane carbon RRLs at frequencies of 68 and 80 MHz were found only in two: towards the Galactic centre and M16 (Anantharamaiah et al., 1988).

The observation of carbon RRLs towards Cas A do not permit yet to select one of the two alternative models for physical conditions in the origination region. It is not still clear where exactly in the "HI envelope-molecular cloud" complex the carbon RRLs originate. The temperature of such a complex decreases from periphery to the centre. Atomic hydrogen prevails in the outer most rarified layers, its spin temperature is <100K (Mebold and Hills,1975). The temperature measured in inner part of the complex by CO, NH3 equals 15-20 K (Troland et al.,1985). The density on the contrary increases from periphery to the cenrte. If "hot" model is valid then RRLs originate within HI envelope where hydrogen is in atomic state with the density of $N_H \cong 500$ cm^{-3}.

If "cold" model is valid then RRLs originate in more condensed layers with the gas density of $(1-3)\cdot 10^3$ cm^{-3} where considerable part of hydrogen is in molecular state. The outer UV radiation of $\lambda > 912$ Å is able to penetrate into such a cloud and ionize carbon. For densities of $N_H = 10^3$ cm^{-3} the penetration depth amounts to 2 pc (by half ionization level); when the density grows the depth falls sharply (Viala and Walmsley, 1976).

It is not excluded that the both models are not alternative at all: the RRL origination region consists of both outer warmer and rarified zone and the ajacent more condensed and colder part. It principally must be so since the interface between the HI envelope and the molecular cloud is not the CII region inner boundary. The question is in the quantative contribution to the intensity of the obsesved RRLs from both parts. Further RRL observations particulary in dm range where the expected intensities of RRL for both models must differ greatly and further development of the RRL theory must contribute to the solution of the problem.

3.3 Hydrogen Ionization Rate Towards Cas A from RRL Observations

Having assumed that hydrogen RRLs originate in the same gas volume as 21cm line Shaver (1976) evaluated the hydrogen ionization rate towards Cas A with the upper limit of H252α line. He used the dependence of ξ_H on the ratio of optical depths of the mentioned lines (Shaver, 1977):

$$\xi_H = 5.7 \cdot 10^{-15} \Phi_2 \left(\frac{\tau_{Hn}}{\tau_{HI}} \right) \cdot \left(\frac{\nu}{100 MHz} \right) \cdot \left(\frac{T_e}{T_s} \right) \cdot \left(\frac{T_e}{(bn\beta n)_H} \right), \quad s^{-1} \qquad (10)$$

where τ_{Hn} and τ_{HI} are the optical depths of hydrogen RRL and of 21cm line correspondingly, T_s-is the spin temperature of hydrogen and Φ_2 is a quantor with a small dependence on temperature (Spitzer, 1981). For accepted T_e=50 K, N_e=0.05 cm^{-3} values Shaver (1976) obtained $\xi_H < 6.7 \cdot 10^{-17} s^{-1}$ and $\xi_H < 3.3 \cdot 10^{-17} s^{-1}$ for the features of 21cm line profile with velocities V_{LSR}=-38 km s^{-1} and V_{LSR}=-48 km s^{-1} respectively. Similar evaluation of hydrogen ionization rate were obtained in subsequent measurements using this method by the upper limits of H300α line (Casse and Shaver, 1977) and H308α line (Payne et al., 1989).

For determination of ξ_H by expression (10) it is however nesesary to postulate the main ISM parameters: T_e and N_e. The values of $bn\beta n$ are in considerable dependence on the magnitude of the electron density. So changing of the adopted value N_e=0.05 cm^{-3} (Shaver, 1976; Payne et al., 1989) to more real value N_e=0.15 cm^{-3} increases the obtained upper limit of hydrogen ionization rate by five times as compared with the above mentioned one: $\xi_H < (1.7-3.5) 10^{-16} s^{-1}$.

The detection of carbon RRLs towards Cas A made it possible to estimate the hydrogen ionization rate by new method: by the ratio of intensities of carbon and hydrogen lines. The case already does not implies postulation of N_e since the electron density in the RRL origination region or, to be exact, the value $N_e T_e^{0.62}$ is determined by direct measurements of the width of carbon RRL. Having assumed that the level population of carbon is hydrogenic and all carbon is ionized: $N_{c+} = N_c$ and $N_c/N_H = 3.3 \cdot 10^{-4}$ Sorochenko and Smirnov (1987) obtained:

$$\xi_H = \frac{6.8 \cdot 10^{-15} \Phi_2 (N_e T_e^{0.62})}{T_e^{1.12}} \frac{\tau_{Hn}}{\tau_{Cn}} \qquad (11)$$

where τ_{Cn} is the optical depth in carbon lines. For T_e=20 K (Φ_2=3.9) and $N_e T_e^{0.62}$=1.73 cm^{-3}K$^{0.62}$ from the measured intensity of C486α line and the upper limit of H486α line it follows that $\xi_H < 4.3 \cdot 10^{-16} s^{-1}$, the measurements of 631α lines yeild yet lower estimation $\xi_H < 1.3 \cdot 10^{-16} s^{-1}$.

At the same time for ξ_H determination by (11) the assumption of hydrogenic population of carbon levels is essential. If the assumption is ignored the measured ratio of the RRL optical depths must be corrected according to:

$$\left(\frac{\tau_{Hn}}{\tau_{Cn}} \right) = \left(\frac{\tau_{Hn}}{\tau_{Cn}} \right)_{meas} \cdot \frac{(bn\beta n)_C}{(bn\beta n)_H}, \qquad (12)$$

where $(b_n\beta_n)_C$ and $(b_n\beta_n)_H$ are the coefficients accounting for variations of optical depth of carbon and hydrogen lines as a result of deviations from LTE unequal for both of them. Application of $b_n\beta_n$ calculations made with an account for DR (Walmsley and Wilson, 1982a) shows that for the model with T_e=50 K, N_e=0.15 cm^{-3} the ξ_H values obtained with 486α lines must be increased by 1.3 times to ξ_H=5.6· 10-16 s^{-1}. For measurements of 631α lines, where divergency of values $b_n\beta_n$ for hydrogen and carbon is the most essential, the obtained ξ_H limit must be increased by an order of the magnitude.

Both above considered methods for determination of ξ_H are mutually complementary. For "hot" ISM with T_e>50 K higher sensitivity for determination of ξ_H gives the comparison of the upper limit of hydrogen RRL with the optical depth in 21 cm line The measurements of carbon line give an opportunity to use for it the real but not *a priori* data on electron density. In "cold" ISM with T_e=20 K lower ξ_H thresholds are obtained from the comparison of the intensities of carbon and hydrogen RRLs. By assembling these two methods one can make a reliable estimation of the hydrogen ionization rate ξ_H<10^{-16} s^{-1} in Perseus arm towards Cas A where RRLs originate.

References

1. Anantharamaiah, K.R., Erickson, W.C., and Radhakrishnan, V. 1985, Nature, **315**, 647.
2. Anantharamaiah, K.R., Payne, H.E., and Erickson, W.C. 1988, M.N.R.A.S., **235**, 151.
3. Blake, D.H., Crutcher, R.M., and Watson, W.D. 1980, Nature, **287**, 707.
4. Brown, D.H. 1980, in Radio Recombination Lines, Ed. P.A. Shaver, Reidel. Publ. Comp., p. 127.
5. Casse, J.L., and Shaver, P.A. 1977, Astr. Ap., **61**, 805.
6. Ershov, A.A., Lekht, E.E., Rudnitskii, G.M., and Sorochenko, R.L. 1982, Sov. Astr. Lett., **8**, 374.
7. Ershov, A.A., Ilyasov, Yu.P., Lekht, E.E., Smirnov, G.T., Solodkov. V.T., and Sorochenko, R.L. 1984, Sov. Astr. Lett., **10**, 348.
8. Ershov, A.A., Lekht, E.E., Smirnov, G.T., and Sorochenko, R.L. 1987, Sov. Astr. Lett., **13**, 8.
9. Federman, S.R., Glassgold, A.E., and Kwan, A.E. 1979, Ap. J., **277**, 446.
10. Goss, W.M., Kalberla, P.M.W., and Dickel, H.R. 1984, Astr. Ap., **139**, 317.
11. Heiles, C., and Stevens, M. 1986, Ap.J., **301**, 331.
12. Konovalenko, A.A., and Sodin, L.G. 1979, Sov. Astr. Lett., **5**, 663.
13. Konovalenko, A.A., and Sodin, L.G. 1980, Nature, **283**, 360.
14. Konovalenko, A.A., and Sodin, L.G. 1981, Nature, **294**, 135.
15. Konovalenko, A.A. 1984, Sov. Astr. Lett., **10**, 353.
16. Lekht, E.E., Smirnov, G.T., and Sorochenko, R.L. 1989, Sov. Astr. Lett., **15**, 396.
17. Lekht, E.E., Smirnov, G.T., and Sorochenko, R.L. 1990,

Sov. Astr. Lett., to be published.
18. Mebold, U., and Hills, D.L. 1975, Astr. Ap., **42**, 187.
19. Payne, H.E., Anantharamaiah, K.R., and Erickson, W.C. 1989, Ap. J., **341**, 890.
20. Shaver, P.A. 1975, Pramana, **5**, 1.
21. Shaver, P.A., Pedlar, A., and Davies, R.D. 1976, M.N.R.A.S., **177**, 45.
22. Sorochenko, R.L., and Smirnov, G.T. 1987, Sov. Astr. Lett., **13**, 77.
23. Spitzer, L. 1978, Physical Processes in the Interstellar Medium, New York: J.Wiley.
24. Ungerechts, H., and Walmsley, C.M. 1978, Techn. Bericht. N45, Max-Planck Institut fur Radioastronomie, Bonn.
25. Troland, T.H., Crutcher, R.M., and Heiles, C.E. 1985, Ap. J., **298**, 808.
26. Viala, Y.P., and Walmsley, C.M. 1976, Astr. Ap., **50**, 1.
27. Walmsley, C.M., and Watson, W.D. 1982a, Ap. J., **260**, 317.
28. Walmsley, C.M., and Watson, W.D. 1982b, Ap.J.Lett., **255**, L123.
29. Watson, W.D., Western, L.R., and Christensen, R.B. 1980, Ap. J., **240**, 956.

Interferometric Observations of Carbon Recombination Lines Towards Cassiopeia A at 332 MHz

H. E. Payne[1], K. R. Anantharamaiah[2], W. C. Erickson[1,3]
[1]*Space Telescope Science Institute, 3700 San Martin Drive, Baltimore, MD 21218*
[2]*Raman Research Institute, Sadashivanagar, Bangalore 560 080, India*
[3]*Astronomy Program, University of Maryland, College Park, MD*

ABSTRACT: VLA maps of Cas A in the carbon recombination line C270α are compared with maps of neutral hydrogen 21 cm absorption and various molecular species to see whether there is any similarity. Single dish observations of line strength as a function of principal quantum number n have suggested models in which the carbon lines originate in molecular clouds colder than 20 K or in neutral clouds as warm as 50–100 K. The general features of the warm model are discussed. The recombination line maps more closely resemble the neutral hydrogen maps, suggesting that the warm model is the correct one.

1 Introduction

The contributions to this volume by Konovalenko and by Smirnov have summarized the results of the low frequency recombination line data towards Cassiopeia A. The outstanding unanswered question is "What are the physical conditions in the carbon line region?" The alternatives seem to be (1) regions with electron temperatures T_e below 20 K (Ershov *et al.* 1984) in which the hydrogen may be predominantly molecular, or (2) regions with T_e of 50 K or warmer and a significant component of atomic hydrogen (Payne, Anantharamaiah, and Erickson 1989, hereafter PAE). In either case the density is high, and the absence of hydrogen recombination lines implies that the fractional ionization of hydrogen is very low.

Observations at high angular resolution have shown that the distributions of molecular and atomic gas across the face of Cas A are quite different. We have mapped the carbon recombination line emission at 332 MHz across Cas A to see whether its spatial distribution more closely resembles that of the molecular or the atomic gas, thereby indicating the physical conditions in the carbon line regions.

2 The models

The two alternative models, which we shall call the "cold" and "warm" models, arose from analysis of the dependence of the line width and line strength on principal quantum number n. The line broadening by the Stark effect is less effective when the electrons are moving slowly, and so, to compensate, the electron density n_e must be higher if the electron temperature is lower. The observed line broadening with increasing n fixes the relationship between electron density and electron temperature.

The highly excited levels of the carbon atoms are not populated according to thermodynamic

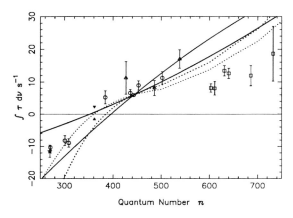

Figure 1: Integrated optical depth in the Perseus arm carbon recombination lines towards Cas A, as a function of principal quantum number n (taken from PAE, except the VLA result, marked by a star). The solid lines are predictions of the warm model (thick line: T_e=100 K, n_e=0.1 cm^{-3}; thin line: T_e=50 K, n_e=0.1 cm^{-3}), and dashed lines are for cold models (thick line: T_e=20 K, n_e=0.27 cm^{-3}; thin line: T_e=16 K, n_e=0.4 cm^{-3}).

equilibrium, and the level populations must be calculated from theory. The observed line optical depth depends sensitively on the relative departures from thermodynamic equilibrium in neighboring levels. By choosing an electron temperature, and therefore an electron density, you can calculate the departures from equilibrium and predict the shape of the dependence of line strength on n. By varying the emission measure, you can adjust the amplitude of this dependence and obtain a comparison to the data. Then, you have to decide whether the best fit model makes sense: could a cloud of the required density reasonably be maintained at the required temperature? The data obtained towards Cas A are shown in Figure 1, taken from PAE. Points marked by circles are from PAE, triangles are from Ershov *et al.* (1984) and Ershov *et al.* (1987), squares are from Konovalenko (1984), and the result from the VLA spectrum is marked with a star.

Besides the temperature, the difference between the cold and warm models is a new mechanism affecting the level populations. This mechanism operates only if the temperature is high enough for there to be a significant number of electrons with about 92 K of energy (Watson, Western, and Christiansen 1980). In a process like dielectronic recombination, these electrons can excite the ground state fine-structure transition of singly ionized carbon, which requires 92 K of energy, while simultaneously being trapped in a bound state with large n. In dielectronic recombination, the atom is stabilized when a photon is emitted and the inner electron returns to the ground state, but in this case that process is too slow. The atom is much more likely to auto-ionize when the inner electron returns to the ground state. This process can be inhibited if a collision knocks the outer electron into a state with a large orbital angular momentum; the atom will likely be maintained until the orbital angular momentum of the outer electron, as a result of repeated collisions, random-walks back to a small value, at which point the atom auto-ionizes. The effect of this mechanism on the level populations can dramatically change the dependence of line strength on n (Walmsley and Watson 1984). The grid of models calculated by Walmsley and Watson (1984) is too sparse to allow interpolation to arbitrary electron temperatures and densities for comparison with the data, but trends apparent in the models suggest the possibility of agreement with the data.

In addition, there is the complicating factor of the radiation field due to the galactic non-thermal background. This radiation provides an additional line broadening mechanism whose magnitude may be comparable to that of Stark broadening. The level populations are also affected due to effects on the rates of radiative transitions between levels. A radiation temperature and dilution factor for this background must be assumed.

Given the limitations of this procedure, and the data available to Ershov *et al.* 1984, the cold model was preferred because it gave the better fit to the data. The deduced emission measure could easily be accommodated in thin shells around the molecular clouds, where the carbon is ionized by the diffuse interstellar ultraviolet radiation field, or as embedded regions ionized by B stars. In the B star case, the temperature expected in the ionized carbon region could be as low as 20 K (Brown 1979), although higher temperatures seem more likely.

PAE extended the observations to higher frequency and to higher velocity resolution. The crossover between absorption and emission lines was observed, and the line was resolved into two features. The similarity between the recombination line spectrum and the neutral hydrogen 21 cm absorption spectrum led us to consider the possibility that both originate in the same regions, in the spirit of Shaver (1976). The neutral hydrogen data seem to imply temperatures in the 50–100 K range, where the dielectronic-like process operates efficiently.

The interesting thing about the models which assume that the carbon lines originate in the same regions as the 21 cm absorption lines is that many physical parameters can be specified. The thermal balance check must satisfy quite a few constraints. The thermal balance in neutral interstellar clouds was considered by Draine (1978). The 21 cm data show that there is a lot of cold neutral hydrogen, and yet no hydrogen recombination lines are observed. The hydrogen ionization rate must therefore be so low that the primary heat sources are photoelectric emission from dust grains, and, if the density is high, the formation of molecular hydrogen. By assuming an interstellar ultraviolet radiation field, the heating rate can be estimated. On the other hand, as long as the carbon is in the form of C^+ rather than CO, the primary cooling channel is the 156 μm fine structure line from the levels directly involved in the dielectronic-like mechanism. The cooling rate depends on the electron temperature and density. But model fitting gives the electron temperature and density, along with the relative populations of the fine structure levels. Cooling is reduced if the carbon is depleted, but even the depletion can be estimated from the observations. Draine's (1978) equilibrium models are parameterized by the ratio of pressure to hydrogen ionization rate, a quantity for which a lower limit can be derived from the observations. There does appear to be a model in thermal balance at 50 K that can explain the carbon recombination line and 21 cm absorption results.

Still, the theoretical calculations of the level populations are so sparse that the inferred temperature and density are not very firm. Uncertainties are introduced as a result of assumptions about the non-thermal background, the UV radiation field, and the properties of grains. On the other hand, as Shaver (1976) pointed out, the 21 cm optical depth and recombination line optical depth have a similar dependence on temperature and density along the line of sight. Quantities depending on the ratio of optical depths should be fairly insensitive to inhomogeneities along the line of sight. Similarly, PAE show that some derived quantities do not depend upon the fraction of Cas A covered by the clouds. Although the true optical depth is higher than the observed optical depth if only part of the source is covered, the density, pressure, and ionization rate—the key parameters for the thermal balance—are not affected.

In summary, model fitting does not now allow a choice between cold and warm models. Both models fail at high n, overestimating the observed optical depth by about a factor of two. At low n, where the lines are in emission, the point at $n=272$ is the only point that may be hard to reconcile with the $T_e = 16$ K cold model. The warm models are especially sensitive to the model parameters in this region, but the scarcity of models makes it impossible to decide whether any warm model fits the data.

3 The data

Recombination line maps of Cas A were made at the VLA in the C270α line at 332.419 MHz. Data from observations in C and D arrays (the two most compact arrays) were combined. The data have a velocity resolution of 1.4 km s^{-1} and an angular resolution of about 2 arcminutes. Maps were made of the line emission in each channel, and then each map was converted to

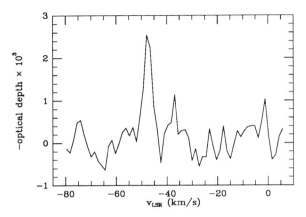

Figure 2: VLA spectrum of carbon recombination line optical depth towards Cas A. The features near -47, -39, and $0\,\mathrm{km\,s^{-1}}$ are all seen in the single dish spectrum Obtained by summing the flux in each map and dividing by the total continuum flux, this spectrum should be directly comparable to the single dish spectrum.

optical depth by dividing by the continuum map. Regions where the continuum flux was less than 10% of the peak were blanked. For comparison with the single-dish spectrum, the total flux in each line map was divided by the total continuum flux. The resulting spectrum is shown in Figure 2. Features are apparent at -47, -37, and $-1\,\mathrm{km\,s^{-1}}$. Full details of the observing are given in Erickson, Anantharamaiah, and Payne (1990).

For comparison with the neutral hydrogen absorption, we used the data of Greisen (1973) and Schwarz et al. (1986), which have angular resolutions of 1.5 and 1.0 arcminutes, respectively. The problem with the 21 cm data is that the $-47\,\mathrm{km\,s^{-1}}$ feature is heavily saturated, and a meaningful comparison can only be done in the line wings. In the recombination line maps, the $-47\,\mathrm{km\,s^{-1}}$ feature is really the only one where the signal-to-noise ratio is high enough to allow a comparison of individual channel maps.

For comparison with molecular data, only the H_2CO maps of de Jager et al. (1978) have a similar angular resolution. The CO maps of Troland, Crutcher, and Heiles (1985), and the NH observations of Batrla, Walmsley, and Wilson (1984) have a higher angular resolution (1.1 and 0.7 arcmin, respectively). We also made use of the OH absorption maps of Bieging and Crutcher (1986), in spite of their much higher angular resolution (0.1 arcmin).

Only for the $-47\,\mathrm{km\,s^{-1}}$ feature is the signal to noise ratio high enough to really consider individual channel maps. As described in Erickson, Anantharamaiah, and Payne (1990), as you examine the maps from -51 to $-45\,\mathrm{km\,s^{-1}}$, the carbon line emission very clearly moves from the eastern and northeastern part of the source to the southwestern part of the source, a trend also seen in the 21 cm data. The two maps in which the recombination line flux is highest have been averaged together to produce Figure 3. There does appear to be emission from the northern half of the source. At these velocities, the 21 cm data are highly saturated, except along the northeastern edge of the source. At the same velocities, the OH is confined to clumps in the southwestern and southeastern portions of the source, as is the formaldehyde and the carbon monoxide, with no detectable lines in the northern half of the source. The CO signal peaks at a position 5 arcmin south-southeast of the edge of the source. The distribution of the recombination lines does seem to more closely resemble that of the neutral hydrogen than those of the molecular tracers.

Figure 4 is the average of the three recombination line maps showing the $-39\,\mathrm{km\,s^{-1}}$ feature. The feature running from the center of the source to the western edge is seen in both the neutral hydrogen and molecular data. The CO data show a peak in the emission just off the west southwestern edge of the source. The extension to another concentration at the northeastern edge of the source is also seen in the neutral hydrogen data, but not in the molecular data.

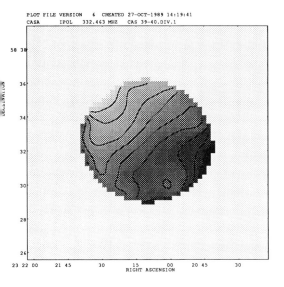

Figure 3: VLA map of carbon recombination line optical depth across Cas A, shown both in grey scale and contour map representations. Darker corresponds to higher optical depths. This is the average of maps at -48.0 and $-46.6\,\mathrm{km\,s^{-1}}$. At these velocities the molecular material is concentrated in the southwestern and southeastern parts of the source, while the neutral hydrogen absorption is saturated everywhere except in the northeastern corner of the source.

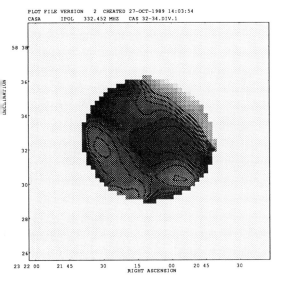

Figure 4: Same as figure 3, except the average of three maps with velocities between -39.8 and $-37.0\,\mathrm{km\,s^{-1}}$. The feature extending horizontally from the center of the source to the western edge is seen in all tracers, but the feature at the northeastern edge of the source is seen only in the neutral hydrogen data.

4 Summary

In spite of the limited signal to noise ratio and angular resolution of the recombination line data, the spatial distribution of the lines is observed to more closely resemble that of the neutral hydrogen than those of the molecular tracers. This lends support to the warm models proposed by PAE. If a warm model is correct, then the recombination lines come from regions with neutral hydrogen densities of 350–650 cm^{-3}, whereas the observed OH and CO lines come from regions with equivalent densities ($2n_{H_2}$) of a few thousand (Bieging and Crutcher 1986, Troland, Crutcher, and Heiles 1985), and the H$_2$CO lines come from regions with densities in the tens of thousands (Goss, Kalberla, and Dickel 1984). Carbon lines may originate in the neutral outer parts of molecular clumps—the most massive neutral clumps identified by Goss, Kalberla, and Dickel (1984) are in regions that have recombination lines. These neutral clumps appear to have typical densities comparable to those predicted by the warm models, with denser, molecular cores. But the lines must also originate in more widely dispersed neutral gas which has no associated molecular material.

REFERENCES

Batrla, W., Walmsley, C. M., and Wilson, T. L. 1984, *Astr. Ap.*, **136**, 127.
Bieging, J. H., and Crutcher, R. M. 1986, *Ap. J.*, **310**, 853.
Brown, R. L. 1979, in *Radio Recombination Lines*, ed. P. A. Shaver (Dordrecht: Reidel), p. 127.
de Jager, G., Graham, D. A., Wielebinski, R., Booth, R. S., and Gruber, G. M. 1978, *Astr. Ap.*, **64**, 17.
Draine, B. T. 1978, *Ap. J. Suppl.*, **36**, 595.
Erickson, W. C., Anantharamaiah, K. R., and Payne, H. E. 1990, in preparation.
Ershov, A. A., Ilyasov, Yu. P., Lekht, E. E., Smirnov, G. T., Solodkov, V. T., and Sorochenko, R. L. 1984, *Soviet Astr. Letters*, **10**, 348.
Ershov, A. A., Lekht, E. E., Smirnov, G. T., and Sorochenko, R. L. 1987, *Soviet Astr. Letters*, **13**, 8.
Greisen, E. W. 1973, *Ap. J.*, **184**, 363.
Konovalenko, A. A. 1984 *Soviet Astr. Letters*, **10**, 353.
Payne, H. E., Anantharamaiah, K. R., and Erickson, W. C. 1989, *Ap. J.*, **341**, 890 (PAE).
Schwarz, U. J., Troland, T. H., Albinson, J. S., Bregman, J. D., Goss, W. M., and Heiles, C. 1986, *Ap. J.*, **301**, 320.
Shaver, P. A. 1976, *Astr. Ap.*, **49**, 149.
Troland, T. H., Crutcher, R. M., and Heiles, C. 1985, *Ap. J.*, **298**, 808.
Walmsley, C. M., and Watson, W. D. 1982, *Ap. J.*, **260**, 317.
Watson, W. D., Western, L. R., and Christensen, R. B. 1980, *Ap. J.*, **240**, 956.

DETECTION OF CARBON RECOMBINATION LINES AT DECAMETER WAVELENGTHS IN SOME GALACTIC OBJECTS

A.A. GOLYNKIN and A.A. KONOVALENKO
Institute of Radio Astronomy Academy of
Sciences of the Ukrainian SSR
4 Krasnoznamennaja str.
Kharkov 310002 USSR

ABSTRACT. The recombination lines of highly excited carbon atoms near 25 MHz were detected in the directions of DR 21, S140 and L 1407. The line intensity is about 10^{-3} of the background level, with the width varying between 1 and 8 KHz and the line of sight radial velocity close to zero. Evaluating the physical parameters of the medium responsible for line formation we have obtained the electron number density in DR 21 and S 140 below 1 cm^{-3}. The estimate for the cloud L 1407 is, with allowance for the data on the background continuum absorbtion, 0.05 to 0.1 cm^{-3} with an electron temperature of 10 to 20 K. The highly sensetive techniques of decameter band radio spectroscopy suggest way of detecting low density cold interstellar plasmas with ionized carbon in various galactic objects.

1. INTRODUCTION

The observations of extremely low frequency carbon recombination lines performed mainly in the direction of Cassiopeia A have demonstrated new possibilities for the study of low density interstellar medium. That is why observations of a potentially wide class of interstellar objects at the frequencies of such lines are of considerable importance. Obviously, the majority of such objects are only observable against the background of the distributed galactic radiation. This makes the problem of detecting and investigating the absorption recombination lines even more complex, mainly because of the antenna beam dilution which further diminishes the already weak effect. By now, these low frequency lines have been detected apart from in Cassiopeia A, in relatively few objects by Konovalenko (1984), Anantharamaiah et al. (1989). Meanwhile, the technical potential of the UTR-2 radio telescope (speci-

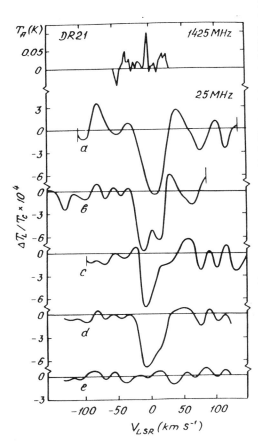

Figure 1. Cnα lines in the directions of DR21. Top of the figure: the C166α line from Pankonin et al. (1977). Curves (a),(b),(c) and (d) are recombination lines near 25 MHz.

fically, its large effective collecting area, the relatively high angular resolution and the possibility of continuous source tracking) make the idea of a comprehensive galactic survey in decameter wave recombination lines of carbon very attractive. As has been shown the objects of prime interest are diffuse HI clouds, dark dust clouds and partially ionized regions around emission nebulae. Our investigations are aimed at evaluating the possibilities of high sensetivity decameter band radio spectroscopy and assessing the different components in the interstellar medium detectability, as well as at establishing the parameters of low density plasmas which cannot be observed by other methods known in astrophysics. In this paper, we present the results of observations of DR 21, S 140 and L 1407 for which we have succeeded in providing sufficiently long integration times (about 100 h) and high sensetivity and finally detected radio recombination lines (RRLs) at decameter wavelength.

2. OBSERVATIONAL RESULTS

The observations were carried out with the radio telescope UTR-2 near 25 MHz (the principal quantum number of the excited atomic state is n=640). The antenna beam tracked the object during ±4 h off the meridian. The correlation type spectrum analyzer employed enabled observing five recombination lines at a time. The line frequencies and the corresponding tuning bands of the receivers were chosen from the range n=637 to 646 according to the condition of minimal interference. During further processing, the spectra were everaged. The total number of the spectral channels was 160, with the bandwidth about each line 20 kHz(V_r= -120 to 120 km/s). In all cases, the antenna temperatures owing to galactic background(nearly 50 000K) were greatly above the internal noise temperature. The conditions, parameters and modes of observations are listed in Table 1.

TABLE 1. Observation parameters

Source	Coordinates (α, δ)	Obser. time (y,m)	Ant. mode	Res. KHz	Int. time hr	3σ - level $T_L/T_C \cdot 10^4$
DR21	20.37.13 42.09.00	1987 2.3	NS+ EW	1.4	330	1.5
S140	22.17.36 63.04.00	1987.7 1989.2	NS EW	2.8 2.8	45 104	2.0 1.5
L1407	04.26.00 54.10.00	1987.2 1987.12	NS+ EW	1.4 1.4	226 63	1.5 2.5

2.1. DR 21

The emission lines C158α and C166α were earlier recorded for this object by Pankonin et. al.(1977). In our measurements, we employed special criteria and test procedures directed towards increasing the identification reliability. Application of the test is illustrated in Fig.1 showing the carbon lines that we have detected (the line at top of the Figure is C166 from Pankonin et al. (1977). Spectrum (b) (integration time T=93 h) was obtained with a 4 kHz shift relative to spectrum (a) (T=77 h). Spectrum (c) (T=160- was measured with a 1 kHz orbital Doppler shift. Spectrum(d) is a result of overaging of the three upper spectra (total time T=330h), while spectrum (e) shows the level of fluctuations for the case when the antenna has been replaced by a noise generator(the integration time T=280 h is the same order).

2.2. S 140

The RRLs C166α and C158α reportedly were detected in this region by Knapp et. al. (1976), however later investigations of Falgarone (1980) characterized by a better angular resolution did not confirm the result. However, a search for decameter band RRLs in this object is of considerable interest. Fig. 2(a) shows the spectrum obtained with the NS antenna and Fig. 2(b) such from the EW antenna whose beamwidth is twice as large. In its upper part, the Figure carries the C166α line according to paper of Knapp (1976). As can be seen, the lines near 25 MHz are indeed observed however they are characterized by a considerable width, comparable with the analysis bandwidth. Hence, application of the frequency shift criteria becomes complicated. For a more reliable identification, observations with a broader band of analysis (up to 50 kHz) would be usefull.

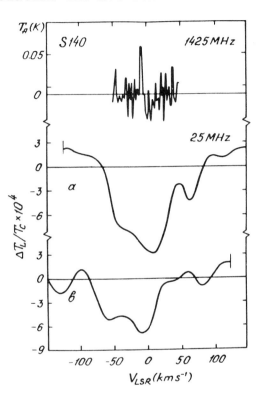

Figure 2. Cnα lines in the direction S 140.

2.3. L 1407

One of the first objects of investigation among the numerous dark clouds was L 1407. The special interest to this object is the absorption of the galactic background that was detected for this location by Abramenkov (1985) during low frequency measurements in the continuum. However, the continuum absorption does not provide sufficient material for establishing the nature of the absorption gas

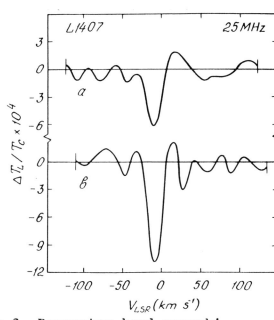

Figure 3. Decameter band recombination lines of carbon in the direction of L 1407.

i.e. determining unumbiguously whether it is a HII region or a low temperature weakly ionized medium. The choice can be done through spectral observation at decameter wave lengths, since the corresponding RRLs cannot be recorded in hot HII regions. As can be seen in Fig. 3(a), a relatively narrow absorption line owing to highly excited atoms of ionized carbon has been detected. The results is quite reliable being confirmed, among other things, by its high reproducibility. Spectrum (b) was measured at a different time. Parameters of the RRLs detected in the above mentioned objects at 25 MHz are listed in Table 2.

TABLE 2. Line parameters

Source	$\Delta T_L/T_C$ $\times 10^4$	V_{LSR} km/s	$\Delta \nu_L$ kHz	ΔV_L km/s
DR 21	-7 ± 3	0 ± 8	3.5 ± 1.0	42 ± 12
S 140 (NS)	-10 ± 3	-6 ± 17	8 ± 2	96 ± 24
S 140 (EW)	-5 ± 2	-36 ± 17	8 ± 2	96 ± 24
L 1407	-8 ± 4	-7 ± 6	1.4 ± 0.5	17 ± 6

3. DISCUSSION

The measured data can be used to evaluate the physical parameters of the line forming medium. Note, however, if the amount of apriori information is be different for different objects, hence the quality of the estimates are going to be different too. A particular difficulty will be presented by the lack, in certain cases, of any knowledge on the spatial structure of the objects and the corresponding uncertainty in the relation between the solid angles occupied by the antenna beam and the source.

3.1. DR 21

The radial velocity of the lines detected is close to zero, wich is in agreement with the velocities shown by high frequency RRLs. In paper of Valles (1987) the region in question was analyzed in the lines $C100\alpha$ and $C125\alpha$, with allowance for the data on $C158\alpha$ and $C166\alpha$. The resulting physical model is as follows: electron number density $N_e = 300$ cm-3, electron temperature $T_e = 30$ K and effective size $S = 0.01$ pc. The high frequency RRLs of carbon are formed in extremely thin layers adjacent to compact HII regions. Making use of the calculated decameter line widths (Konovalenko(This issue)) whose collisional component can be qualitatively described as

$$\Delta \nu_p \sim N_e T_e^{0.5} n^{5.2}. \qquad (1)$$

It can be easily shown that the number density suggested cannot explain the observations of carbon RRLs at frequencies below 30 MHz.

To estimate the medium parameters that could be responsible for the decameter lines observed, we have made the following assumptions:

a) the region is uniform, isotropic and reasonably compact (less than the antenna beam in its cross-section, i.e. 20'),

b) the electron temperature is $T_e = 20$ K,

c) the electron number density can be estimated as $N_e \sim 0.5$ cm^{-3}, according to the measured line width and with the Doppler and radiative broadening neglected. The observed intensity of the low frequency lines can be expressed as

$$\Delta T_L / T_c = - \tau_L^* b_n \beta_n m \Omega_s / \Omega_a, \qquad (2)$$

where $\tau_L^* = f(N_e, T_e, S)$ is the equilibrium optical depth of the recombination line whose explicit form can be found e.g. in Shaver (1975): $b_n \beta_n$ are the departure factors (the dielectron-like recombination mechanism at $T_e=20$ K can be neglected): Ω_s and Ω_a are the solid angles of the source and the antenna beam, respectively, and m is a factor alowing for the temperature ratio of the background and the equipment noise, as well as for the galactic background brightness temperature in front of the objects and behind.

For $n \sim 640$, we have with the 2.5 kpc distance to DR 21, actual experimental conditions and the above assumption $b_n \beta_n \sim 0.9$(Shaver(1975)); $\tau_L^* \sim 3.9 \cdot 10^{-2} S$ (S in pc); m=0.7; the angular size of the cloud $\theta=1.5$ S(S in pc, θ in minutes of arc); the solid angle $\Omega_s = 2.2$ S^2 (min of arc)^2, and the effective solid angle of the antenna beam $\Omega_a=10^4$(min of arc)^2. By substituting these values and the measured results into equation (2) we can find S>6 pc and $\theta > 10'$. This estimate is obviously different from the size of the region responsible for high frequency RRLs.

The authors are fully aware of the fact that some of the assumption introduced in the analysis, especially (a) and (b), may be not true. Nevertheless, the principal conclusion concerning the observability of decameter band RRLs in various component of the interstellar medium, including the vicinity of emission nebulae, remains valid. Such lines indicate the presence of partially ionized cold matter with extremely low electron dencities. Equally valid is the implication that the physical model of the line forming region is markedly different from that responcible for high frequency RRLs.

3.2. S 140

The radial velocities measured with different antennas are not the same, although this may be caused by errors in determining the parameters of as broadlines. Yet the probability of actually observing different areas of the medium cannot be excluded, since the "knife" antenna patterns have different orientations relative to the galactic plane (while their axes have the same orientation). Note radial velocities of 0 to -36 km/s to be quite usual for l=105°. In case the observed lines are formed by the same isotropic region then its size most probably is less than 20', since the line intensity yielded by the NS antenna is twice as high as that from the EW array. The estimated upper level to the electron density following from the line width is Ne 1cm^{-3} Following the same pattern of analysis as for DR 21, it can be shown that the line forming region responsible for decameter lines might be even more compact than in the case of DR 21, i.e. the lower estimate of its angular size might prove less then 10'.

3.3. L 1407

To interprete the results obtained for the dust cloud L 1407, we have assumed the absorption at the line frequencies and in the continuum to occur in the same gaseous volume. Thus equations (1) and (2) are supplemented by the equation

$$\Delta T_c / T_c = 1 - exp(-\tau_c) \approx \tau_c \quad (\tau_c < 1), \tag{3}$$

where τ_c is the optical depth for the continuum (Shaver (1975), i.e.

$$\tau_c = 0.0314 \, N_e^2 \, S \, \nu^{-2} \, T_e^{-3/2} \left[1.5 \ln T_e - \ln(20.2\nu) \right], \tag{4}$$

ν - in GHz, T_e in K. In contrast to the two preceding cases, the lines of L 1407 are narrow and hence practically unresolved. Therefore, it proves more convenient to operate with the integrated line intensity,

$$I_L = \int \Delta T_L / T_c \, d\nu = -2 \cdot 10^6 N_e^2 S \, T_e^{-5/2} b_n \beta_n \, m \, \Omega_s / \Omega_a. \tag{5}$$

Making use of our experimental data (i.e. m=0.8 and I_L=1 s^{-1}) and the results of paper of Abramenkov (1985) τ_C=1.5·10^{-2}) and combining equations (4) and (5), we arrive at

$$\Omega_a / \Omega_s = 1.5 \cdot 10^3 b_n \beta_n \, T_e^{-1} (1.5 \ln T_e - 0.68)^{-1}. \tag{6}$$

If we assume for the angular size of the absorbent cloud the value of 20' to 40' following from the maps of continuum from Abramenkov (1985), then the ratio of Ω_a/Ω_s falls into the range 10 to 20. This estimate for the region size also has been confirmed by preliminary spectral measurements performed in the pattern multiplication mode of two antennas when the radio telescope is insensetive to source components of larger extent than 30'.

By evaluating the amount of line broadening due to collisions with the aid of equation(1) we can arrive at an upper estimate for the electron density namely N_e<0.1 cm^{-3}. Thus, equation (6) has been used to set limits to the range of most probable parameter value are T_e=10 to 20 K and N_e=0.05 to 0.1cm^{-3}. The fact that low frequency RRLs of carbon have been recorded near the dust cloud L 1407 (in wich object neither low frequency nor high frequency RRLs were observed before) is very significant, suggesting a way of detecting cool gas with a content of **ionized carbon.**

4. CONCLUSION

The above described experiments have resulted in the detection of decametric RRLs of carbon in three galactic objects, thus demonstrating a possibility in prinsiple of study in weakly ionized, cold components of the interstellar medium. As follows from the results obtained for DR 21, S 140 and L 1407, ionized carbon with electron densities of 0.05 to 1cm^{-3} can be present in significant volumes of the cold interstellar gas, while hydrogen atoms in the volume remain practically neutral. In spite of the low intensity of the lines and intrinistic difficulties of the radio spectrometry at extremely low frequencies, high sensetivity measurements are quite real. This gives us grounds for believing that positive results are achievable with other galactic objects, too. The medium parameters could be estimated to better accuracies and structural models of the objects refined if data on spatial distribution of line absorption were available. For such measurements, the angular resolution of the UTR-2 radio telescope might prove sufficient.

REFERENCES

Abramenkov, E.A. (1985) 'Low frequency radio observations of HII regions:The galactic disk region 147°< l <153°, Sov. Astron.J., 62, 1057-1064.

Anantharamaiah, K.R., Payne, H.E., Erickson W.C. (1988)'Detection of carbon recombination lines below 100 MHz towards the Galactic Centre and M16', Mon. Not. R. Astr. Soc. 235, 151-160.

Falgarone, E. (1980)' Carbon and sulfure ionized regions in dark clouds', in P.A. Shaver (ed.), Radio Recombination Lines, pp. 141-147.

Knapp, G.R., Brown R.L., Kuiper, T.B.K. and KakarR.K. (1976)'Carbon recombination line observations of the Sharpless 140 region', Astrophysical J. 204, 781-788.

Konovalenko, A.A.(1984)'Detection of decameter wave radio lines of excited carbon in some galactic objects',Sov. Astron. J. Lett., 10, 912-917.

Konovalenko, A.A. 'Review of decameter wave recombination lines: problems and methods', this issue.

Pankonin, V, Thomasson, P. and Barsuhn, J.A.(1977) 'Survey of radio recombination lines from HI regions and associated HII regions', Astron. and Astrophys., 54, 335-344.

Shaver, P.A.(1975)'Theoretical intencities of low frequency recombination lines', Pramana 5, 1-28.

Valles, I.(1987)'The neutral interface ajoining HII regions', Astrophys. and Space Sci. 127, 339-351.

RRLs FROM THE LOCAL INTERSTELLAR MEDIUM

N.G.BOCHKAREV
Sternberg State Astronomical Institute
Universitetskij prosp. 13
119899 Moscow V-234
U.S.S.R.

ABSTRACT. Great value $\zeta = 2 \cdot 10^{-16}-10^{-13} s^{-1}$ and uncertainty for ionization rate of the local interstellar medium (LISM) results in expediency of RRL measurement to determine ζ from observations of cold part of LISM: nearby atomic clouds (at first Sancini's-van Woerden H I filament), diffuse low-mass molecular clouds and IR cirruses. From distribution of ionization inside the clouds there is a chance to make out principal heating mechanism: subcosmic rays (CR) or electromagnetic radiation. Excess of the intensity of RRL in 10-cm range above the relic radiation is suggested to be at least $10^{-(2 \div 4)}$.

1. INTRODUCTION

LISM is named a region with radius of $\simeq 150$ pc near the Sun connected with large scale bubble observed from inside (Cox and Reinolds, 1987; Bochkarev, 1987b, 1990). LISM is characterized by low average density, hot (T = 10^6 K) coronal gas, and by old SNR (North Polar Spur) and has complicated structure.
 LISM is a unique superbubble, nearby location of which gives a chance to study it in detailes. Inside LISM new classes of objects are detected such as low density H II regions (Zhidkov, 1970, Gry et al, 1983, Reynolds, 1984a,b) and low-mass molecular clouds (Blitz et al, 1984; Magnani et al, 1985; see also Bochkarev, 1990) which correspond often to IR cirruses (Low et al., 1984; de Vries, 1986; Deul,1988).
 Optical methods permit to measure extended ($\geqslant 1°$) warm (T $\simeq 10^4$ K) H II regions with emission measure EM as small as 0.1-1 pc cm^{-6} (Kutyrev and Reynolds, 1989). Soft X-ray observations give a chance to investigate hot (T= 10^6-10^7 K) gas with EM=10^{-2}-10^{-3} pc cm^{-6} (Rocchia et al.,1984; McCammon, 1984). For cold gas, similar sensitivity (EM = 10^{-1}-10^{-3} pc cm^{-6}) is achieved for RRLs (e.g. Ershov et al., 1984, 1987; Payne et al.,1989).
 RRLs are a sensitive indicator of the ISM ionization rate ζ (Payne et al., 1984, 1989; Sorochenko and Smirnov, 1987). Radiation of coronal gas and hot white dwarfs, and also CR generated in North Polar Spur can make anomalous ζ in LISM. The most part of the CR penetrate from SNR into surrounding ISM (Ammosov et al.,1989) and contribute to ζ. Therefore it is important to use RRL diagnostic possibilities to study physical conditions in LISM.

2. LOCAL INTERSTELLAR MEDIUM

Superposition of observational data are discussed by two colloquia (Kondo et al., 1984; Gry and Wamsteker, 1986) and in reviews (Cox and Reynolds, 1987; Bochkarev, 1987b, 1990) confirms Weaver's suggestion that the Sun is near the inner edge of a giant bubble (cavern). It has a diameter of about 350 pc and was formed around an old stellar association Sco-Cen by collective action of stellar winds, SNRs, and WR-stars (Bochkarev, 1987a,b).

Most part of the cavern volume is filled by coronal gas with $T=1.0\ 10^6 K$ and $EM = 0.001-0.01$ pc cm^{-6}. This gas is surrounded by an H I envelope observed as H I filaments extending as far as $\simeq 180°$ on the sky, which were studied by Helies and Jenkins (1976), Cleary et al.(1979) and others. North Polar Spur inside LISM contains gas with $T = (3-5)\ 10^6$ K and $EM = 0.001-0.01$ pc cm^{-6}. About one third of the cavern is filled by low density H II regions ($T = 10^4 K$ electron number density $n = 0.1-0.3$ cm^{-3}, sizes $l = 10-30$ pc). There are many atomic-dust (Knude, 1979) and molecular clouds. Among the latter, there are giant ones such as the molecular complex Oph-Cen and low-mass clouds studied by Blitz et al.(1984), Magnani et al.(1985) and others - see Table 1 and discussion in Bochkarev's (1990) monograph. The gas temperature in duffuse molecular clouds is 8-10 K, the dust one is 20-30 K. As marked in Sect.1, many such clouds correspond to IR cirruses (Table 2). Cold gas occupies 2% of the LISM volume and molecular gas - 0.03% (Bochkarev, 1990).

The Sun is embedded in warm ($T = 8000 \pm 1000$ K) corona of a H I cloud, the center of which is Sancini's - van Woerden (1970) H I filament at the distance 10-20 pc (Olano and Poppel, 1981; Crutcher, 1982). The filament covers 4x20° on the sky (Table 3). Around Sancini's-van Woerden, filament some regularity of the interstellar polarization (so-called Tinbergen (1980) patch of polarization) is found for stars nearer than 35 pc. Bochkarev (1987a,b) shows that the gas producing the polarization forms a horseshoe-like structure surrounding the Sun from three sides.

Important and discussed question is the value of ionization rate ζ in LISM. Consideration of physical conditions in matter around the heliosphere ($10^{-4}-3$ pc around the Sun) shows that n_e and degree of gas ionization are very uncertain, with n_e between 0.01 cm^{-3} and $\simeq 0.5$ cm^{-3} (Reynolds, 1986, 1988; Frisch et al., 1987; Bochkarev, 1990) while H I number density $n(H\ I)=$ 0.05-0.3 cm^{-3}. Reynolds (1986, 1988) found that among observed sources of ζ the greatest is radiation of hot ($T \simeq 60\ 000$ K) nearby white dwarfs for which data collected by Paresce (1984). Together with other observed sources $\zeta =$ $=2.5\ 10^{-16}$ s^{-1}, but the uncertainty is very great: extrapolation of the spectrum of X-ray background (Bloch et al., 1986; McCammon et al., 1984) until 30-40 eV yields $\zeta = (2-10)\ 10^{-16}$ s^{-1} and addition of other possible sources can increase ζ until 10^{-13} s^{-1}, which is an upper limit of ζ following from faint H$_\alpha$ emission (Reynolds, 1988; Bochkarev, 1990). There are several indications that EM of warm gas near the Sun can be close to upper limit given by H$_\alpha$ observations, i.e. ζ is not far from 10^{-13} s^{-1}. Thus near the Sun

$$2.5\ 10^{-16}\ s^{-1} \leq \zeta \leq 10^{-13}\ s^{-1}. \tag{1}$$

Bochkarev (1987a,b) notices that a gas ionized by Antares when it was O-star (less than 10^6 years ago) has not finished recombining. It is an additional source of n_e. Important source of ζ can be CR generated by SNRs LISM.

Table 1. High-latitude diffuse molecular clouds with estimated distances (Bochkarev, 1990).

Cloud	Coordinates (deg) l	b	Distance (pc)	Av in cloud center (mag)
MBM55	89.2	-40.9	≤175	0.5
MBM55A	105.1	-39.9	≤275	0.5
L1228	112.	+22.	≈150	-
MBM26	136.4	+32.6	175 ± 50	0.6
MBM32	146.4	+39.6	≤275	0.5
MBM7	150.4	-38.1	125 ± 50	0.9
MBM12 = L1457/8	159.4	-34.3	65 ± 5	0.4
MBM16	171.7	-37.7	60 ÷ 95	1.9
Pleiades	175.	-18.	140 ± 15	1.6
MBM18 = L1569	189.1	-36.0	≤175	1.1
MBM20 = L1642	210.9	-36.5	70 ÷ 125	1.4
No.113	337.8	-23.0	≤90	-
No.126	355.5	-21.1	≈100	-

MBM - Magnani, Blitz and Mundy (1985)

Table 2. Main IR cirruses (Deul, 1988)

Name	Coordinates (deg) l	b	Sizes (deg)	$\langle V_r \rangle$ km/s	N_H 10^{20} cm^{-2}
D	43	-53	6 x 6	0	4 ÷ 10
in Hercules	46	+24	7 x 6	+10	6 ÷ 13
Pintcher	54	+15	20 x 12	-60, +25	3 ÷ 18
Coathanger	90	-37	15 x 15	0	4 ÷ 10
Angle	247	+72	10 x 7	-30, 0	2 ÷ 5
Helene	248	+15	9 x 7.5	0	5 ÷ 17
B	275	+75	2 x 3	-18	1.5 ÷ 5
X	276	+73	24 x 10	-8	

Table 3. H I clouds nearby to the Sun (Bochkarev, 1990)

Name	Coordinates (deg) l	b	Distance pc	logN_H cm^{-2}	$\langle n_H \rangle$ cm^{-3}	T K	V_{LSR} km/s
Sancini's - - van Woerden	from 343 to 358	+16 +33	10 - 20	20.	30	≤200	-13
to star HD 2151	305	-40	6	18.6	18	-	-
to α Ophiuchi	from 30 to 55	+25 +10	10 - 17	-	-	70÷ ÷500	-8

3. POTENTIALITIES OF RRL ANALYSIS

It is possible to measure RRL in absorption or emission if its central intensity is only $3 \cdot 10^{-4}$–10^{-3} part of continuum (Konovalenko and Sodin, 1980; Ariskin et al., 1982; Ershov et al., 1984; Payne et al., 1989). Accuracy of the same order (10^{-4}) is achieved for measurement of individual space fluctuations of radio background (BG), e.g. in search for Zel'dovich-Sunyaev effect.

Let us neglect deviations from LTE for estimations of RRL central intensity because correction is usually less than a factor of two. For transition between hydrogen levels n+1 and n, LTE brightness temperature in the RRL center is approximately (Sorochenko, 1979)

$$T_{Hn\alpha} \approx 80 \text{ EM } \lambda/(T_e^{3/2} T_D^{1/2}), \text{ K} \tag{2}$$

where EM is in $pc \text{ cm}^{-6}$, λ is wavelength in cm, T_e is electron temperature in T_D is HWHI for line profile in K. Thermal and turbulent movements and Stark effect for low-frequency lines contribute to T_D. For n < 200 Stark effect is usually neglected. Typical turbulent velocity is about the thermal one. Therefore we can use for estimations

$$T_{Hn\alpha} \approx 50 \text{ EM } \lambda / T_e^2, \text{ K}. \tag{3}$$

RRL is observed together with BG continuum. Outside H II regions, brightness temperature of non-thermal BG $T_B \sim \lambda^{2.8}$ and for $\lambda \lesssim 10$–15 cm the principal one is the relic radiation ($T_B = 3$ K). Thus, for $\lambda = 10$ cm

$$\eta = T_{Hn\alpha}/T_{BG} \approx 100 \text{ EM } / T_e^2 \tag{4}$$

and on the level η gas with

$$EM \approx 0.01 \; \eta \; T_e^2, \; pc \text{ cm}^{-6} \tag{5}$$

can be found.

Value $\eta = 10^{-3}$–10^{-4} corresponds to EM = 100–1000 $pc \text{ cm}^{-6}$ for warm ($=10^4$ K) gas which is 3 orders of magnitude greater than by optical methods (Sect.1). For T = 10–100 K gas with EM $\approx$ 0.001–0.01 $pc \text{ cm}^{-6}$ can be measured. Sensitivity of observations of decametric carbon RRLs is of the same order of magnitude (Sect.1). Therefore cold matter embedded in coronal gas such as low-mass molecular clouds (Table 1), IR cirruses (Table 2), and HI filaments around local cavern together with those mentioned in Table 3, are the most proper objects for diagnostic of LISM by means of RRLs.

4. PENETRATION OF IONIZING RADIATION INTO COLD GAS AND INTENSITY OF RRL

Warm gas around the heliosphere has T = 8000 K and EM $\lesssim$ 1 $pc \text{ cm}^{-6}$. According to (4), RRL contrast for the gas is $\eta \lesssim 2 \cdot 10^{-6}$, which is far below the possibilities of observations. The problem gets easier if we take into account that the emission comes from all the sky. It can permit to achieve lower limit of detection similar to searching for relic radiation fluctuations.

For colder gas, as is marked in Tables 1-3, it is necessary to know how ionizing radiation passes through the cloud gas. For CR with E = 1÷100 MeV/nucleon, which can be the main source of ISM ionization (Spitzer and Tomasco, 1968; Bochkarev, 1972 and others), the problem of penetration of CR into clouds was discussed by Cesarsky and Volk (1978), Volk (1983) and for purpose of interpretation of decametric carbon RRLs toward Cas A - by Bochkarev (1989). In one case it can be free penetration of CR into the clouds, in others particles with E < 50 MeV can be reflected from the cloud edges. Intensification of CR inside molecular clouds is also important.

EUV and soft X-ray radiation is mostly absorbed in the low-energy part, because cross-section $\sigma \sim E^{-\beta}$, $\beta \simeq 3$. As can be readily seen for a flat spectrum (in photon per energy unit) in the range $E_{min} < E < E_{max}$, ionization rate decreases with increasing gas column density N as

$$\zeta \simeq N_0/N \tag{6}$$

where N_0 is determined the through optical depth for photoabsorption

$$\tau(E_{min}) = N_0 \sigma(E_{min}) = 1. \tag{7}$$

Therefore

$$N_0 = \sigma_0^{-1} \simeq 1.5 \, 10^{17} \, cm^{-2}. \tag{8}$$

if $E_{min} < \chi_H$ - hydrogen ionization potential, and $N_0 > \sigma_0^{-1}$ if $E_{min} > \chi_H$. Equation (6) is correct as far as $\tau(E_{max}) < 1$.

We use eq.(4), the relationship of degree of ionization of ISM heated by hard electromagnetic radiation or CR, according to Bochkarev (1972), and equation of ionization equilibrium $\zeta n(H\,I) = \alpha(T) n_e^2$, where α is the coefficient of recombination. For observations made normally to gas layers, we get bearing in mind (6) an estimation

$$\eta = T_{Hn\alpha}/T_{BG} < 10^{-2} - 10^{-4} \tag{9}$$

for heating by the electromagnetic radiation in the case $E_{min} < \chi_H$ and $E_{max} \gg \chi_H$. This is an approximate lower limit estimation of η.

For the same ζ outside the clouds, η can be by orders of magnitude larger when heating by CR. It results from greater track lengths N of CR in the ISM, e.g. for E = 2 MeV N = 10^{21} cm^{-2}. However, trajectory of CR in ISM is complicated, so CR do not penetrate into ISM so deeply.

5. CONCLUSION

Observations of RRLs both in emission in the centimetre range and in absorption in the decameter range can give a chance to specify the upper limit of ionization rate in the LISM. Nearby location of the observed structures resulting in large angular sizes (degrees or tens of degrees) can permit to analyse the distribution of physical conditions inside cold gas in LISM and to set limits on the type and spectrum of dominating ionizing agent. Nevertheless, in most suggestions the problem of RRL measurement from LISM is difficult.

REFERENCES

Ammosov,A.E., Berezhko,E.G., Elshin,V.K. 1989: Sov.Astron., in press
Ariskin,V.I. et al 1982: Astron.Zh. V.59. P.38. (Sov.Astron. V.26. No.1)
Blitz,L., Magnani,L.,Mundy,L. 1984; Ap.J. V.282. P.L9
Bloch,J.J., Jahoda,K., Juda,M. et al 1986: Ap.J. V.308. P.L59
Bochkarev,N.G. 1972; Astron.Zh. V.49. P.756 (Sov.Astron.V.16,P.619)
Bochkarev,N.G. 1987a: Astron.Zh. V.63. P.38 (Sov.Astron.V.63,No.1)
Bochkarev,N.G. 1987b: Astrophys.Space Sci. V.138. P.229
Bochkarev,N.G. 1989: Astron.Nachr. V.310. P.399
Bochkarev,N.G. 1990: Local Interstellar Matter. Moscow: Nauka
Cesarsky,C.J., Volk,H.J. 1978: A.Ap. V.70. P.36
Cleary,M.N.,Heiles,C., Haslam,C.G.T. 1979: A.Ap.Suppl. V.36. P.95
Cox,D.P., Reynolds,R.J. 1987: Ann.Rev.A.Ap. V.25. P.303
Crutcher,R.M. 1982: Ap.J. V.254. P.82.
de Vries,C.P. 1986: Adv.Space Res. V.6. No.2. P.35
Deul,E.R. 1988: Interstellar Dust and Gas in the Milky Way and M33.
   Sterrewacht Leiden
Ershov,A.A. et al 1984: Pis'ma Astron.Zh.V.10.P.833(Sov.Astr.Lett.V.10.No.6)
Ershov,A.A., et al 1987: Pis'ma Astron.Zh.V.13.P.19(Sov.Astr.Lett.V.13.No.1)
Frisch,P.C., York,D.G., Fowler,J.R. 1987: Ap.J. V.320. P.842
Gry C., Wamsteker W. eds. 1986: Proc. COSPAR Symp.Physical Processes
   in the Local Interstellar Medium: Adv. Space Res. V.6. No.2
Gry,C., York,D.G., Vidal-Madjar,A. 1983: A.Ap. V.124. P.99
Heiles,C., Jenkins,E.B. 1976: A.Ap. V.46. P.333
Knude,J. 1979: A.Ap.Suppl. V.38. P.407
Kondo,Y. et al eds.1984:The Local Interstellar Medium.IAU Coll.81.NASA CP-2345
Konovalenko,A.A., Sodin,L.G. 1980: Nature V.283. P.360
Kutyrev,A.S., Reynolds,R.J. 1989: Wisconsin Ap. Prepr. No.321
Low,F.J., Bointema,D.A., Gautier,T.N. et al 1984: Ap.J. V.278. P.L19
Magnani,L., Blitz,L., Mandy,L. 1985: Ap.J. V.295. P.402
McCammon,D. 1984: in Local Interstellar Medium. NASA CP-2345. P.195
McCammon,D., Burrows,D.N., Sanders,W.T., Kraushaar,W.L. 1983:Ap.J.V.269.P.107
Olano,C.A., Poppel,W.G.L. 1981; A.Ap. V.95. P.316
Paresce F. 1984: in Local Interstellar Medium. NASA CP-2345. P.169
Payne,H.E., Salpeter,E.E., Terzian,Y. 1984: AJ V.89. P.668
Payne,H.E., Anantharamaiah,K.R., Erickson,W.C. 1989; Ap.J. (15 June)
Reynolds,R.J. 1984a: in Local Interstellar Medium. NASA CP-2345. P.97
Reynolds,R.J. 1984b: Ap.J. V.252. P.191
Reynolds,R.J. 1986: AJ V.92. P.653
Reynolds,R.J. 1988: rep. on V Meeting on Cosmic Gas Dynamics (Moscow, 1988)
Rocchia,R., Arnaud,M., Blondel,C. et al 1984: A.Ap. V.130.P.53
Sancini,R., van Woerden,H. 1970: A.Ap. V.5. P.135
Sorochenko,R.L. 1979: in Spectral Research of Cosmic and Atmospheric
   Radiation. Ed. A.G.Kislyakov. Inst. Apply Physics. P.5
Sorochenko,R.L., Smirnov,T.G. 1986: Prepr. FIAN No.321
Spitzer,L., Tomasko, M.G. 1968: Ap.J. V.152. P.971
Tinbergen,J. 1982: A.Ap. V.105. P.53.
Volk,H.J. 1983: Space Sci.Rev. V.36. P.3
Weaver,H. 1979: Large-Scale Characteristics of the Galaxy.IAU Symp.84.P.295
Zhidkov,V.F. 1970: Sov. Astron. Circ. No.578. P.5; No.655. P.4

RECOMBINATION LINES AND GALACTIC STRUCTURE

FELIX J. LOCKMAN
National Radio Astronomy Observatory[1]
Edgemont Rd.
Charlottesville, VA 22903
USA

ABSTRACT. Radio recombination lines provide information that is crucial in understanding the distribution of ionized gas in the Galaxy. The new Green Bank HII region survey, which detected > 450 nebulae, is used in conjunction with southern hemisphere surveys to investigate the radial distribution of HII regions, the question of an inner boundary of star formation in the galactic disk, and star formation in the 3-kpc arm.

I. The Use of Recombination Lines

From the beginning, it was clear that the discovery of radio recombination lines would have a significant impact on the study of galactic structure. The first two recombination line surveys were published within three years of the discovery (Dieter 1967; Mezger and Höglund 1967), and although they detected only a few tens of nebulae, that was sufficient to show many of the basic features of their galactic morphology. Since then, recombination line surveys have detected hundreds of galactic HII regions and have been the major source of information on the kinematics and distribution of ionized gas in the Galaxy.

From the point of view of galactic structure, there are two main pieces of information that come from recombination line observations of a continuum source: first, the line's very existence, which identifies the radio source as an HII region, and second, its velocity, which can be used to derive its distance and location in the Galaxy through a kinematic model. But there is more: (1) The distance and continuum flux give the luminosity of the HII region and thus information about its exciting stars; (2) The velocity of a nebula can establish its association with other objects, like molecular clouds or supernovae. Sometimes the most interesting discovery about an HII region is that it is <u>not</u> connected with something nearby, as was the case with the HII region toward the first millisecond pulsar (Heiles *et al.* 1983). (3) Recombination line measurements allow us to measure physical properties (like the He abundance or ionization state) in special places in the Galaxy, like the 3-kpc arm or the nucleus; (4) Continuum sources that do not have detectable lines can be of interest in themselves, and recombination line surveys produce lists of such objects.

---

[1] The National Radio Astronomy Observatory is operated by Associated Universities Inc. under agreement with the National Science Foundation.

This aspect of recombination line surveys has been valuable in identifying galactic supernova remnants. Examples of recent work on galactic structure in which recombination line surveys have played an important role are the papers by Avedisova and Palouš (1989), Digel, Bally and Thaddeus (1990), and by Oskanyan (1989).

This paper gives a brief overview of a few aspects of galactic structure that have been revealed by recombination line surveys. After considering the sensitivity, completeness and selection effects of recombination line surveys in general, the main topics will be: the radial distribution of HII regions in the Galaxy; nebulae in and interior to the 3-kpc arm, and the provocative question of spiral structure. I will draw heavily on data from a survey that has just been completed using the 140-foot telescope of the NRAO at Green Bank (Lockman 1989). There is not space here to discuss the connection between HII regions and molecular clouds except in the most global sense, nor even to mention interesting physical properties, for example, the extremely low temperatures, that some nebulae have (Lockman 1989). Ionized gas in the galactic center is reviewed by Goss and others elsewhere in this volume. A systematic study of one set of objects that do not have detectable recombination lines is given by Helfand *et al.* (1989).

II. Making a Recombination Line Survey

There are two types of recombination line surveys: those which use continuum sources as targets, and those which systematically cover an area of the sky with uniformly spaced observations in the way that HI surveys are made. There are only a few examples of the second type; they have been made almost exclusively at wavelengths longer than ~ 10 cm and are discussed elsewhere (Lockman 1980a; Cersosimo *et al.* 1989; Cersosimo, this volume). Here I will be concerned with the first group, the "discrete-source" surveys. Discrete-source surveys are made by searching a flux-limited sample of radio sources (provided by a radio continuum survey) to see if they have recombination line emission.

II.1 Discrete-source Surveys

Table 1 lists the larger discrete-source surveys along with the number of nebulae detected in each. The total number of nebulae now known is quite large, but most galactic continuum sources still have not been examined for recombination line emission.

The specific hydrogen transition that is observed has an important effect on the results of a survey, for different transitions sample different parts of the HII region population. If the HII regions being searched are all optically thin ($\tau_c \ll 1$), then the recombination line antenna temperature scales approximately as $T_L \propto \nu^{-1}$ if the source size is much larger than the antenna beamwidth, and $T_L \propto \nu^{+1}$ if the source is much smaller than the beamwidth. While galactic continuum sources are found with all angular sizes, the brightest sources tend to be the most compact, so HII region surveys have usually been made at the highest convenient frequency, taking into account the state of available receivers and the vicissitudes of the weather at the specific observatory. All the surveys listed in Table 1 were made in recombination lines near 5 GHz except for the last, the recent 140-foot survey, which was made mainly at 10 GHz.

Since much of this paper uses results from the new Green Bank survey (Lockman 1989), I will discuss its properties briefly. It was made on the 140-foot telescope which, at 10 GHz where most of the lines were observed, has a beamwidth of $3'$. The angular resolution

Table 1

Observers	Approximate Number of HII	Notes
Reifenstein et al. 1970	80	Northern
Wilson et al. 1970	130	Southern
Downes et al. 1980	170	Northern
Caswell and Haynes 1987	320	Southern
Lockman 1989	460	Northern

is thus very close to that of the recent continuum surveys by Altenhoff *et al.* (1978), and Wenkder (1984) which were used to select most of the target continuum sources. About 500 objects were observed and 462 were detected, half of these for the first time. The survey covers $\delta \geq -37°$. It is most complete and uniform in its coverage of longitudes $\ell \leq 60°$, less so for the Cygnus region, and is decidedly nonuniform in the anticenter where there are few nebulae.

II.2 The Problem of the Rosette Nebula

To understand some of the selection effects in discrete-source surveys, it is useful to consider how effective they would be at detecting the Rosette nebula in comparison to Orion A. Both HII regions are associated with large molecular clouds that have a size ~ 100 pc and a mass $\sim 10^5$ $M_\odot$ (Blitz 1978; Maddalena *et al.* 1986). Both have an almost identical radio continuum luminosity. Both are ionized by a cluster of young stars: Orion A has stars as early as O6, the Rosette as early as O5 (Hoffleit 1982; Celnik 1985). Both have an angular size that is large compared to the beams used for recombination line surveys. Both should be included in any census of galactic nebulae. In radio emission, however, they are quite dissimilar: at H85α Orion A has $T_L = 3100$ mK while the Rosette has only $T_L = 14$ mK, a variation of a factor ~ 200. If the Rosette had been in the inner Galaxy, obscured by foreground dust so that it had no visible optical emission, it probably would not have been detected in the new Green Bank survey, or in any other discrete-source survey to date. In contrast, the set of nebulae detected in the new Green Bank survey probably includes all of the "Orion A" type within about 20 kpc of the Sun at $\delta > -37°$.

Low surface brightness nebulae, like the Rosette, are underrepresented in most recombination line surveys even though they may be as significant, in all other ways, as nebulae which are readily detected. As a part of the Green Bank survey I made some recombination line measurements at 2.7 GHz of low surface-brightness continuum sources that were fairly extended ($\gtrsim 10'$), to see if they might be faint HII regions like the Rosette. As discussed above, it is expected that for these objects $T_L \propto \nu^{-1}$, and in fact, a number were detected that would have been below the sensitivity threshold of the higher frequency surveys. The 2.7 GHz observations were not uniform or complete, but even so, ~ 40 low surface-brightness, moderate size nebulae were detected.

It is not known how abundant these faint nebulae are. They may contribute a significant fraction of the low-frequency recombination lines observed in "blank" areas of the plane (Lockman 1976; Anantharamaiah 1985, Cersosimo this volume). In all discussions of the properties of galactic nebulae we should be aware that a complete class may be missing because the surveys are unavoidably selective.

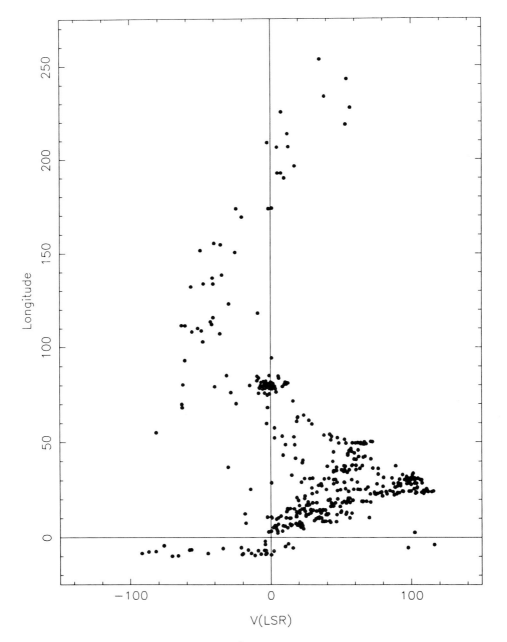

Figure 1: The velocity-longitude diagram for nebulae detected in the new Green Bank recombination line survey (Lockman 1989). Directions with $|\ell| < 2°$ were not included in the survey. This figure contains > 450 nebulae with $\delta \geq -37°$.

III. The Distribution of HII Regions

III.1 Velocity-Longitude Diagrams

The 462 nebulae measured in the Green Bank survey are shown in velocity-longitude coordinates in Figure 1. The new survey did not cover directions within 2° of the galactic center, so this area will not be considered in any subsequent discussion. The new survey, combined with data from the Parkes Survey by Caswell and Haynes (1987), form a data set that contains almost 750 HII regions, the largest number ever available for study. Their distribution in velocity and longitude is shown in Figure 2.

Clearly, galactic HII regions are most prevalent in the inner Galaxy, at $R < R_0$, (i.e., $|\ell| \leq 90°$), but their numbers do not increase all the way into the galactic center – they are distributed in a broad ring. This has been known since the continuum survey of Westerhout (1958), but is especially evident here. We will return to Figure 2 in the discussion of spiral structure.

III.2 Radial Distribution

A galactocentric distance, R, was calculated from the velocity of each HII region using the Burton and Gordon (1978) rotation curve scaled to $R_0 = 8.5$ kpc, except for a few HII regions associated with the 3-kpc arm (W31) that were assumed to be at $R = 3.4$ kpc (see §IV) and a few optical nebulae in the anticenter that were placed at their spectrophotometric distances (from, e.g., Avedisova and Kondratenko 1984). Of course, in many directions kinematic distance estimates are not very accurate, and the errors tend to be systematic and correlated, producing spurious groupings (see, e.g., Burton 1972, Lockman 1979). Kinematic distance estimates are also poor at low longitudes, so most of the discussion that follows will be restricted to nebulae at $|\ell| > 10°$.

The radial distribution of HII regions in the North (10° – 180°) is compared with that in the South (180° – 350°) in Figure 3. The difference in the number of nebulae between North and South is most probably an artifact of the greater sensitivity of the northern survey. The difference in shape of N(R), however, seems to be real, and is quite interesting. Southern nebulae are distributed more broadly, and the peak in the distribution occurs farther from the galactic center, than the northern nebulae. These characteristics are qualitatively shared by the H_2 distribution traced by CO emission (Bronfman et al. 1988).

The decrease in N(R) at R < 4 kpc seen in all radial distributions is real and not a product of any selection effect, although since we require that $|\ell| > 10°$, the figures do not include any objects at R < 1.5 kpc. It is somewhat troubling, though, that N(R) at R $\leq$ 4 kpc is extremely sensitive to the treatment of just a few nebulae, specifically W31 and others that may be in the 3-kpc arm. If W31 were omitted from this analysis by, e.g., only analyzing nebulae that are at $|\ell| > 11°$, the surface density at 3 kpc in the North would be reduced by a factor ~ 3. In the innermost parts of the galactic disk the kinematics and distribution of nebulae are still mysterious and need much further study.

The surface density distribution of the combined northern and southern nebulae is given in Figure 4. Hodge and Kennicutt (1983) found that such distributions fall into one of three classes in spiral galaxies: continually decreasing outward, oscillating and decreasing outward, or ring-shaped with a deep central minimum. Our Galaxy is clearly in the third class, which is populated mostly by early spirals of type Sb and Sbc. Many galaxies show a nearly exponential decrease in the surface density of nebulae in the outer parts of their disks. The data for our Galaxy are not especially well fit by such a model because of the

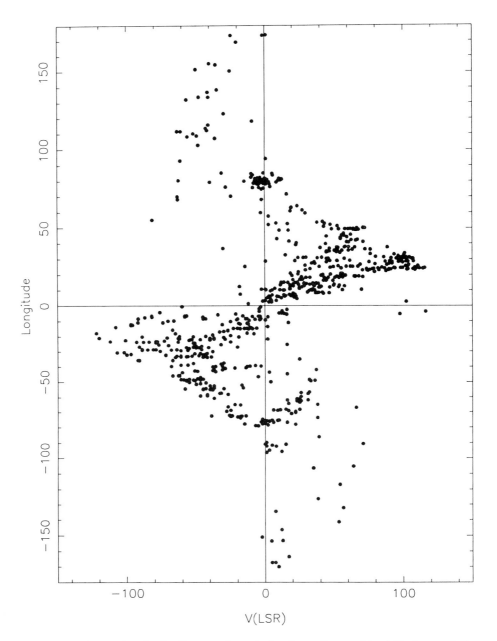

Figure 2: A velocity-longitude diagram for nebulae from the combined Parkes and Green Bank surveys (Caswell and Haynes 1987; Lockman 1989). This is the most complete sample of radio HII regions assembled to date. It contains almost 750 HII regions and is incomplete only at very low longitudes ($|\ell| < 2°$).

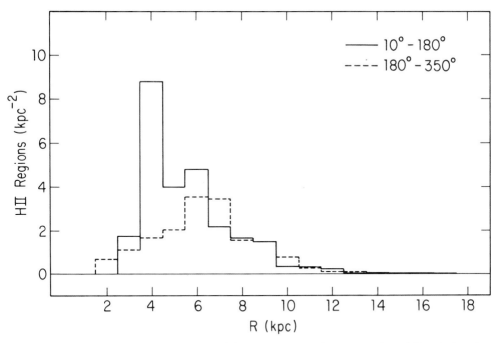

Figure 3: Surface density of nebulae, N(R), in the North (10° − 180°) and South (180° − 350°) from the combined Green Bank and Parkes surveys. The figure does not include any nebulae at R < 1.5 kpc because of the longitude limit. The relative amplitude of the distributions is a function of the sensitivity of the surveys, but the difference in the shape of the distributions is real.

peaks at R=4 and 6 kpc; between 6 and 12 kpc N(R) decreases with an exponential scale length of approximately 2 kpc.

On the small scale, HII regions are observed to be intimately connected with molecular clouds, although star-formation rates on the largest scale are not especially well correlated with the mean molecular gas density in a galaxy (Kennicutt 1989). A comparison of the distribution of molecular gas (derived from CO observations by Bronfman *et al.* 1988) and HII regions from the combined surveys is given in Figure 5. The distribution of the two species is quite similar, a fact which was not at all obvious in earlier versions of this comparison. The ratio of surface densities of ionized to molecular gas appears to vary with R, being lower at small and large radii. Kennicutt (1989) has discussed ways in which this measure of star formation efficiency might be related to the global properties of a galaxy.

IV. The 3-kpc Arm and the Inner Boundary of Star Formation

The 3-kpc arm is a coherent aggregation of atomic and molecular clouds that extends over at least 30° in the galactic plane. Where it passes in front of the galactic center it has a

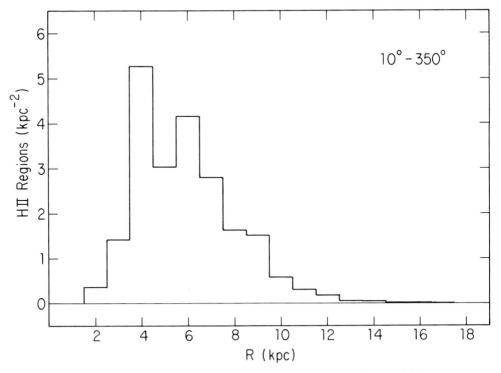

Figure 4: Surface density of all radio HII regions at R > 1.5 kpc and $|\ell| > 10°$.

velocity of -50 km s^{-1}. Its gaseous mass is estimated to be $\sim 2 \times 10^7$ $M_\odot$. Bania (1980) has summarized its observed properties, and Oort (1977) reviews various models for its origin. The question of star formation in the 3-kpc arm was first considered a decade ago (Lockman 1980b), and it was concluded that there was little evidence for HII regions, or any other extreme Population-I species, in the numerous molecular clouds of the arm. The recent data from the HII region surveys by Downes et al. (1980), Caswell and Haynes (1987) and the new Green Bank survey now show that this early conclusion is wrong.

Figure 6 shows an expanded velocity-longitude diagram of the HII regions in the inner Galaxy, together with the approximate locus of the 3-kpc arm. There are at least six nebulae at negative longitudes that may be in the 3-kpc arm and several at positive longitudes. In addition, Cersosimo (1990) has detected H166α emission from extended nebulae at five longitudes between 338°5 and 348° that appear to be in the arm. Just because an object overlaps with the 3-kpc arm in longitude and velocity, however, does not mean that it is necessarily associated with the arm. The nebula at 7.47+0.06, for example, which has a velocity of -18 km s^{-1}, is exactly on the locus of the arm, but HI absorption spectra show that it lies beyond the tangent point, and thus most probably is not in the 3-kpc arm (Garwood and Dickey 1989). Nonetheless, there do seem to be nebulae in the arm, and the earlier conclusion that star formation was totally absent arose from incomplete observations and the patchiness of the 3-kpc arm at positive longitudes. This topic is discussed further by Cersosimo (1990).

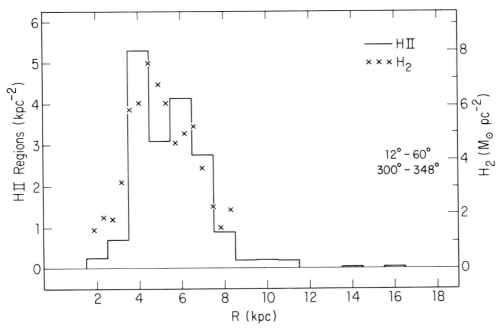

Figure 5: The surface density of radio HII regions compared with that of H_2 (Bronfman et al. 1988). The distributions are quite similar except that there seems to be more molecular gas per HII region at the largest and smallest radii. The comparison is restricted to longitudes $12° - 60°$ and $300° - 348°$ over which there are complete CO surveys.

The distance to the 3-kpc arm is derived solely from its apparent projection tangent to the line-of-sight somewhere near $\ell \lesssim 340°$. In the molecular clouds traced by CO emission it is seen to perhaps 336° (Dame et al. 1987) which requires that it be located at $R > 0.41 R_0$. This arm has often been identified (theoretically) with a resonance phenomenon and (observationally) with the rings of gas and HII regions that are often found at the inner termination of spiral arms (e.g., Lin 1971). On such a view, there should be little star formation interior to the arm, except, perhaps, in the galactic nucleus. Figure 6 shows, however, that at negative longitudes there are a number of nebulae whose velocities place them near $R = 2.5$ kpc, as shown by the fiducial line. These HII regions appear to be located <u>interior</u> to the 3-kpc arm, and in some cases there is additional evidence to support this conclusion. The nebula at 350.13+0.09, for example, has a velocity that places it at $R = 2.7$ kpc, and the 3-kpc arm is clearly seen in HI absorption in front of the nebula (Garwood and Dickey 1989) in agreement with the kinematic interpretation. Also, there are nebulae that lie in well-defined structures interior to the 3-kpc arm (Lockman 1981; Cersosimo 1990). Whatever the nature of the 3-kpc arm, it is not the inner boundary of star formation in the galactic disk.

Finally, Figure 6 shows a clear asymmetry between the positive and negative longitudes in the number of nebulae at high $|V|$ and thus small R. While there are > 10 near the $R = 2.5$ kpc line in the fourth longitude quadrant, there are only one or two in the first quadrant. Star formation appears to extend much further inward at negative longitudes

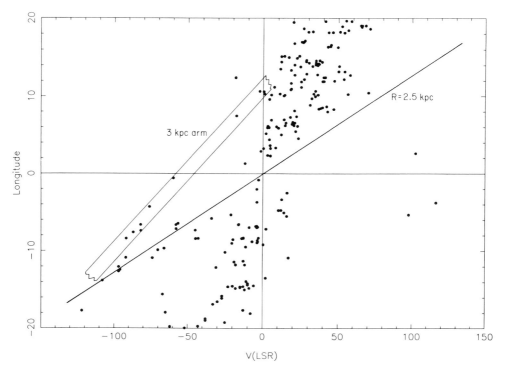

Figure 6: The velocity-longitude diagram of radio HII regions in the inner 40° of the Galaxy (it is not complete in the inner ±2°). The approximate locus of the 3-kpc arm is outlined, and there is a line marking the expected location in velocity and longitude of nebulae that are 2.5 kpc from the galactic center. There appear to be HII regions both in and interior to the 3-kpc arm, especially at negative longitudes.

than at positive longitudes. Given that the 3-kpc arm is also most prominently defined (in HII regions) at negative longitudes, it is tempting to speculate that these two phenomena are somehow connected.

V. Spiral Structure.

If Figure 2, the complete velocity-longitude diagram, is studied with relaxed eyes, many of the HII regions will seem to lie in bands, e.g., from $(v, \ell) = (0, 0)$ to $(v, \ell) = (100, 30°)$ and to $(v, \ell) = (70, 50°)$. This is the distribution that nebulae in a small pitch-angle spiral arm would have (Lockman 1979), and I believe that this is the strongest evidence for spiral structure in the inner Galaxy. Even so, it is not as persuasive as one would like, for the nebulae at negative longitudes appear more scattered on this diagram and do not fit such a simple scheme. A true "grand design" spiral should extend in a coherent pattern at least half way around the Galaxy, but it is not apparent that our HII regions do.

VI. Final Notes

In this paper I have mentioned briefly some of the complications and consequences of recombination line surveys as they affect our understanding of galactic structure. For some studies, such as the comparative radial distribution of HII regions and molecules, and of star formation in the 3-kpc arm, the relatively few objects detected in the early surveys led to erroneous conclusions. This again emphasizes the need to obtain and use large samples of data in any analysis of the properties of the Galaxy. As of now, probably fewer than half of all galactic HII regions have been measured in a recombination line, so there is still much to be done, and we can look forward to many more surprises in the next 25 years.

References

Altenhoff, W.J., Downes, D., Pauls, T., and Schraml, J. 1978, Astr. Ap. Suppl., **35**, 23.
Anantharamaiah, K.R. 1985, J. Astrophys. Astr., **6**, 203-226.
Avedisova, V.S., and Kondratenko, G.I. 1984, Nauchnye Informacii Astronomicheskogo Soveta Akademii Nauk SSSR., issue 56, p. 59.
Avedisova, V.S. and Palouš, J. 1989, Bull. Astron. Inst. Czechosl., **40**, 42-52.
Bania, T.M. 1980, Ap.J., **242**, 95-111.
Blitz, L. 1978, Ph.D. Thesis, Columbia Univ., NASA Technical Memorandum 79708.
Bronfman, L., Cohen, R.S., Alvarez, H., May, J., and Thaddeus, P. 1988, Ap.J., **324**, 248-266.
Burton, W.B. 1972, Astr. Ap., **19**, 51.
Burton, W.B. and Gordon, M.A. 1978, Astron. Ap., **63**,, 7-27.
Caswell, J.L. and Haynes, R.F. 1987, Astr. Ap., **171**, 261.
Celnik, W.E. 1985, Astr. Ap., **144**, 171.
Cersosimo, J.C., Azcárate, I.N., Hart, L., and Colomb, F.R., 1989, Astron. Astrop., **208**, 239-246.
Cersosimo, J.C. 1990, Ap.J. (Letters), (in press).
Dame, T.M. et al. 1987, Ap.J., **322**, 706-720.
Dieter, N.H. 1967, Ap.J., **150**, 435-451.
Digel, S., Bally, J. and Thaddeus, P. 1990, Ap.J. (Letters), (in press).
Downes, D., Wilson, T.L., Bieging, J. and Wink, J. 1980, Astr. Ap. Suppl., **40**, 379.
Garwood, R.W. and Dickey, J.M. 1989, Ap.J., **338**, 841-861.
Heiles, C., Kulkarni, S.R., Stevens, M.A., Backer, D.C., Davis, M.M., and Goss, W.M. 1983, Ap.J. (Letters), **273**, L75-L79.
Helfand, D.J., Velusamy, T., Becker, R.H. and Lockman, F.J. 1989, Ap.J., **341**, 151.
Hoffleit, D. 1982, *The Bright Star Catalogue*, (Yale Univ. Observatory: New Haven).
Hodge, P.W., and Kennicutt, R.C. 1983, Ap.J., **267**, 563.
Kennicutt, R.C. 1989, Ap.J., **344**, 685.
Lin, C.C. 1971, in *Highlights of Astronomy*, Vol. 2, ed. C. de Jager, (Dordrecht: Reidel). p. 88.
Lockman, F.J., 1976, Ap.J., **209**, 429.
Lockman, F.J., 1979, Ap.J., **232**, 761-781.

Lockman, F.J. 1980a, in *Radio Recombination Lines*, ed. P.A. Shaver, (Dordrecht: Reidel), p. 185.
Lockman, F.J. 1980b, Ap.J., **241**, 200-207.
Lockman, F.J. 1981, Ap.J., **245**, 459-464.
Lockman, F.J. 1989, Ap.J. Suppl., **71**, 469-479.
Mezger, P.G., and Höglund, B. 1967, Ap.J., **147**, 490.
Maddalena, R.J., Morris, M., Moscowitz, J., and Thaddeus, P. 1986, Ap.J., **303**, 375-391.
Oort J.H. 1977, Ann. Rev. Astr. Ap., **15**, 295.
Oskanyan, A.V. 1989, Astrofizika, **30**, 128-139.
Reifenstein, E.C., Wilson, T.L., Burke, B.F., Mezger, P.G. and Altenhoff, W.J. 1970, Astr. Ap., **4**, 357-377.
Wendker, H.J. 1984, Astr. Ap. Suppl., **58**, 291-316.
Westerhout, G. 1958, Bull. Astr. Inst. Netherlands, **14**, 215-260.
Wilson, T.L., Mezger, P.G., Gardner, F.F. and Milne, D.K. 1970, Astr. Ap., **6**, 364-384.

STUDY OF THE H166α RECOMBINATION LINE IN THE SOUTHERN MILKY WAY

J. C. CERSOSIMO [1]
NAIC - Arecibo Observatory [2]
P.O.Box 995
Arecibo P.R. 00613

ABSTRACT. We review the observations of recombination line emissions with low angular resolution made at the Instituto Argentino de Radioastronomía, and new detections of diffuse ionized gas located in the 3-kpc arm are shown. The observations suggest that the H166α line-emitting gas is associated with Population I material, but it is not always connected to active star formation regions. Around $l = 335°$ the lines show evidence of interarm ionized gas, probably connecting giant complexes of HII regions located in Scutum-Crux and Norma arms. If a non-LTE model is suitable for this ionized gas the analysis of the line emission, along with low frequency continuum observations, suggests a low electron temperature ($T_e < 500$ K) and a low emission measure ($E < 200$ pc cm^{-6}).

1. Introduction

In order to study the large scale distribution of the ionized hydrogen, Hart and Pedlar (1976) and Lockman (1976) observed the H166α line (1424.734 MHz) in the first quadrant of the Galaxy, with uniform coverage of the galactic plane of these lines. In addition, to complete the search in the inner galaxy an extensive radio recombination line (RRL) survey along the South galactic equatorial plane (fourth quadrant) has been made by Hart et al. (1982), and later completed and discussed by Cersosimo et al. (1989). The observations of RRLs at 1.4 GHz in the south quadrant has been carried out using a new receiver at the Instituto Argentino de Radioastronomía (IAR) was mounted, a description of the equipment used is given by Cersosimo et al. (1989). The IAR radio telescope beam is 34', and the spectral resolution of the observations is 15.9 km s^{-1}. The survey gave the first adequate description of the south galactic distribution of these RRL.

---

[1] Fellow of CONICET of Argentina. Visiting astronomer at Arecibo Observatory.
[2] The Arecibo Observatory is part of the National Astronomy and Ionosphere Center, which is operated by Cornell University, under agreement with the National Science Foundation.

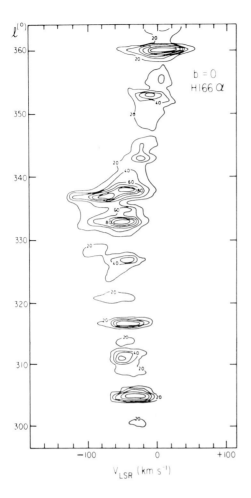

Figure 1: Velocity-longitude distribution at $b = 0°$ for H166α line antenna temperature (in mK). Contour are labeled with the antenna temperature of the line in milli-Kelvins. Outermost contour is larger than 3 times the rms noise. The data were taken each degree from longitude $300°$ to $4°$.

The line is detected in almost all positions between longitudes $300°$ and $360°$; the strongest emission originates between longitudes $330°$ and $340°$. In order to study in detail this zone, we observed each half degree between longitudes $330°$ and $340°$ along the galactic plane. Also as part of the Southern Survey a new search has been made between longitudes $345°$ and $349°$ obtaining new spectra with an integration time of 10 hr; our main objective was to discover new line emissions at high velocity located in the 3 kpc arm. The lines were detected in directions of $l = 345°.5$, $347°$, and $348°$. Furthemore using a spectral resolution of 2 km s^{-1} an intense line at high velocity was detected at $l = 347°.6$, $b = 0°.2$. In this talk we will review the main results of the southern survey, discuss the emission around $l = 335°$ and present new detections of RRL associated with the 3-kpc arm.

2. Results of the Survey

The aim of this section is to show the results obtained in the fourth quadrant. The observations of the H166α transition were taken every one degree from longitudes 300° to 4° along the galactic plane (Cersosimo et al. 1989). In Figure 1 we show the antenna temperature versus velocity where the general characteristics and quality of the data are illustrated. The lowest level contours are drawn every 20 mK which is above 3σ. The lines are broad and in many positions there are multiple components. The profiles show no emission of the He166α and C166α lines. In most the map of the line emission is traced by small scale HII regions actually standard HII regions like RCW 74 or NGC 6334 that are relatively dense and have definable boundaries. Probably a smoother component connecting classical HII regions is visible between longitude 330° and 345°. The H166α emission is widely distributed at radial velocities in several positions of the fourth quadrant. Not including the profile in the direction of the galactic center, which has the most intense and broad line, the emissions are all at negative velocities. There is no significant emission at $V = 0$ km s^{-1} (LSR), except for the profiles at $l = 349°$ to 355°. The absence of substantial emission around zero velocity indicates that the RRL do not come from regions close to the Sun. The H166α emission shows the highest emission at longitudes about 30° away from the galactic center.

In Figure 2 we show a more complete view of the emission from the plane between longitudes 330° and 340° where the more intense emission was observed. The spectra were taken at an interval of 0°.5. The emissions have the highest negative velocities and the widest profiles between $l = 331°$ and 334°, and between 336° and 339°. The emission has outstanding power, with the largest peak and the greatest velocity extent of the South quadrant. A similar feature has been found at $25° < l < 32°$ in the North quadrant (Lockman 1976; Hart and Pedlar 1976). At longitudes between $334° < l < 336°$ we can distinguish weak emissions, almost with constant line temperature connecting the complex located around $l = 333°$ and $l = 337°$.

Another view of the emission in the fourth quadrant is shown in Figure 3 where the integrated strength of the line ($\sum T_i \times v_i$) is plotted against longitude. For comparison, the results found in the first quadrant (Lockman 1976) are shown. In Figure 3 we can see the intense emission originated in the 330-340 region. The counterpart in the North is located at $23° < l < 31°$, that is 3° farther away from the galactic center than the one main southern emission. Some relatively intense emission was detected at $l < 329°$ in the South, which has no counterpart in the North at $l > 31°$. These emissions are due to southern HII complexes around $l = 305°$, $l = 311°$ and $l = 317°$. Assuming a spiral pattern (Georgelin and Georgelin 1976) the peak of the emission at $l = 305°$ is located in the Sagitarius-Carina arm, at $l = 311°$ it is located in the Scutum-Crux arm, and at $l = 317°$ it is located in the far Scutum-Crux arm. The emission at zero velocity from $l = 349°$ to $l = 355°$ is located in the Sagitarius-Carina arm.

The comparisons of the H166α results with the HI antenna temperatures in ve-

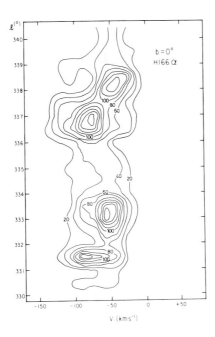

Figure 2: Same as in fig. 1 except that the data were taken each half a degree from longitude 330° to 340°.

locity and longitude for both quadrants (see Cersosimo et al. 1989, and Lockman 1976), show that the HI line is more extensive in velocity than H166α line. Generally, the intensity of HI does not show large variations in longitude that are evident in the H166α line. There is a substantial amount of HI at positive velocities probably originating at $R/R_o > 1$, which has no counterpart in the 1.4 GHz RRL. On the other hand, the H166α emission is correlated with the CO emission. Broadly speaking both components are confined to galactocentric radii $0.5 < R/R_o < 0.9$ (Cersosimo et al. 1989).

3. The 330°-340° Region

3.1. H166α LINE AND STAR FORMING REGIONS

Compact HII regions represent a very early stage in the life of an OB star and the object must be in state of rapid expansion (Mezger et al. 1967). As their size increases their density and central emission measure decreases. Expansion of the HII regions continue until they achieve pressure equilibrium with the surrounding neutral gas, or until the ionizing O stars move off the main sequence. Once the electron densities have decreased to some 10 cm^{-3}, they form extended low density HII regions with low surface brightness which are difficult to observe at high radio frequencies. However, the free-free emission from a diffuse region is easily observable

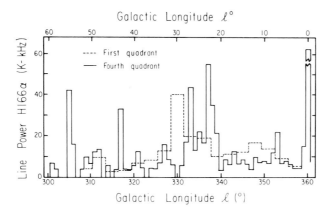

Figure 3: Total power of the H166α line as a function of galactic longitude in both fourth (*full-lines*) and first quadrants (*dashed-lines*).

with low angular resolution at decimeter wavelengths, and probably RRL emission is observable in these low brightness regions.

The map of Figure 4 (top) shows the H166α line emission contours at 40 mK intervals in the region. The crosses mark the location of 5 GHz RRL from Caswell and Haynes (1987), and the dots mark H_2O masers from Braz and Epchtein (1983) which probably indicate star forming regions. The H166α line shows four maxima of emission at the positions $l = 331°$ to $332°$, $l = 332°.5$ to $334°$, $l = 336°$ to $338°$ and $l = 337°.5$ to $339°$. All of these positions are correlated with emission at H110α, which came from sources with $N_e \geq 100$ cm^{-3}, and with H_2O maser sources.

Continuum observations at 1.4 GHz were made using the IAR radio telescope, the data reduction are described by Cersosimo (1990). The results are shown in the lower panel of Figure 4 where the measurements of the continuum brightness temperature are in units of antenna temperature. The observations are referenced to the point $l = 300°$, $b = -5°$ which has an antenna temperature, $T_A = 0.95$ K. The map shows five maxima which are correlated with optical features of the RCW (Rodgers Campbell and Whiteoak 1960) catalogue. The continuum emission is also well correlated with the main emission of the RRL along the galactic plane. Generally there exists a clear relationship between strong H166α line emission, star forming regions and strong continuum emission. In these directions the emission arises from very complex regions where star forming is located in the Scutum-Crux and Norma arms.

The continuum emission detected in the direction of RCW 108, at 1°.5 south of the plane, is associated with the young galactic cluster NGC 6193. RRL from the gas associated with the cluster has been observed by Cersosimo (1982) using a velocity resolution of 2 km s^{-1}. The parameters obtained suggests that the H166α line is formed in a diffuse thermal gas with $N_e \sim 10$ cm^{-3}, probably ionized by the

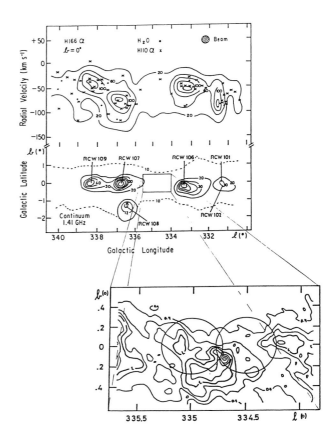

Figure 4: Top: contours of antenna temperature each 40 mK of H166α line in the velocity-longitude plot. The H₂O maser sources are marked with *dots*, and the 5 GHz RRL positions are marked with *crosses* (upper panel). Continuum antenna temperature contours at 1415 MHz (lower panel). Bottom: Continuum at 5 GHz. Circles indicate the 34' sample of 30-m antenna.

stars of the cluster. The ionized gas around the cluster is like an envelope with a linear diameter of 50 pc.

The region around $l = 335°$ presents weak H166α RRL emission. The main characteristics are the continuity of the emission through 2 degrees along the galactic plane, which is free of high density HII regions observable at 5 GHz and evidence of star forming regions. The continuum emission observations show a minimum between the two maxima located at $l \sim 333°$ and $l \sim 337°$. This region is discussed in the next section.

3.2. THE REGION AROUND $l = 335°$

In this direction the (l, V) plot of Figure 4 (top) shows the peak temperatures of the line to have values between 20 to 60 mK spread in velocity from -120 to -25 km s^{-1}. The peaks of the emission have a velocity gradient of approximatly

25 km s^{-1} per degree along the galactic plane (see Figure 2). The profiles observed around the position $l = 335°$ have substantially greater widths, and thus the excess broadening must be due to velocity effects of galactic rotation and turbulence. If the ionization is distributed along the line of sight, the broad velocity extent and the velocity gradient along the galactic plane, both, can be caused by the differential rotation of the Galaxy. The gas connects the regions G332.5-334 and G336-338 which are locted in the Scutum-Crux and Norma arms respectively (Georgelin and Georgelin 1976). By examining the continuum emission at 5 GHz (Haynes et al. 1979) we can see an extended low brightness emission region correlated with the absorption feature at 29.9 MHz (Jones and Finlay 1974). This low frequency is well suited for observing details of the absorption of galactic radiation by thermal electrons. Then the extended low brightness emission at 5 GHz can be mostly thermal. Figure 4 (bottom) shows the 5 GHz continuum map and the HPBW of our antenna superposed at the position where the H166α line has been detected. In this map we see the extended low brightness source and an intense feature at the position G334.684-0.107, where Caswell and Haynes (1987) detected the H110α line at $V = -32$ km s^{-1}. This feature is diluted and dimmed by our beam antenna, and we think that its contribution to our spectra is not important. Then the contribution to our observations at 1.4 GHz is not associated with discrete HII regions around $l = 335°$. The constancy of the RRL emission over this galactic region suggests that the gas may actually be widespread rather than confined to relatively small angular structure. This is supported by the lack of H$_2$O masers in Figure 4 (top), and also by the lack of early type stars (Humphreys 1970).

It has been suggested (Gottesman and Gordon 1970; Gordon and Gottesman 1971; Cesarsky and Cesarsky 1971) that RRL emission from the galactic plane and away from catalogued HII regions can be attributed to partially-ionized cool ($T_e <$ 500 K) interstellar gas. In such a case for the interpretation of the emission the LTE theory is not valid. A convenient way is to use non-LTE models along with results of continuum observations, then the electron temperature and the electron density can be deduced. By examining the continuum maps of Haslam et al. (1982) and of Jones and Finlay (1974) we find emission at 408 MHz and absorption at 29.9 MHz, respectively. To explore the electron temperature and the emission measure, we used the continuum results assuming the optical depths: $\tau_{(408)} = 0.01$ and $\tau_{(30)} = 1.5$. In Figure 5 are shown the constraints on the emission measure and the electron temperature of the ionized gas around $l = 335°$, derived from the continuum optical depth at 408 and 30 MHz. In order to explain the line emission around $l = 335°$ we use a plasma model illuminated by a nonthermal radiation field. We use the equation gived in the literature,

$$T_L = b_n \tau_L^* T_e (1 - \beta \frac{T_o}{T_e}), \tag{1}$$

where T_e is the electron temperature, τ_L^* is the optical depth of the line, T_0 is the temperature of the background which from our data is assumed to be 15 K. The

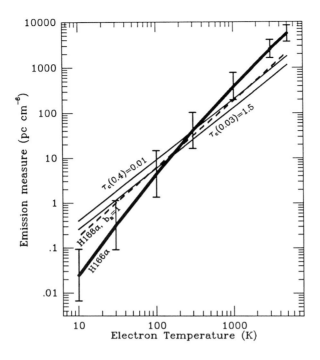

Figure 5: Emission measure plotted against the electron temperature. The *thin* lines represent the constant optical depth obtained from observations at 0.03 and 0.4 GHz in direction of $l = 335°$. The *dashed* line is derived from the power H166α line assuming a LTE model. And the *thick* curve is derived from a non-LTE model using the equation (1).

b_n and β_n factors were interpolated from tables gived by Salem and Brocklehurst (1979). In Figure 5 the thick line shows the run of the emission measure and the electron temperature for the H166α line emission assuming a path length of 3 kpc. The error bars are estimated errors due to errors of interpolation of the coefficients, uncertainties in the background continuum temperature, and assuming an error of 30% in the path length.

The models suggest an upper limit of the electron temperature, $T_e < 1000$ K, and an upper limit of the emission measure of 200 pc cm^{-6}, then with the path length of 3000 pc the electron density is $N_e < 0.2$ cm^{-3}. This upper limit suggests a dispersion measure of 600 pc cm^{-3}. This value is large compared with the values derived from pulsar dispersion in the directions of the galactic plane (Hulse and Taylor 1975). A reasonable value is $DM = 200$ pc cm^{-3}, which implies a mean electron density of .07 cm^{-3}. Finally using this electron density we need an electron temperature no more than 200 K for explain the H166α line emission and the continuum observations. This model is not unique, we also computed equation (1) assuming $b_n=1$ and $\beta=0$, the results are shown in the Figure 5 with a dashed line. The LTE line emission model is in agreement with the continuum observations for the temperature range between 100 and 2000 K and emission measure between 50 and 700 pc cm^{-6}. In sum, we show evidence for the existence of moderate density ionized gas that is, at the hottest, about 2000 K.

4. Ionized Gas In The 3-kpc Arm

Observations with high sensitivity were made between $345°$ and $349°$ longitude. The observations allow us to see a significant detection around $V = -90$ km s^{-1} in the directions $l = 345°.5$, $347°$ and $348°$, which can be located in the inner zone of the galaxy at distances $R_G < 4$ kpc.

The recombination lines can usually be characterized as Gaussian profiles. The results of the H166α line are summarized in Table 1. In addition we include the emissions detected at high velocity in the fourth quadrant at longitudes $350°$, $352°$ and $356°$ by Lockman (1980a), and the emission at $l = 338°.5$ shown in the map of Figure 2. In Table 1 column (1) gives the galactic longitude. Columns (2), (3) and (4) give the peak temperature, the width, and the central radial velocity of the lines respectively. According to the analyses by Cersosimo (1990) the emissions of Table 1 arise from gas with low electron density (< 10cm^{-3}). In Figure 6 are plotted with filled dots the data of Table 1 in the velocity-longitude diagram. The emissions of the H166α line are seen as a band located in the 3-kpc arm, which is marked with solid lines.

TABLE 1: Parameters of the H166α Line

Galactic Longitude (degree)	Line Temperature (mK)	Line Width (km s^{-1})	Radial Velocity (km s^{-1})
338.5	38	20±8	-112±9
345.5	11	22±10	-120±9
347.0	14	32±12	-85±9
347.6	100	32±2	-98±4
348.0	13	36±12	-96±9
350.0[1]	≤15	-	-110
352.0[1]	≤15	-	-90
356.0[1]	≤15	-	-80

(1) From Lockman 1980.

For comparison, in Figure 6, the triangle mark dense HII regions detected in 5 GHz discrete-source (Caswell and Haynes 1987), which kinematically, are located in the inner zone of the Galaxy, at $R_G < 3.4$ kpc (assuming $R_\odot = 8.5$ kpc). Almost half of the dense HII regions are in the 3-kpc arm. It is shown in Figure 6 that in some cases the H166α RRL is not associated with RRL emission at 5 GHz, which indicates that the low density gas could not be associated with dense HII regions. Excluding the source at $l = 347°.6$, the average line emission has temperature of ~ 15 mK, it appears on angular scales ≤ 20', and does not show continuity along of the arm.

The intense line detected at $347°.5$ is associated with line emission at 5 GHz.

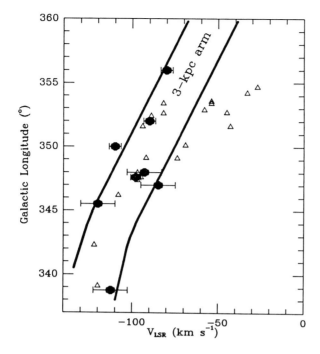

Figure 6: Low density HII regions detected at H166α line (*dots*) compared with dense HII regions (*triangle*), in the inner region of the Galaxy. The low density HII regions are located in the locus of the 3-kpc arm.

The continuum source has an angular size of 9×5' and the emission measure is $E = 4.3 \times 10^4$ pc cm^{-6} (Reifenstein et al. 1977). The mean velocity of the profile is almost coincident with the H109α profile. Furthemore the ionized gas is correlated with a gap of HI (Kerr et al. 1981) which is suggesting the existence of an interactive mechanism between stars and the interstellar medium, like supernova remnant or open cluster.

The ionized gas located at $l = 338°.5$ and with a radial velocity ~ -120 km s^{-1} (see Figure 2), is located where the HI emission, which define the 3-kpc arm, appears to merge gradually with the terminal velocity ridge of HI (Kerr et al. 1981).

5. Conclusions And Final Comments

We have presented observations of H166α recombination lines of the fourth quadrant. Generally the emission is traced by small scale HII regions, and the main emission came from between longitude 330° and 340°. In this region, which have large feature of continuum emission, the RRL is probably originated in relatively dense HII regions and in its associated envelopes, the young open clusters and the OB associations must be responsable of the ionizing radiation.

At $l = 335°$ we detected weak emission not associated with traditional HII regions and not connected to active star forming regions. Also weak emission at high velocity between longitude $345°$ and $349°$ has been detected, which is thought to be associated with the 3-kpc arm. Like the emission detected at $l = 36°$ (Lockman 1980b) these diffuse materials have not been detected except in the H166α line.

With a 30-m single dish at 1.4 GHz the observations are sensitive to low density gas at distances farther than 5 kpc to the Sun (Cersosimo 1990), then as suggested by Anantharamaiah (1986), large low-density envelopes of the normal HII regions can explain the emissivity of distant regions. From the analyses of the H166α RRL and continuum observations, using LTE and non-LTE models we obtain for the ionized gas a low emission measure $E < 1000$ pc cm^{-6} and a low electron temperature $T_e < 2000K$. If the non-LTE model is suitable the analysis suggests an electron temperature of $T_e \leq 500K$ and an emission measure of $E \leq 200$ pc cm^{-6}.

Apparently the emission around $l = 335°$, like those found by Lockman (1980b) at $l = 36°$, is not coming from traditional HII regions, the recombination line 5 GHz survey (Downes et al. 1980 and Caswell and Hynes 1987) with a *rms* noise level of 10 mK, does not show dense HII regions associated with these emission. However it is probable that only a small fraction of galactic HII regions have been observed. Hughes and Mac Leod (1989) identified 1255 infrared HII regions from the IRAS Point Source Catalog to an estimated 89% confidence level. The interesting aspect of this result is the fact that the IRAS HII regions are toward the galactic plane. A latitude-longitude plot (Hughes and Mac Leod 1989) shows that most of the HII regions lie within $\pm 3°$ of the galactic plane. Of course the regions around $l = 335°$ and $l = 36°$ are highly populated with these small regions. By observing RRL we are probably detecting numerous smaller HII regions and the recombination lines can arise in extended outer envelopes of these small regions. Always we have irregularities in the interstellar electrons which are responsable of the phenomenon of interstellar scintillation (ISS). In fact the ISS dilates images of compact extragalactic sources, broadens pulsar pulse profiles and causes pulsar signals to vary in time and radio frequency. By considering the observational aspects it is probably that irregularities of the electron distributions can be created by the presence of small HII regions, more than a distributed electron component. However, more data such as the optical depth at 1.4 GHz of these HII regions are desirable.

The author wants to express his sincere thanks to the staff of the National Astronomy and Ionophere Center and to the Scientific Organizing Committee for their help, also to T. Phillips by the critical reading manuscript. The present paper was prepared at the Arecibo Observatory supported by the CONICET of Argentina.

REFERENCES

Anantharamaiah, K.R.: 1986, J. Ap. Astr. **7**, 131.
Braz, M.A., Epchtein, N.: 1983, Astr. Ap. Suppl., **54**, 167.

Caswell,J.L., Haynes, R.F.: 1987, Astr. Ap. **171**, 261.
Cersosimo, J.C.: 1982, Ap. Lett. **22**, 157.
Cersosimo, J.C.: 1990, Ap. J., January 20.
Cersosimo, J.C., Azcárate, I.N., Hart, L., Colomb, F.R.: 1989, Astr. Ap. **208**, 239.
Cesarsky, C.J., Cesarsky, D.A.: 1971, Ap. J. **169**, 293.
Downes, D., Wilson, T.L., Bieging, J., Wink, J.: 1980, Astr. Ap. Suppl. **40**, 379.
Georgelin, Y.M., Georgelin, Y.P.: 1976, Astr. Ap. **49**, 57.
Gordon, M.A., Gottesman, S.T.: 1971, Ap. J., **168**, 361.
Gottesman, S.T., Gordon, M.A.: 1970, Ap. J., **162**, L93.
Hart, L., Azcárate, I.N., Cersosimo, J.C., Colomb, F.R.: 1982, Proc. Leiden Workshop, *Survey of Southern Galaxy.* Eds: W.B. Burton, and F.P. Israel, Dordretch, p. 43.
Hart, L., Pedlar, A.: 1976, M.N.R.A.S., **176**, 547.
Haslam, C.G.T., Salter, C.J., Stoffel, H., Wilson, W.L.: 1982. Astr. Ap. Suppl. **48**, 1.
Haynes, R.F., Caswell, J.L., Simons, L.W.: 1979, Australian J. of Phys., Astrophys. Suppl., N° 48.
Hughes, V.A., MacLeod G.C.: 1989, A. J. **97**, 786.
Hulse, R.A., Tylor, J.H.: 1975, Ap. J. Lett., **201**, L55.
Humphreys, R.M.: 1970, A. J. **75**, 602.
Jones, B.B., Finlay, E.A.: 1974, Australian. J. Phys. **27**, 687.
Kerr, F.J., Bowers, P.F., Henderson, P.D.: 1981, Astr. Ap. Suppl. **44**, 63.
Lockman, F.J.: 1976, Ap. J., **209**, 429.
Lockman, F.J.: 1980a, Ap. J., **241**, 200.
Lockman, F.J.: 1980b, Proc. Ottawa Workshop, *Radio Recombination Lines.* Ed: P.A. Shaver, p.185.
Mezger, P.G., Altenjoff, W., Schraml, J., Burke, B.F., Reifenstein, E. C., Wilson, T.L.: 1967, Ap. J. (Letter) **150**, L137.
Pedlar, A.: 1980, M.N.R.A.S., **192**, 179.
Reifeistein III, E.C., Wilson, T.L., Burke, B.F., Mezger, P.G., Altenhoff, W.J.: 1970, Astr. Ap. **4**, 357.
Rodgers, A.W., Campbell, C.T., Whiteoak, J.B.: 1960, M.N.R.A.S., **121**, 103.
Salem, M. Brocklehurst, M.: 1979, Ap. J. Suppl. **39**, 633.

RADIO RECOMBINATION LINE IMAGING OF SGR A

W. M. Goss
NRAO*, Socorro, New Mexico
USA

J. H. van Gorkom
Columbia University
New York, New York
and
NRAO*, Socorro, New Mexico
USA

D. A. Roberts
NRAO*, Socorro, New Mexico
and
University of Oklahoma
Norman, Oklahoma
USA

J. P. Leahy[+]
NRAO*, Socorro, New Mexico
USA

Abstract:
Recent Very Large Array observations of Sgr A West have been carried out at 8.3 and 14.7 GHz. The kinematics of the central 1-2 pc have been delineated. The electron temperature increase in the central bar towards Sgr A* found at 14.7 GHz by Schwarz et. al (1989) has been confirmed at 8.3 GHz. Zeeman observations of the H92 α line at 8.3 GHz have been carried out; the resulting upper limit to the magnetic field is about 15m Gauss.

Introduction:

In his ground breaking paper, Kardashev (1959) has said (translation taken from Soviet Astronomy, 1960): "It would be particularly interesting to attempt to discover lines from the mass of ionized gas found at the

---

* Operated by Associated Universities Inc. under cooperative agreement with the National Science Foundation.

[+]Present address: Nuffield Radio Astronomy Laboratories, Jodrell Bank, Macclesfield, Chesire, England

galactic center. This would provide information on its temperature, velocity and ionization which are not well known". In the intervening 30 years these predictions have certainly been fulfilled.

The early radio recombination line (RRL) detections towards Sgr A are now known to arise from stimulated emission due to HII regions along the line of sight. The first detection of RRL's from Sgr A itself was reported by Pauls et. al (1974) who observed broad lines ($\Delta V \sim 200$ kms^{-1}) at 5 GHz in the H109 α line. It is now recognized that these line widths are produced by the large velocity gradients in Sgr A west. The 12.8μm Ne$^+$ fine structure line has been observed by many groups with high angular resolution in recent years. The comparison of these IR results with the radio images of RRL's with comparable angular resolutions has been illuminating.

Previous reviews of Sgr A RRL observations have been presented by Pauls (1980) and van Gorkom et. al (1985).

In this paper we will not discuss observations of the fascinating RRL studies of other galactic center thermal sources, such as the thermal arc at $\ell=0°1$, b=0°08 and the source G0.15 - 0.05 in the linear arc. Various papers by F. Yusef-Zadeh and collaborators (e.g. Yusef-Zadeh et. al 1987 and Yusef-Zadeh et.al, 1989) have described these VLA observations.

Sgr A West: pre 1989 data source.

Sgr A West is a radio source of $\approx$ 20 Jy which surrounds Sgr A*, presumably the nucleus of the Galaxy. Ekers et. al (1983), and Lo and Claussen (1983) discovered a multi-arm spiral structure within Sgr A West. Since this structure has an inverted spectrum between 1.4 and 5 GHz, the source is an optically thick HII region. It is from this HII region that the RRL lines arise.

Sgr A West is a "giant" HII region; however there are many HII regions in the galaxy with higher luminosities which have a larger number of exciting stars e.g. Sgr B2 and W49. The unique aspects of Sgr A West are the location, shape, kinematics and ionization. From the RRL velocities, the mass distribution in the central 1-2 pc of the galaxy can be derived. We assume a distance of 10 kpc; 20" arc is thus 1 pc.

The first images of recombination line emission from Sgr A West were made with the WSRT by Bregman and Schwarz (1982) in the H110 α line. These observations showed that the emission was related to the spiral features, thus confirming the thermal origin of the continuum radiation. The first VLA observations, made in the H76 α line with 4"5 arc resolution, (van Gorkom et. al, 1985) revealed that each of the spiral features had a large scale systematic velocity gradient. The VLA results lead to a fundamental change in the way in which the [Ne II] gas velocities had previously been interpreted; instead of "clumps" of gas in a cloudy distribution, a large scale coherent velocity field was indicated.

More sensitive images of the H76 α line were analyzed by Schwarz et. al (1989). This time the VLA was used in the D array with a resolution of 4"5x7"5 arc (α x δ). The rms sensitivity was about 2 mJy beam^{-1} and the velocity range covered was -340 to 393 kms^{-1}.

The H76 α results are presented in Fig. 1 which shows an intensity-hue image of the H76 α RRL. An almost complete ionized ring is detected; this ring forms the inner boundary of the molecular ring observed in many molecules (e.g. HCN, Güsten et. al 1987). There are three long curved "filaments" to the north, east and west which cross the ring and themselves have systematic velocity gradients. The typical velocity widths of the profiles are 45 ± 15 kms^{-1}. In Fig. 2 the outline of the various features are shown with the velocities superimposed. Note that in the RRL emission, the "southern" arm does not bend into the center but continues straight to the north; this feature has also been called the western arc by some authors.

A simple kinematic model of circular motion has been fitted to the velocities. The circular velocity at a radius of 1.5 pc is 105 kms^{-1} with an inclination of 61° and a position angle of the major axis of 19°. The ionized ring in circular motion implies a total mass within this radius of 3.8×10^6 M$_\odot$. Fig. 3 shows the various estimated masses within a given radius based on observations of H76 α, [Ne II], [OI], CO and HI.

A surprising result found from the H76 α data is the systematic decrease in the line to continuum ratios along the eastern bar. Schwarz et. al have interpreted this as an increase in electron temperature towards Sgr A* (see following section).

Schwarz et. al have also compared their observations in detail with observations of the H66 α line obtained with the 100m antenna at Effelsberg with a resolution of 40" arc (Mezger and Wink 1986) and the extensive set of [NeII] observations of Serabyn and Lacy (1985) with a 6" arc resolution. The H76 α data agrees with these observations when the data are compared at similiar resolutions.

Sgr A West: New results

In 1989, Sgr A was observed in a 2x8 hour period in the H92 α line at 8.309383 GHz using the new 8-9 GHz system at the VLA (Roberts, Leahy, van Gorkom and Goss, in preparation). In the B/C array the beam width is essentially circular with a resolution of 2".4 arc. Both senses of circular polarization were observed. The velocity coverage was -13 to 213 kms^{-1} and thus only the velocity range of the northern part of Sgr A West is covered; the velocity resolution is 7.05 kms^{-1}. A major goal was the detection of the Zeeman effect in the northern arm. The rms noise in this data is 0.4 mJy beam^{-1}.

In Fig. 4 the continuum image is shown; two representative profiles are shown in Fig. 5. The profile in Fig. 5a was observed in the direction of Sgr A* and undoubtedly arises from stimulated emission in the intervening gas at a velocity of ~57 kms^{-1}. In Fig 5b the position of the profile (6" arc from Sgr A*) is 2" arc south of IRS 10 in the northern arm.

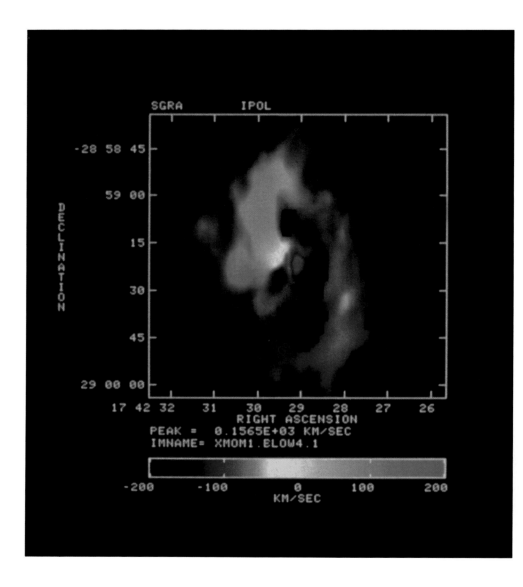

Fig. 1. Pseudo-color representation of the H76 α emission from Sgr A West (Schwarz et. al, 1989). The intensity represents the strength of the recombination line emission and the hue (color) represents the velocity. The velocity scale (l.s.r.) is shown at the bottom. Note the ring of ionized emission. The angular resolution of these VLA observations is 4".5x7".5 arc and the velocity resolution is 32 kms^{-1}.

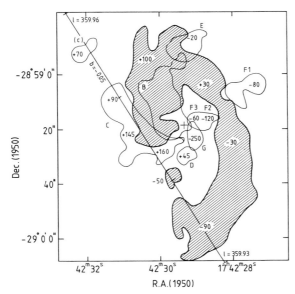

Fig. 2. Outline of the H76 α emission from Schwarz et. al, 1989. The ionized ring (their feature A) is shaded. Typical velocities are indicated in kms^{-1}. The cross indicates the position of Sgr A*. The line represents a line of constant galactic latitude (0°.05).

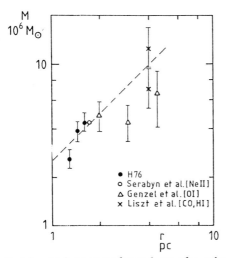

Fig. 3. Estimated masses based on the circular velocity model of Schwarz et. al (1989). The H76 α data are indicated by the filled circles, the [NeII] data (open circles) are taken from Serabyn and Lacy (1985), the [OI] data (triangles) are taken from Genzel et. al (1985) and the CO and HI data (crosses) are from Lizst et. al (1983, 1985).

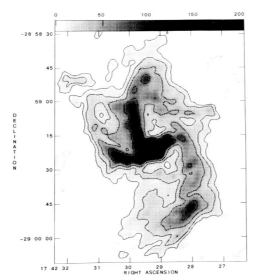

Fig. 4. Continuum image of Sgr A West at 8.3 GHz obtained with the VLA with an angular resolution of 2″.4 arc. The contour intervals are -8, 8, 16, 32, 64, 128, 256, 512 mJy beam^{-1}. The peak intensity in the image is 920 mJy.

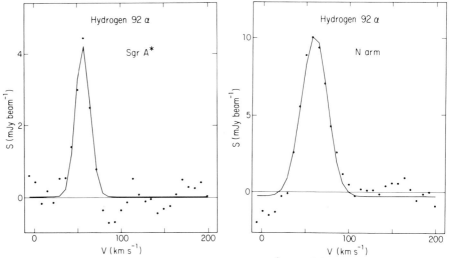

Fig. 5. H92 α (8.3 GHz) profiles at two positions in Sgr A West. Velocity is with respect to the l.s.r. The solid line shows the fitted Gaussian and the points are the original data. (a) Profile in the direction of Sgr A*; velocity is 57 kms^{-1} and the velocity width (FWHM) is 19 kms^{-1}. (b) Profile 6" arc north of Sgr A* in the northern arm, close to the position of IRS 10 (Serabyn and Lacy, 1985). The velocity is ~60 kms^{-1} and the velocity width is 35 kms^{-1}.

A new determination of the electron temperature in the bar to the east of Sgr A* has been made. The results of Pedlar et. al (1989) at .33, 1.5 and 5 GHz indicate that this continuum radiation is completely thermal. Local thermodynamic equilibrium (LTE) is assumed following Schwarz et. al. The electron temperature as a function of projected distance from Sgr A* is shown in Fig. 6 for both the H76 α and H92 α data. The straight line shows the 1/r dependence determined from the H76 α data. The excellent agreement over this almost two to one range in frequency has three implications: (1) there can be no over-estimate of the thermal continuum due to a substantial non-thermal component which would in turn lead to apparently increased electron temperatures; (2) there can be no substantial apparent increase in the electron temperature due to continuum opacity effects ($\propto \nu^{-2.1}$) and (3) pressure broadening can play no role in reducing the line to continuum ratios since the line broadening would be substantially larger at 8.3 GHz. In conclusion, the increase in electron temperature toward Sgr A* proposed by Schwarz et. al has been confirmed.

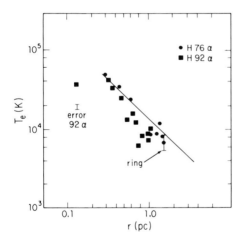

Fig. 6. Distribution of electron temperature with distance from Sgr A* along the eastern arm (see Schwarz et. al, 1989; their Fig. 8). The H76 α data are the filled circles. The data from the ionized ring (feature A, Fig. 2) are shown with an error bar. The H92 α data are shown as filled squares (the typical error is shown). The fitted line with a slope of -1 (taken from the H76 α data) indicates a 1/r decrease of T_e with distance from Sgr A*.

In several positions in the northern portion of Sgr A West, the 3 σ upper limit to the magnetic field based on the Zeeman effect is about 15 m Gauss. This value is comparable to the > 10 m Gauss fields proposed by Aitken (1989). However these values are an order of magnitude larger than the fields in the adjacent molecular ring determined from HI (Schwarz and Lasenby, 1990) and OH (Killeen et. al, 1990).

In the first observations of the broad RRL's from Sgr A West (Pauls et. al 1974), a possible identification of the He 109 α line was proposed. Later observations showed that this line was, in fact, a lower velocity H recombination line. Pauls (1980) does state that the ratio He/H line is less than 10%. In 1985, Goss and van Gorkom (unpublished) observed the H76 α and He76 α line with a resolution of 5 x 10" arc (α x δ) using the VLA in the D array. The rms noise is 1 mJy beam^{-1} and the total velocity coverage is about 500 kms^{-1} (25 MHz) with a velocity resolution of 32 kms^{-1}. Two sample spectra are shown in Fig. 7. The upper limit for any He line is y^+ (He$^+$/H$^+$) < .05 (2σ). This value is quite consistent with the effective temperature of 35,000 K derived by Lacey et. al (1989) based on observations of Br α, [Ne II] and [Ar III]; with this efective temperature the helium is not completely ionized. Also Serabyn and Lac y (1985) have suggested a value of 35,000 K based on a number of arguments.

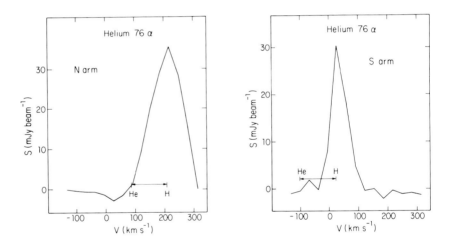

Fig. 7. H76 and He76 α profiles observed at two positions in Sgr A West. The VLA in the D array with an angular resoltuion of 5" x 10" arc was used with a velocity resolution of 32 kms^{-1}. The velocity scale (l.s.r.) refers to the helium line. The difference in velocity of the two lines is 122 kms^{-1}. The theoretical location of the corresponding helium line is indicated. The upper limit to the He76 α line is ≲ 5% (2σ) of the H76 α line. (a) A position in the northern arm where the H line has a positive velocity and (b) a position in the southern arm at a negative velocity.

In the direction of Sgr A* there is no evidence for the broad He line (up to ± 700 kms^{-1}) detected at 2.06µm by Hall et. al (1982) and Geballe et. al (1984). A search for this line has been made by van Gorkom (unpublished) with the VLA in the A array at 2, 6 and 20cm. The results are consistent with the interpretation of Hall et. al (1982), that the line arises in a very dense region. Pressure broadening would decrease the peak line intensity to below the detection limits at radio wavelengths.

References

Aitken, D. K.: 1989, Proc. IAU Symp. 136, ed. M. Morris, Kluwer, Dordrecht, p. 457
Bregman, J. D., Schwarz, U. J.: 1982, Astron. Astrophys. 112, L6
Ekers, R. D., van Gorkom, J. H., Schwarz, U. J., Goss, W. M.: 1983, Astron. Astrophys. 122, 143
Geballe, T. R., Krisciunas, K. L., Lee, T. J., Gatley, I., Wade, R., Duncan, W. D., Garden, R., Becklin, E. E.: 1984, Astrophys. J. 284, 118
Genzel, R., Watson, D. M., Crawford, M. K., Townes, C. H.: 1985, Astrophys. J. 297, 766
Güsten, R., Genzel, R., Wright, M.C.H., Jaffe, D. T., Stutzki, J. Harris, A. I.: 1987, Astrophys. J. 318, 124
Hall, D. N. B., Kleinmann, S. G., Scoville, N. Z.: 1982, Astrophys. J. 262, L53
Kardashev, N. S.: 1959, Astron. Zh. 36, 838 (English translation, 1960, Soviet Astronomy - A.J. 3 813)
Lacy, J. H., Achtermann, J. M., Bruce, D. E.: 1989, Proc. IAU Symp. 136, ed. M. Morris, Kluwer, Dordrecht, p. 523
Liszt, H. S., Burton, W. B., van der Hulst, J. M., Ondrechen, M.: 1983, Astron. Astrophys. 126, 341
Liszt, H. S., Burton, W. B., van der Hulst, J. M.: 1985, Astron. Astrophys. 142, 237
Lo, K. Y., Claussen, M. J.: 1983, Nature 306, 647
Killeen, N., Lo, K. Y., Crutcher, R. M.: 1990, in press
Mezger, P. G., Wink, J. E.: 1986, Astron. Astrophys. 157, 252
Pauls, T., Mezger, P. G., Churchwell, E.: 1974 Astron. Astrophys. 34, 327
Pauls, T.: 1980, Radio Recombination Lines, ed. P. A. Shaver, Reidel, Dordrecht, p. 159
Pedlar, A., Anantharamaiah, K. R., Ekers, R. D., Goss, W. M., van Gorkom, J. H., Schwarz, U. J., Zhao, J. H.: 1989, Astrophys. J. 342, 769
Schwarz, U. J., Bregman, J. D., van Gorkom, J. H.: 1989, Astron. Astrophys. 215, 33
Schwarz, U. J., Lasenby J. 1990 Proc. IAU Symp. 140, eds. R. Beck, P. P. Kronberg, R. Wielebinski, Kluwer, Dordrecht, in press
Serabyn, E., Lacy, J. H.: 1985, Astrophys. J. 293, 445
van Gorkom, J. H., Schwarz, U. J., Bregman, J. P.: 1985, Proc. IAU Symp. 106, eds. H. von Woerden, R. J. Allen, W. Burton, Reidel, Dordrecht, p. 371
Yusef-Zadeh, F., Morris, M., van Gorkom, J. H.: 1987, The Galactic Center, ed. D. C. Backer, AIP, New york, p. 190
Yusef-Zadeh, F., Morris, M., van Gorkom, J. H.: 1989, Proc. IAU Symp. 136, ed. M. Morris, Kluwer, Dordrecht, p. 275

LIMITS ON THE TEMPERATURE AND FILLING FACTOR OF THE WARM IONIZED MEDIUM TOWARDS THE GALACTIC CENTRE

K.R. Anantharamaiah[1], H.E. Payne[2], D. Bhattacharya[1]
[1] *Raman Research Institute, Bangalore 560 080, INDIA*
[2] *Space Telescope Science Institute, Baltimore, MD 21218, USA*

ABSTRACT. We have detected a hydrogen recombination line, near 145 MHz, in the direction of the galactic centre with an optical depth of $-2.7 \pm 0.3 \times 10^{-4}$. This is shown to arise in a nearby HII region with an electron density of 7 cm^{-3}. The absence of any significant contribution to this line from the warm ionized medium (WIM) implies that it must have a temperature > 6700K and a filling factor >25%. If the temperature of the WIM is 8000K, then it must have a filling factor >60%.

1 Introduction

The warm ionized medium (WIM), which was first detected through the dispersion of pulsar signals is one of the main components of the interstellar medium (ISM). This gas has also been seen as diffuse faint Hα emission in the galactic plane (e.g. Reynolds 1987). From analysis of dispersion measures of pulsars (e.g. Vivekanand and Narayan 1982), the line-of-sight average electron density, $\langle n_e \rangle$, of this medium is known to be ~ 0.03 cm^{-3}. Other properties of this gas like temperature and filling factor are not well determined (see review by Kulkarni and Heiles 1988). In early theoretical models (Field, Goldsmith, and Habing 1969), this component was thought to be a uniformly distributed partially ionized gas, with a filling factor close to unity. However, with the discovery of the coronal gas (Rogerson *et al* 1973), and its possible ubiquity in the ISM, the WIM is now thought to be made of clumps of ionized gas with higher electron density and correspondingly smaller filling factor, so as to maintain the average electron density of 0.03 cm^{-3}. For example, in the model of Mckee and Ostriker (1977), this warm gas has a local electron density $n_e = 0.17$ cm^{-3}, filling factor $f_w = 0.23$, and a temperature $T_e = 8000$K. However, there is considerable controversy regarding these values (Cowie and Songaila 1986, Kulkarni and Heiles 1988, Reynolds 1989).

The sensitivity of low frequency recombination lines to low density ionized gas, coupled with the high non-thermal galactic background, which aids stimulated emission, offers the possibility of detecting the WIM in radio recombination lines. Several attempts have been made, but so far there is no detection (Shaver 1975, 1976, Shaver, Pedlar, and Davies 1976, Hart and Pedlar 1980). Anantharamaiah and Bhattacharya (1986, hearafter AB86) used the then available upper limit to the H351α line towards the galactic centre to obtain constraints on the WIM of $T_e \geq 4500$K and $f_w \geq 0.2$. They also predicted that if the WIM has a temperature and filling factor close to the values proposed by Mckee and Ostriker (1977), then recombination lines should be detectable from this gas in the frequency range 100–150 MHz towards the galactic centre. The direction of galactic centre is advantageous because the absence of differential galactic rotation allows line emission over long path lenghts to accumulate near zero velocity.

In this paper, we report the detection of hydrogen recombination lines near 145 MHz, with an optical depth of -2.7×10^{-4} towards the galactic centre. But as we show in Section 3, this line cannot be attibuted to the WIM. Using this result and the arguments of AB86, we obtain strong constraints on the temperature and filling factor of this component.

2 Observation and Result

Observations were made using the 140-ft telescope of the National Radio Astronomy Observatory in Green Bank[†]. We used the 110–250 MHz broadband crossed dipole feed at the prime focus. At our frequency the beam is $3°.3$. Two transitions, H353α (148.86 MHz) and H359α (141.54 MHz), were observed simultaneously in two orthogonal polarizations by splitting the 1024 channel autocorrelator into four independent spectrometers of 256 channels each. Frequency resolution was 1.2 kHz. Reference spectra were measured by switching to a noise source adjusted to match the on-source system temperature. Spectra were measured in units of the ratio of line to continuum temperature (T_L/T_C). After careful editing of the data, to remove interference, we combined the data of both transitions and both polarizations to obtain the spectrum shown in Figure 1. The effective integration time for this spectrum is 66 hours. The feature near $V_{lsr} = 4$ km s^{-1} is real since it could be seen in the averaged data of each of the four spectrometers. From a gaussian fit to the profile in Figure 1, we obtain the line parameters $T_L/T_C = 2.7 \pm 0.3 \times 10^{-4}$, $V_{lsr} = 4.2 \pm 1.8$ km s^{-1}, and $\Delta V = 41 \pm 4$ km s^{-1}. As the continuum temperature

---

[†] National Radio Astronomy Observatory is operated by the Associated Universities Inc. under a co-operative agreement with the NSF.

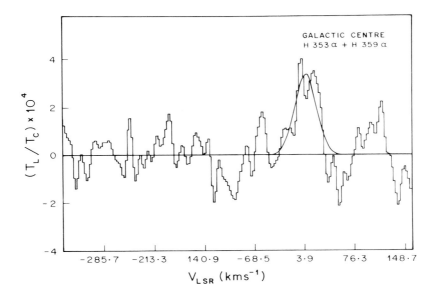

Figure 1: Hydrogen recombination line observed towards the galactic centre. The profile is smoothed to a resolution of 10.4 km s^{-1}.

is dominated by the non-thermal background, the line to continuum ratio gives the line optical depth, i.e. $\tau_L = -T_L/T_C$.

3 Discussion

The first question is "Is the detected line from the WIM?". The direction of the galactic centre is schematically shown in Figure 2. Although the ionized clouds of WIM could be present along the entire line of sight, confusion can occur due to the presence of a nearby low density ($n_e \sim 5$–15 cm^{-3}) HII region, which is known from recombination line observations in the 300 to 400 MHz range (Pedlar *et al* 1978, AB86). At even higher frequencies, non-zero velocity clouds near the galactic centre begin to dominate the recombination line spectrum. Typical spectra observed near 327 and 1420 MHz are shown in Figure 3. The strong H272α feature has a velocity of 2.1±0.5 km s^{-1} (AB86) and the line is due to the nearby HII region. The slight positive velocity of the feature in Figure 1 strongly suggests that the present observation may also have detected the recombination line from this HII region. A model fit to the observed intensities at different frequencies confirms this as shown in Figure 4. In fact the present observation fixes the electron density of this HII region at ~ 7 cm^{-3}.

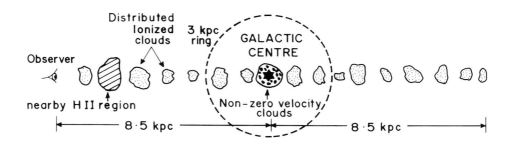

Figure 2: Schematic representation of known ionized components along the line of sight towards the galactic centre.

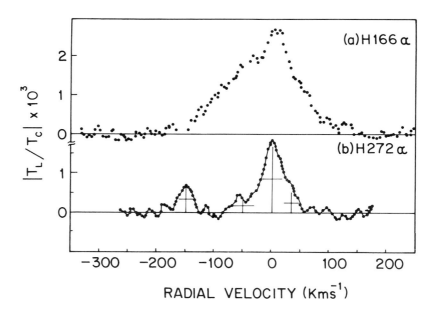

Figure 3: Higher frequency lines towards the galactic centre observed by AB86 (H272α) and Kesteven and Pedlar (1977) (H166α).

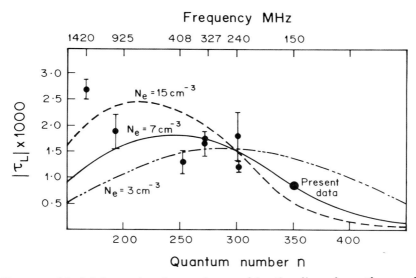

Figure 4: Model fit to the observed recombination lines from the nearby HII region. The parameters are cloud size $2°$, EM $= 2200$ pc cm^{-6}, $T_e = 7500$K. The data points are from Kestevan and Pedlar (1977), Casse and Shaver (1977), Pedlar *et al* (1978), and AB86. The optical depths were computed from the published line and continuum temperatures after taking into account the beam size and assuming that 10% of the observed intensity is due to spontaneous emission. Higher frequency lines have contribution from ionized gas near the galactic centre.

Therefore, the WIM makes little or no contribution to the observed line optical depth. Conservatively, we can assume that the optical depth due to WIM is less than half of the observed value (i.e. $|\tau_L| \leq 1.35 \times 10^{-4}$), and realistically we may have $|\tau_L| \leq 1.0 \times 10^{-4}$. These limits can now be used to obtain constraints on the WIM.

If the local density of the distributed clouds, which constitue the WIM is n_e, then their filling factor $f_w = \langle n_e \rangle / n_e$, where $\langle n_e \rangle = 0.03$ cm^{-3}. The effective emission measure of the WIM is EM $= n_e^2 f_w L = n_e \langle n_e \rangle L$, where L is the total pathlength over which the line intensity accumulates. This unique correspondance between n_e and EM together with the known expression for the line optical depth (e.g. Shaver 1976), allows us to obtain a relation between n_e and T_e which can produce a give line optical depth. We have computed this relation for two optical depth limits obtained above. We used the departure coefficients from Salem and Brocklehurst

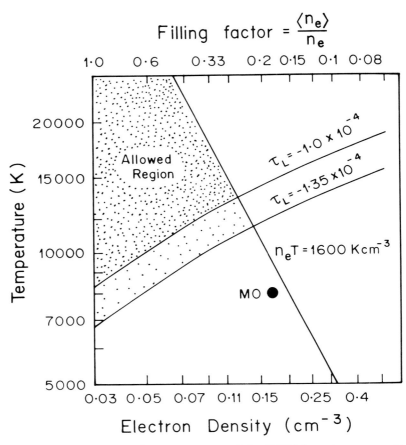

Figure 5: Constraints on the parameters of the WIM based on recombination lines (marked τ_L) and interstellar pressure assumed to be $nT = 3200$ K cm^{-3}. The parameters suggested by Mckee and Ostriker (1977) is indicated by a filled circle.

(1979) and assumed L = 8.5 kpc since the region behind the galactic centre is likely to be blocked out due to the high continuum optical depth of the non-zero velocity clouds near the galactic centre. The results are shown in Figure 5. If L>8.5 kpc, then the lines maked τ_L move upward.

The allowed T_e, n_e, and f_w of the WIM are above the lines marked τ_L. According to Figure 5, our conservative upper limit to $|\tau_L|$ of 1.35×10^{-4}, imply that the temperature of the WIM is $\geq$6700K. The present upper limit also rules out the Mckee and Ostriker (1977) parameters for the WIM. Since the WIM is expected to be in pressure equilibrium with the ISM, the line marked n_eT in Fig 5 imposes another constraint. The allowed values are to the left of this line. The minimum filling factor of the WIM is therefore 25% if its temperature is ~11000K. Lower temperatures imply larger filling factors. The electron temperature and filling factor of the WIM are well constrained in Figure 5. If the temperature of the WIM is 8000K, as usually adopted (Kulkarni and Heiles 1988), then our upper limit implies a filling factor for this gas of >60%. It is possible that the ISM is closer to the early model of Field, Goldsmith, and Habing (1969) than to the model of Mckee and Ostriker (1977).

A word of caution. In our discussion, we have assumed that the measured line to continuum ratio with a $3°.3$ beam directly gives a limit on the line optical depth of the WIM. This may be justified since the WIM is expected to fill the beam and the non-thermal background towards the galactic centre overwhelms the system temperature at this frequency. But, because of the presence of other ionized gas in this direction and their optical depth effects, a more careful analysis is warranted in the light of the significance of the above result.

REFERENCES

Anantharamaiah, K.R., Bhattacharya, D., 1986, *J. Ap. Astr.* **7** 141 (AB86).
Casse, J.L., Shaver, P.A., 1977, *Astr. Ap.*, **61**, 805.
Cowie, L.L., Songaila, A., 1986, *Ann. Rev. Astr. Ap.*, **24**,499
Field, G.B., Goldsmith, D.W., Habing, H.J., 1969, *Ap.J.*, **155**, 149.
Hart, L., Pedlar, A., 1980, *MNRAS*, **193**, 781.
Kulkarni, S.R., Heiles, C., 1988, in *Galactic and Extragalactic Radio Astronomy*, ed. G.L. Verschuur and K.I. Kellerman, (Springer, New York) p95.
McKee, C.F., Ostriker, J.P., 1977, *Ap. J.*, **218**, 148.
Pedlar, A., Davies, R.D., Hart, L., Shaver, P.A., 1978, *MNRAS*, **182**, 473.
Reynolds, R.J., 1987, *Ap. J.*, **323**, 118.
Reynolds, R.J., 1989, *Ap. J.*, **345**, 811
Rogerson, J.B., Spitzer, L., Drake, J.F., Dressler, K., Jenkins, E.B., Morton, D.C., York, D.G., 1973, *Ap. J.*, **181**, L97.
Salem, M., Brocklehurst, M., 1979, *Ap. J. Suppl.*, **39**, 633.
Shaver, P.A., 1975, *Astr. Ap.*, **43**, 465.
Shaver, P.A., 1976, *Astr. Ap.*, **49**, 149.
Shaver, P.A., Pedlar, A., Davies, R.D., 1976, *MNRAS*, **177**, 45.
Vivekanand, M., Narayan, R., 1982, *J. Ap. Astr.*, **3**, 399.

VLA OBSERVATIONS OF RECOMBINATION LINES FROM THE STAR BURST GALAXY NGC253

K.R. Anantharamaiah
Raman Research Institute, Bangalore 560 080, INDIA

W.M. Goss
NRAO, Very Large Array, Socorro, NM 87801, USA

ABSTRACT. NGC253 was observed in H166α, H110α, and H92α recombination-lines using the C and D configurations of VLA[1]. Lines were detected in the nuclear region, which also dominates the continuum emission. The spatial location of the H166α line is different from that of the other two. We suggest that the higher frequency lines arise in high density and high emission measure gas, and the lower frequency line in a region of low density and lower emission measure . The observed velocity distribution indicate unusual motions of the ionized gas near the nucleus.

1 Introduction

Beyond the Magellanic clouds, radio recombination lines have been detected, till to-date, in only two extragalactic sources, M82 and NGC 253. The first detections in these sources were reported by Shaver, Churchwell, and Rots (1977) and Seaquist and Bell (1977). Subsequent searches towards several galaxies and quasars have yeilded no detection (Churchwell and Shaver 1979, Bell et al 1984). Both M82 and NGC253 are relatively nearby and active galaxies, which are considered as proto-types of starburst galaxies. Interferometric studies of recombination lines from M82 have been carried out using both the Very Large Array (VLA) and the Westerbork Synthesis Radio Telescope (Seaquist, Bell, and Bignell 1985, Roelfsema 1986). Here we report VLA observations of recombination lines from NGC253 at wavelengths

---

[1] The VLA is a part of the National Radio Astronomy Observatory which is operated by the Associated Universities Inc., under a cooperative agreement with the NSF

of 20, 6 and 3.6 cm. The angular resolutions are from 45″ at 20 cm to 3″ at 3.6 cm. At a distance of 2.7 Mpc, 1″ corresponds to ~13 pc.

2 Observations and Data Reduction

Observations were carried out in several sessions during the period May 87 to Aug 88, using the C and D configurtions of the VLA. Three transitions, H166α (1424.734 MHz), H110α (4874.158 MHz), and H92α (8309.384 MHz), were observed using 31 spectral line channels centered at a heliocentric velocity of 200 km s^{-1}. The velocity resolutions were 20.5, 24.0, and 28.1 km s^{-1} for the three transitions respectively. The amplitude scale was determined by observations of 3C48, assuming its flux density to be 15.75 Jy at 1424 MHz, 5.37 Jy at 4874 MHz, and 3.19 Jy at 8309 MHz. Instrumental and atmospheric phase variations were corrected using frequent observations of the calibrator 0023-263. The bandpass calibrators were 3C48 at 20 cm and 3C84 at 6 and 3.6 cms.

Further processing of the data was carried out using the Astronomical Image Processing System (AIPS). The data corresponding to each of the transitions were processed in nearly identical manner. Line images were formed after applying phase corrections obtained from self-calibration of the continuum channel and then subtracting the continuum 'clean' components from the visibility data. Residual continuum in the line images were removed by subtracting an average image of outer channels where no line was expected (and none found). Where necessary, the line images were deconvolved using 'CLEAN'. Both natural and uniform weighting of the visibility data was tried to obtain either good signal to noise ratio or better angular resolution. The uniformly weighted images were useful only at 3.6 cm where the signal to noise ratio was large. Because of the low signal to noise ratio at 20 cm, the set of line images were hanning smoothed, which resulted in a spectral resolution of 41 km s^{-1}.

3 Results

Recombination lines were detected at all the three frequencies. In Figures 1a, 1b, and 1c, we show the continuum images obtained using the average of the central three quarters of the observed band. The continuum emission is dominated by the central region at all the three frequencies.The spectral indicies of the core region, determined after convolving all the three images in Fig 1 to the resolution at 20 cm (Fig 1a), are −0.53 between 3.6 and 6 cm and −0.36 between 6 and 20 cm.

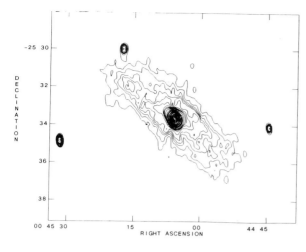

Figure 1a: Continuum at 20 cm. Beam size $22.7'' \times 11.4''$. Contour levels: 3, 5, 7, 9, 11, 15, 20, 25, 40, 60, 80, 100, 150, 200, 300, 400 to 1600 in steps of 200 mJy/beam. The peak flux density is 1600 mJy per beam.

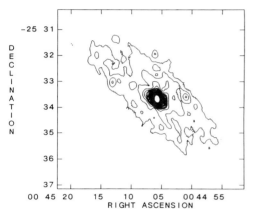

Figure 1b: 6cm Continuum. Beam size $12.5'' \times 8.8''$. Contour levels: 1, 2, 4, 6, 8, 10, 15, 20, 25, 30, 40, 50, 60, 80, 100 to 350 in steps of 50 mJy/beam. Peak flux density is 875 mJy/beam.

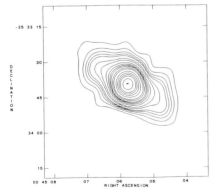

Figure 1c: Continuum at 3.6 cm. Beam size $7.4'' \times 5.2''$. Contour levels: 2, 4, 6, 8, 10, 15, 20, 30, 40, 50, 60, 80, 100, 150, 200, 250, 300, 400, 500 mJy/beam. Peak flux density is 503 mJy/beam.

The break in the spectral index indicates presence of foreground thermal gas. The disk continuum emission is observed only at 20 and 6 cm, indicative of a steeper spectrum.

Line emission was detected only near the bright central region at all the three frequencies. The integrated line profiles are shown in Fig 2. Some relevent parameters of the line and continuum measurements are given in Table 1. The line parameters were obtained from a single component gaussian fit to the profiles shown in Fig 2.

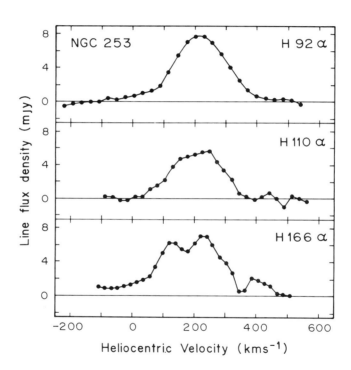

Figure 2: Recombination line profiles towards NGC253, obtained by integrating the line images over identical regions in each channel. This region for each transition was chosen based on the velocity integrated intensity (zeroth moment) shown in Figures 3b, 4b, and 5b.

Table 1: Line and Continuum parameters for NGC253

Line	Resolution (")	$\Delta S_L^\dagger$ (mJy)	$S_C^{\dagger\dagger}$ (mJy)	$\frac{\Delta S_L}{S_C}$ (10^{-3})	V_{Hel} (kms^{-1})	ΔV (kms^{-1})
H166α	45.2 × 27.3[+]	6.18 ± 0.34	1440	4.3	195 ± 36	220 ± 87
H110α	12.5 × 8.8[+]	5.77 ± 0.15	1048	5.5	209 ± 13	185 ± 32
H92α	7.4 × 5.2[+]	7.97 ± 0.09	764	10.4	217 ± 8	189 ± 19
	3.7 × 1.9[++]	6.57 ± 0.05	518	12.7	222 ± 5	200 ± 11

[†] Integrated Line emission. [††] Continuum integrated over line region. [+] Obtained with natural weighting. [++] Uniform weighting.

In Figures 3, 4, and 5, we show the distribution of the continuum in the central region together with the velocity integrated line emission and velocity centroids at the three observed frequencies.

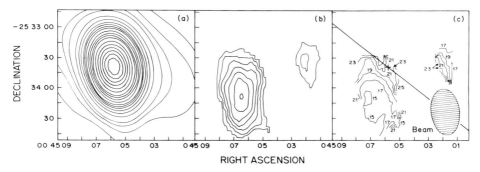

Figure 3: (a) Continuum, (b) integrated line intensity (Zeroth moment), and (c) the velocity field (first moment) in the central region of NGC253 for H166α. Beam size 45.2" × 27.3". Contour levels are (a) 0.6, 0.8, 1 to 4 in steps of 0.5, 5, 6, 7, 8 to 20 in steps of 2, in units of 100 mJy/beam. (b) 0.5, 2 to 12 in steps of 2, in units of 100 Jy/beam m/s. (c) As indicated with 1 Unit = 10 km s^{-1}.

At the two higher frequencies (H110α and H92α), the line emission is centered on the continuum peak and is marginally resolved with the present angular resolution. The centroids and widths of the detected lines also agree within errors (Table 1) indicating that the two lines arise from the same gas. However, at the lowest frequency (H166α), the peak of the line emission is shifted towards south-east from the continuum peak indicating a different origin for this line. There is also a suggestion of unresolved H166α line emission from a blob to the west of the nucleus (see Fig 3).

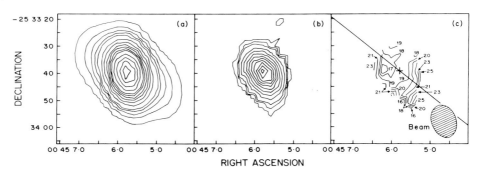

Figure 4: Same as Fig 3 for H110α. Beam size 12.5″ × 8.8″. Contour levels:(a) 20 to 100 in steps of 20, 100 to 400 in steps of 50, 400 to 800 insteps of 100 mJy/beam. (b) 0.2, 0.6, 1 to 10 in steps of 1, in units of 100 Jy/beam m/s. (c) As indicated with 1 unit = 10 km s^{-1}.

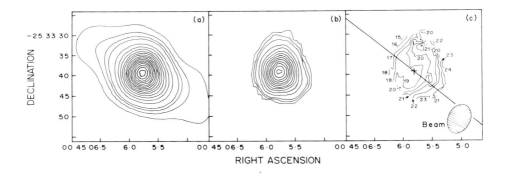

Figure 5: Same as Fig 3 for H92α. Beam size 7.4″ × 5.2″. Contour levels: (a) 10, 20, 30, 40, 70 to 490 in steps of 30 mJy/beam. (b) 0.2, 0.6, 1 to 12 in steps of 1, in units of 100 Jy/beam m/s. (c) As indicated with 1unit = 10 km s^{-1}.

The centre velocity and width of the H110α line (Table 1) is consistent, within errors, with the H112α single-dish measurements of Mebold *et al* (1980). However, our integrated flux density is a factor of two lower, which can be accounted for only if there is an additional contribution to the single-dish measurement from a broader region, which may be resolved out in the interferometric data. A similar difference between single-dish and interferometric data for M82 has been noted by Roelfsema (1986), who rules out the possibility of resolution effects. At this point, these differences are some what puzzling.

4 Discussion

4.1 LINE INTENSITIES

Some constraints can be obtained on the properties of the line emitting region based on the intensity of the recombination lines observed here. From the spatial distribution seen in Figures 3, 4, and 5, it appears that the line emitting region detected in H92α and H110α is not detected in H166α and vice versa. If we refer to these two distinct regions as A and B, then the 3σ upper limit to H166α emission from region A is $\sim$2 mJy. For region B, the 3σ limit to H110α is 1.2 mJy and to H92α is 0.7 mJy. These measurements would be consistent with a model in which, region A is of high density, high emission measure, and lies in the direction of the nuclear source. The H166α from this region is diminished due to pressure broadening and continuum optical depth. Region B is of low density and lies to the south-east of the continuum peak. The H166α from region B is dominated by stimulated emission, which is not very effective at higher frequencies. This is illustrated in the models depicted in Figures 6a and 6b. At an electron density of 10^4 cm^{-3} (Fig 6a), the H92α is due to 75% spontaneous emission and 25% stimulated emission, and for H110α the contributions are 65% and 35%. On the other hand, at an electron density of 100 cm^{-3} (Fig 6b), the H166α is 99% stimulated emission. The parameters of the model are chosen for illustration and the present observations cannot uniquely determine the exact values. Presence of both higher and lower denisty components are required to explain the observations. Further observations at higher and lower frequencies are required to determine their properties. The illustrated models indicate volume filling facotrs of 10^{-4} for the high density gas and 10^{-1} for the low density gas.

4.2 LINE WIDTHS

The FWHM of the observed lines are 180-200 km s^{-1}, which are large compared to typical widths of recombination lines from galactic HII regions (20-50 km s^{-1}). Therefore it is unlikely that the observed lines towards NGC253 arise in a single HII region. A number of HII regions with both systematic and peculiar motions

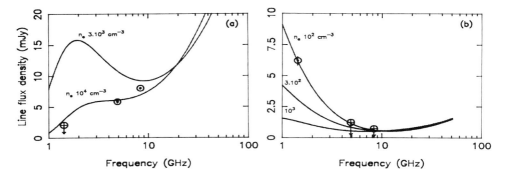

Figure 6: (a) Variation of recombination line intensity of gas with higher densities which are indicated. Other model parameters are EM $= 1.2 \times 10^6$ pc cm^{-6}, $T_e = 5000$K, Background continuum 764 mJy at $\nu = 8309$ MHz, spectral index $\alpha = -0.6$. The departure coefficients are from Salem and Brocklehurst (1979). (b) Lower density models. EM $= 1.1 \times 10^6$ pc cm^{-6}, $S_\nu = 1.45$ Jy at $\nu = 1424$ MHz, $\alpha = -0.6$. In both cases the background continuum and the foreground line regions are both assumed to be 10″ in size. Measured intensities are indicated by circles with error bars and upper limits by circles with a cross.

located in the nuclear region can account for the large line widths. It can be noted that in the Galaxy, large line widths are observed only towards SgrA West, which is located near the galactic centre.

4.3 VELOCITY FIELD

The central velocities of the observed lines are near 200 km s^{-1}, which is lower than the systemic velocity of the galaxy of 250 km s^{-1}, suggesting an outflow. The distribution of velocity with postion, shown in Figures 3c, 4c, and 5c, shows an interesting behaviour. The velocity gradient seen in H110α (Fig 4c) is roughly consistent with normal rotation around the centre of the galaxy. The same is true of the outer iso-velocity contours of H92α seen in Fig 5c. However, in Fig 5c, the region near the nucleus shows a gradient in velocity perpendicular to the major axis (i.e. along the minor axis). This suggests motion of the gas along the minor axis, which is unusual. The angular resolution in Fig 5c is insufficient to delineate the region of this unusual motion. In Fig 7, we show higher resolution H92α images obtained using uniform weighting of the visibility data. Fig 7c clearly confirms that the velocity gradient is along the minor axis.

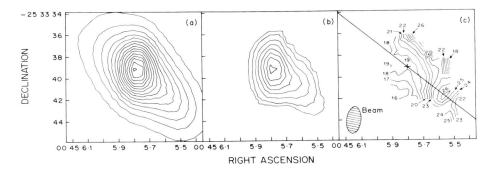

Figure 7: Same as Fig 3, for H92α obtained with uniform weighting of visibility data. Beam size $3.7'' \times 1.9''$. Contour levels: (a) 20 to 300 in steps of 20 mJy/beam. (b) 0.5, 1 to 7 in steps of 1, in units of 100 Jy/beam m/s. (c) As indicated with 1unit = 10 km s^{-1}.

Optical observations (Demoulin and Burbidge 1970, Ulrich 1978) have also indicated peculiar motions of the ionized gas in the nuclear region of NGC253. This has been interpreted as outflow from the nucleus (Ulrich 1978). The optical data do not show a systematic gradient in velocity as does the H92α data in Fig 7c. Clearly, higher angular resolution H92α observations can help model the motion of the ionized gas near the nucleus, which may not be easily accessible at optical wavelengths due to 3–7 magnitudes of visual extinction (Waller et al 1988).

REFERENCES

Bell, M.B., Seaquist, E.R., Mebold, U., Reif, K., and Shaver, P.A., 1984, *Astron. Astrophys.* **130**, 1.
Churchwell, E. and Shaver, P.A., 1979, *Astron. Astrophys.*, **77**, 316.
Demoulin, M.H. and Burbidge, E.M., 1970, *Astrophys. J.*, **159**, 799.
Mebold, U., Shaver, P.A., Bell, M.A., and Seaquist, E.R., 1980, *Astron. Astrophys.*, **82**, 272.
Roelfsema, P.R., 1986, *Ph. D. Thesis*, University of Groningen, The Netherlands.
Salem, M. and Brocklehurst, M., 1979, *Astrophys. J. Suppl*, **39**, 633.
Seaquist, E.R. and Bell, M.B., 1977, *Astron. Astrophys.*, **60**, L1.
Seaquist, E.R., Bell, M.B., and Bignell, C., 1985, *Astrophys. J.*, **294**, 546.
Shaver, P.A., Churchwell, E., and Rots, A.H., 1977, *Astron. Astrophys.*, **55**, 435.
Ulrich, M., 1978, *Astrophys. J.*, **219**, 424.
Waller, W.H., Kleinman, S.G., and Ricker, G.R., 1988, *Astron. J.*, **95**, 1057.

RADIO RECOMBINATION LINES AT 25

A Summary of IAU Colloquium No. 125, Puschino, U.S.S.R.

> P.A. SHAVER
> *European Southern Observatory*
> *Karl-Schwarzschild-Str. 2*
> *D-8046 Garching bei München*
> *Federal Republic of Germany*

It is most fitting that this meeting to celebrate and review 25 years of radio recombination line studies should take place in this country, where both the prediction of their detectability and the first detections themselves were made. Great progress has been made since those early observations. As we have seen in this meeting, radio recombination line observations now span virtually the entire radio spectrum from 1 mm to 20 meters wavelength, and are observed in sources as diverse as dense stellar winds, diffuse interstellar clouds, and distant radio galaxies. Radio recombination lines have become a powerful tool to study ionized gas under a wide range of conditions.

1. Physics of Radio Recombination Lines

The basic theory of radio recombination lines is well understood. As Professor Sorochenko showed in his introductory review, the major issues concerning the role of pressure broadening and potential importance of stimulated emission were already addressed in the late 1960s, and excellent agreement with observation has been achieved for both line widths and intensities. The energy level populations, meanwhile, have been computed using various increasingly sophisticated techniques and comprehensive codes, as summarized by Gulyaev and Beigman.

Laboratory studies on Rydberg atoms have followed a parallel course. After the initial period of exploration in the 1970s, Rydberg atoms are now being used almost routinely as a test ground to study basic physics at the boundary between the quantum and classical regimes. Rempe outlined various applications, including photon counting detectors, one-atom masers, two-photon masers, and the study of chaos.

The remaining questions concerning the astrophysics of radio recombination lines are either highly specific, concerned with limited ranges of physical parameters, or have more to do with the physical conditions under which the lines are emitted. Thus, for example, we heard of the possible importance of line broadening by ions in cold plasmas (Hoang-Binh), of radiative excitation in H II regions around late O-stars (Gordon), and of a dielectronic recombination-like process for carbon in cool H I clouds (Konovalenko, Smirnov, Payne).

Old assumptions which had been made years ago based on the canonical 10^4 K electron temperature and 10^3 cm^{-3} electron density, concerning for example the effectiveness of collisional re-distribution among l-states or the relative importance of various processes, must be reexamined in the context of the very different temperatures and densities which apply for many of the line-emitting regions studied now. And studies continue of possible diagnostic methods, such as the use of suitable pairs of radio recombination lines to determine electron densities and temperatures (Dravskikh, Gordon). Aside from such questions, however, the known theory is being confidently applied to observations to derive the physical conditions of the line-emitting gas and answer various astrophysical questions.

2. The Sun, Stellar Winds, and Planetary Nebulae

One possible source of radio recombination line emission that was not discussed at this meeting is the sun. Dupree (1968) suggested that dielectronic recombination onto coronal ions may produce detectable lines. Attempts to observe such lines have so far failed, perhaps because the lines are too broad and weak due to pressure or radiation broadening, or Zeeman splitting. Hoang-Binh (1982) pointed out that hydrogen recombination lines from the chromosphere may be better prospects; these would be strongest at far-infrared and submillimetre wavelengths. Renewed efforts may now be warranted.

A spectacular new development was the detection, just a year ago, of strong variable recombination line maser emission in the stellar wind source MWC 349, as reported here by Martin-Pintado. This opens up a new domain for radio recombination line research; as the many transitions each sample different layers in the source, these lines will be powerful probes. There are of course many questions to be answered. What are the detailed physical conditions for these masers? How large is the line-emitting region in the different transitions? What causes the strong variations? Is the observed line doubling due to a bipolar outflow? Does this masering occur only in sources with slow massive flows? Is there emission from an adjacent C II region, as suggested for Be stars by Tarafdar? More observations will help to answer these questions — monitoring the variations, studying more transitions, looking at other sources, and of course interferometry. The discovery of this maser emission, occurring as it does at high frequencies where maser emission was thought to be less likely, serves as a reminder that surprises can turn up anywhere in the wide parameter space available to radio recombination lines.

Radio recombination lines have been detected in several planetary nebulae, although as Terzian pointed out, there are over a thousand to go. The radio recombination line emission from planetary nebulae is relatively uncomplicated, as they are generally sharply-defined objects uncontaminated by lower density gas, and pressure broadening, for example, can be clearly seen. Garay showed that the electron temperatures and densities derived on the simplest assumptions from the recombination lines agree very well with those obtained from radio continuum and optical line observations. The filling factor of the gas is apparently close to unity, and Garay has suggested that the electron density derived from the radio recombination lines can be used together with information about the radio continuum to obtain reliable distances from radio data alone. As the extinction can be large, the radio recombination lines are important in that they provide information on the dynamics of planetary nebulae, and at high spectral resolution. The large Doppler line widths often seen in planetary nebulae are apparently due to systematic mass motions (specifically expansion), which can be mapped in detail with interferometers such as the VLA. As many planetary nebulae are apparently ionization-bounded, they should be surrounded by C II regions, and it would be interesting to search for the associated carbon recombination lines.

3. H II Regions and Their Environs

Compact H II regions are bright radio sources, but they are often embedded in extended regions of lower density gas which dominate the integrated radio emission. The study of recombination lines from compact H II regions did therefore not mature until the late 1970s, with the advent of interferometers like the WSRT and the VLA with high sensitivity and dynamic range which are capable of isolating the lines of sight to the compact sources from the "contaminating" lower density gas.

Much work on radio recombination lines from compact H II regions has been done over the past decade, as outlined by Roelfsema, Garay, Churchwell, and Goss. Extremes in the line properties are seen in these compact regions, including pressure broadening, stimulated emission, continuum optical depth effects, and large variations in both the line-to-continuum ratio and the He^+/H^+ ratio. The Doppler line widths can be large, particularly in regions of very high density and emission measure (Garay, Churchwell), probably the result of systematic mass motions (e.g., bow shocks and bipolar outflows, as argued by Churchwell). The electron temperatures determined at suitable high frequencies, and the electron densities determined from pressure broadening, agree with temperatures and densities determined by other methods. In some cases, such as G10.6–0.4 discussed by Churchwell, there is evidence from the frequency dependence of the line widths for multiple components in the line of sight; most compact H II regions probably do not provide as "clean" a laboratory to study radio recombination lines as do isolated planetary nebulae with their sharp boundaries.

The most startling result to come out of recombination line observations of compact H II regions is the very large variation in the He^+/H^+ ratio, as discussed by Goss and Tsivilev. The ratio is usually less — sometimes much less — than the canonical 10%, but these deviations can fairly easily be explained if the helium Strömgren sphere is substantially smaller than the hydrogen Strömgren sphere, as may occur due to ionization by a late O or B star, line blanketing in the stellar atmosphere, or selective absorption of helium-ionizing photons by dust. Much harder to explain are the two confirmed cases in which the ratio is considerably greater than 10%, as high as 30–40%. In one of these, W3A, the high ratio is confined to just one part of the nebula, indicating that the effect is not caused by the exciting star itself. Possible explanations include contamination of the helium recombination line by other elements (carbon), selective absorption of hydrogen-ionizing photons by dust, or, perhaps more likely, a true local abundance anomaly due to contamination by a (possibly extinct) Wolf-Rayet star (W3A also has an anomalously high ^{3}He abundance — Rood *et al.*, 1984).

Extremes such as these are not generally seen in single-dish (low resolution) observations of recombination lines from more typical H II regions. The integrated line emission at any given frequency tends to be dominated by low density gas in the outer regions for which pressure broadening and stimulated emission are not yet important, and the lines have approximately Gaussian shapes and LTE intensities. These complications were gradually appreciated through the 1970s, and it is now understood under what conditions useful information can be derived about the physical parameters of H II regions.

As discussed by Anantharamaiah and Wilson, radio recombination line observations of extended H II regions with moderate angular resolution make it possible to determine the distributions of electron temperature, line width, and radial velocity. Variations which are found in velocity give additional information about the evolution and structure of these H II regions. Maps of the He^+/H^+ ratio, preferably obtained from several different transitions, can in principle be used to estimate the true helium abundance, at least for simple H II regions. Gordon showed that high-frequency recombination lines can provide unique information, although there can be problems due to contamination by molecular lines in the spectra and dust emission in the continuum. Low-frequency recombination lines (~ 0.1–1 GHz) also provide unique information, in particular concerning the outer low-density envelopes of the H II regions, as most recently elaborated by Anantharamaiah (1986).

The cool, partially ionized peripheries of H II regions, at and beyond the hydrogen Strömgren radii, can be studied using the narrow hydrogen recombination lines, and recombination lines of carbon and other heavy elements with low ionization potentials (Anantharamaiah, Roelfsema, Qaiyum). It is found that the H^0 and C II regions cover only parts of the H II regions, so observations with good angular resolution are desirable. The H^0 and C II regions are distinct, and generally have different velocities. Observations such as these, coupled with comparable molecular and far-infrared line observations, can give detailed information on these transition zones on the outskirts of H II regions.

Extended C II regions with little or no associated H II, produced by B stars embedded in dark clouds, were discovered by Brown and Knapp (1974) fifteen years ago. Subsequent work showed that prominent C II regions are not commonplace, and interest in them has lapsed over the past decade, although diffuse C II regions excited by B stars could be relevant in the context of the very low frequency recombination lines described below.

4. Interstellar Clouds and the Diffuse Interstellar Medium

One of the major developments in radio recombination line research over the last decade was the serendipitous discovery by Konovalenko and colleagues of very low frequency carbon recombination lines in absorption. In fact, this discovery had been anticipated by considerations which showed that cold clouds may well dominate the low frequency recombination line spectrum, that lines at the lowest frequencies must ultimately go into absorption, and that the prospects for detectable recombination lines in absorption were greatest for cold gas (or dielectronic recombination in hot gas), for which absorption sets in before pressure broadening. However, previous searches for such lines had failed for a variety of reasons — bandwidths which covered only hydrogen, or observing frequencies where the lines go from emission into absorption and have almost zero intensity — and the lines were finally found serendipitously.

Many such lines have now been observed in the spectrum of Cas A, ranging from the 166α line in emission at 1425 MHz to the 768α line in absorption at 14.7 MHz. The latter is the lowest frequency spectral line so far observed in astronomy, and comes close to the limits ($n \sim 1000$) set by ionospheric opacity and by radiation damping due to the galactic nonthermal radiation field which ultimately causes adjacent lines to merge.

Pressure broadening is clearly present at the lowest frequencies, indicating electron densities of ~ 0.1–0.5 cm^{-3}; the run of line intensities with frequency provides further constraints on the parameters and physical processes involved. The issue discussed by Konovalenko, Smirnov, and Payne is whether the lines arise in "cool" ($T_e \sim 50-100$ K) gas with energy levels populated by a dielectronic-like recombination process, or "cold" ($T_e \sim 20$ K) gas with hydrogenic level populations. The relative line intensities argue for the latter, but the VLA data presented by Payne show the (stimulated) recombination line emission at 327 MHz to be distributed similarly to the H I absorption, for which the former conditions are thought to be more appropriate.

A possible complication is suggested by the fact that the absorption lines in the spectrum of Cas A are blueshifted relative to the emission lines by 2.2 ± 0.4 km s^{-1} (from the data of Payne et al., 1989). This is significant at the 5σ level, and there may be a continuing trend with decreasing frequency, the lowest frequency lines being blueshifted by as much as ~ 5 km s^{-1}. Thus, there may be multiple components in the line of sight with different densities. It is also quite possible that other heavy elements contribute to these lines; the corresponding recombination lines of magnesium and sulfur, for example, are 7.0 and 8.6 km s^{-1} to the blue of the carbon line, and such lines observed in dark clouds become stronger relative to carbon at lower frequencies. Whatever the reason, this velocity shift indicates that, here too, single-component models may not be adequate, and demonstrates the importance of observations with good spectral resolution.

These low-frequency absorption lines have also been observed in several other directions along the galactic plane, including some containing no prominent discrete sources, as summarized by Golynkin. The diffuse galactic nonthermal emission itself serves as the "background" source against which the absorption at the line frequency takes place, and low frequency absorption lines with optical depths of $\sim 10^{-4}$–10^{-3} appear to be ubiquitous along the galactic plane. Problems in interpretation arise here due to the continuum emitting and line absorbing regions being entangled along the line of sight, and uncertainties in beam dilution, but there is no question that these lines provide an important new probe of the interstellar medium.

As shown by Anantharamaiah, low frequency recombination lines can already be used to set strong constraints on the filling factor of H II regions and the cool or cold partially ionized medium, and therefore also on the filling factor of the hot intercloud medium. They are also used to derive or set limits on the interstellar ionization rate. Bochkarev summarized what is known about the very local interstellar medium, and suggested that it may also give rise to detectable radio recombination lines. It appears that this field is just developing, and will be increasingly important in the years to come.

5. Large Scale Properties of the Galaxy

One of the early important applications of radio recombination lines was distinguishing between thermal and nonthermal radio sources along the galactic plane. This work continues, and recent extensive galactic plane recombination line surveys bring the total number of sources so observed to over 800, as Lockman pointed out. There are thousands more, however, so this work is far from complete. Cersosimo reported on distributed centimetre-wavelength recombination line emission along the galactic plane, and highlighted one region where there is significant line emission but no discrete sources. Lockman and Cersosimo both showed that north-south asymmetries in galactic structure are revealed by the distribution of the line emitting regions, and that H II regions appear to be present in the 3-kpc arm.

In Lockman's survey, several more "cool" H II regions were found, whose recombination line widths impose absolute upper limits of just 3000–4000 K for the electron temperatures; it appears that such H II regions are not at all rare. From earlier work it seems that these cool H II regions generally have low densities. There appears to be a global relationship between line width and density, in that high density H II regions often have large Doppler widths due to systematic mass motions, and low density H II regions which are well evolved and relaxed often have low turbulence and small Doppler widths.

The galactic centre region has also been the subject of several radio recombination line studies, as summarized by Goss. Aside from helping to identify the various thermal sources and examine their properties, these spectacular observations have probed the central few parsecs of the Galaxy. As usual the large line widths seen at low angular resolution are found at higher resolution to be due to systematic mass motions; strong velocity gradients are seen, and a solid body rotation model fits the data well. A mass of $2.5 \ 10^6 \ M_\odot$ within the central 1.3 pc is implied. A strong gradient in electron temperature also appears to be present, increasing to tens of thousands of degrees in the inner regions, possible evidence for a central source of ionizing radiation. Limits of the order of 20 milliGauss can be placed on magnetic fields in the vicinity of the galactic centre, from the absence of Zeeman splitting in the recombination lines. Thus, radio recombination lines are now providing unique information about the nearest active galactic nucleus.

6. Extragalactic Radio Recombination Lines

No new sources of extragalactic radio recombination lines have been found over the past decade, and work in this area has focussed instead on detailed studies of the recombination line emission from M82 and NGC 253. As described by Anantharamaiah and Roelfsema, VLA observations of these galaxies have revealed the distribution of the line-emitting gas, the velocity fields of the central parts of these galaxies, and other global properties. One new development has been the detection by Puxley et al. (1989) of spontaneous radio recombination line emission from M82 at millimetre wavelengths; as complications due to stimulated emission no longer play a role, these data are useful in determining the total mass of ionized gas in the galaxy and the star formation rate, parameters which are otherwise difficult to obtain due to obscuration.

Over the past decade there have been improvements in instrumentation — greater dynamic range in interferometric observations, improved receivers at high frequencies, etc. — and these may justify renewed attempts to detect extragalactic radio recombination lines from other sources. The potential rewards could be substantial — redshifts for unidentified radio sources, redshift-independent distances (Shaver, 1978), etc.

There are several possibilities. Spontaneous line emission at high frequencies will almost certainly be detectable from other nearby galaxies. Special cases, such as the molecule-rich gas in front of the strong nucleus in Centaurus A, may provide interesting detections of radio recombination lines in (stimulated) emission or absorption. Ionized gas around the ultra-steep spectrum high redshift radio galaxies may produce detectable radio recombination lines, perhaps enhanced by dielectronic recombination. The gas-rich IRAS quasars at low redshifts may be promising, and the narrow emission line regions of similar (radio-loud) quasars at high redshifts may produce detectable lines by stimulated emission. Finally, absorption systems in the spectra of distant radio-loud quasars may produce detectable lines: the strongly damped systems with high column density may produce stimulated emission or absorption lines in cold or cool partially ionized gas, and even the more typical high excitation heavy-element absorption systems might produce lines enhanced by dielectronic recombination, as suggested by Val'tts (1983). Still further possibilities include stimulated radio recombination lines from protogalaxies at very high redshifts, and perhaps ultimately even the general intergalactic medium and the cosmic microwave background itself.

7. Prospects for Future Research

The major developments in radio recombination line research over the past decade have been the discovery of the very low frequency recombination lines from interstellar clouds, the use of radio interferometers (particularly the VLA) to study the detailed recombination line emission from a variety of sources, and the recent discovery of variable maser emission from a stellar wind source.

What will the next decade bring? Perhaps submillimetre recombination lines will finally be observed in the solar spectrum. The newly discovered phenomenon of masering recombination lines from emission-line stars will certainly be pursued and explored in detail. Studies of radio recombination lines from planetary nebulae and H II regions will undoubtedly continue, particularly using interferometers, and phenomena such as bipolar flows may increasingly be discovered and studied. Work on low frequency recombination lines will mature, and provide increasingly comprehensive information on the interstellar medium. Still higher angular resolution observations of the recombination lines from the galactic centre will be of great interest. New detections of extragalactic radio recombination lines are anticipated, perhaps including some from high-redshift quasars at last. The first detections of radio recombination lines using VLBI techniques may be achieved. Ionized gas is ubiquitous in the universe, and detection of the associated radio recombination lines is limited only by instrumentation, which continues to develop at a rapid pace.

References

Anantharamaiah, K.R. (1986), *J. Astrophys. Astron.* **7**, 131.
Brown, R.L. and Knapp, G.R. (1974), *Astrophys. J.* **189**, 253.
Dupree, A.K. (1968), *Astrophys. J.* **152**, L125.
Hoang-Binh, D. (1982), *Astron. Astrophys.* **112**, L3.
Payne, H.E., Anantharamaiah, K.R. and Erickson, W.C. (1989), *Astrophys. J.* **341**, 890.
Puxley, P.J., Brand, P.W.J.L., Moore, T.J.T., Mountain, C.M., Nakai, N. and Yamashita, T. (1989), *Astrophys. J.* **345**, 163.
Rood, R.T., Bania, T.M. and Wilson, T.L. (1984), *Astrophys. J.* **280**, 629.
Shaver, P.A. (1978), *Astron. Astrophys.* **68**, 97.
Val'tts, I.E. (1983), *Sov. Astron.* **27**, 18.

AUTHOR INDEX

Anantharamaiah, K. R., 123, 203, 259, 267
Apparao, K. M. V., 169
Bachiller, R., 161
Bhattacharya, D., 259
Bochkarev, N. G., 219
Cersosimo, J. C., 237
Churchwell, E., 83
Dewdney, P. E., 123
Dravskikh, A. F., 115
Dravskikh, Z. V., 115
Erickson, W. C., 203
Filges, L., 105
Garay, G., 73, 155
Golynkin, A. A., 209
Gordon, M. A., 93
Goss, W. M., 123, 249, 267
Gulyaev, S. A., 37
Hoang-Binh, D., 51
Konovalenko, A. A., 175, 209
Leahy, J. P., 249
Lockman, F. J., 225
Martín-Pintado, J., 161
Payne, H. E., 203, 259
Rempe, G., 19
Roberts, D. A., 249
Roelfsema, P. R., 59
Shaver, P. A., 277
Smirnov, G. T., 189
Sorochenko, R. L., 1, 189
Steppe, H., 83
Tarafdar, S. P., 169
Terzian, Y., 141
Thum, C., 161
Tsivilev, A. P., 131
van Gorkom, C. M., 249
Walmsley, C. M., 83
Walther, H., 19
Wilson, T. L., 105
Wood, D. O. S., 83

CITATION INDEX

Abramenkov, 212, 216
Achtermann, 256
Ahmad, 42
Aitken, 256
Akabane, 105, 109, 111
Albinson, 206
Altenhoff, 105, 161, 162, 227, 240, 246
Alvarez, 229, 231, 233
Ammosov, 219
Anantharamaiah, 10, 12, 14, 176, 181, 182, 183, 185, 186, 191, 192, 194, 196, 199, 200, 203, 204, 206, 208, 209, 219, 222, 227, 247, 255, 260, 261-263, 280, 281
Anderson, 26
Ansari, 70
Apparao, 169
Ariskin, 10, 212
Arnaud, 219
Arons, 78
Avedisova, 226, 229
Azcárate, 226, 237, 239, 240
Baade, 169, 170
Baars, 84
Bachiller, 8, 141, 143, 145, 146, 153, 162, 164, 167
Backer, 225
Baker, 40
Balick, 106, 145, 151, 159
Ball, 123, 124
Bally, 226
Balona, 170
Bania, 232, 279
Banks, 44
Barcia, 137
Barlow, 161, 162
Barnes, 125
Barsuhn, 134, 137, 210, 211
Barton, 20, 25
Bastien, 110
Batrla, 110, 135, 206
Bayfield, 34
Becker, 165, 226

Becklin, 257
Beigman, 40, 41, 44, 100, 103
Beiting, 20
Bell, 267, 272
Belyaev, 39
Berezhkov, 219
Berulis, 5-8, 14, 76, 97, 108, 147, 150,
Bhattacharya, 260-263
Bica, 137
Bieging, 8, 105-107, 110, 125, 167, 206, 208, 232, 247
Bignell, 145, 159, 267
Blake, 10, 176, 189, 194
Blitz, 219-221, 227
Bloch, 220
Bloemberger, 25
Blondel, 219
Blount, 135
Blümel, 34
Bochkarev, 219-221, 223
Bodenheimer, 78
Bohr, 1
Bointema, 219
Borodzich, 2, 4, 59, 107, 161
Bowers, 246
Brand, 76, 162, 282
Braude, 178
Brault, 52
Braz, 241
Bregman, 206, 249-253, 255
Brocklehurst, 5, 8, 38, 40, 42, 47, 76, 79, 93, 99, 101, 107-109, 118, 152, 164, 165, 242, 274
Bronfman, 229, 231, 233
Brown, 5, 79, 94, 106, 108, 118, 131, 133, 194, 205, 212, 280
Bruce, 256
Brune, 34
Budker, 39
Bujarrabal, 8, 145, 153, 162
Burbridge, 275
Burgess, 42, 43, 101
Burke, 227, 240, 246

Burrows, 220
Burton, 229, 253
Campbell, 241
Cantó, 159, 161, 162
Carral, 76, 77
Casati, 34
Casse, 10, 194, 200, 263
Caswell, 227, 229, 230, 232, 241, 243, 245, 247
Celnik, 227
Cersosimo, 226, 232, 233, 237, 239-241, 245, 247
Cesaroni, 90
Cesarsky, C., 14, 213, 242
Cesarsky, D., 14, 124, 242
Chaisson, 108, 133, 145, 147
Childs, 20
Chirikov, 34
Christensen, 43, 177, 204
Churchwell, 5, 7, 14, 74, 75, 84, 85, 90, 91, 107, 131, 133, 135, 143, 145, 149, 151, 156, 250, 256, 267
Clarke, 40
Claussen, 250
Cleary, 220
Coffeen, 109
Cohen, 167, 229, 231, 233
Colomb, 226, 237, 239, 240
Conklin, 9
Cook, 24
Cooke, 2
Copetti, 137
Costero, 156, 159
Cowie, 259
Cox, 219, 220
Crawford, 253
Crutcher, 10, 125, 176, 189, 194, 199, 206, 208, 220, 256
Cummings, 20, 26, 29, 30
Dachs, 169, 170
Dame, 233
Dauns, 13, 14
Davidovich, 34
Davies, 5, 7, 9, 10, 107, 175, 176, 192, 194, 225, 260, 261, 263

Dehmelt, 26
DeMartini, 26
Demoulin, 275
Dent, 135
Deul, 219, 221
Dickel, H., 135, 198, 199, 208
Dickey, J., 232, 233
Dieter, 6, 225
Digel, 226
Dobiasch, 20, 25
Doherty, 106, 124, 129, 130, 145
Dopita, 108
Downes, 136, 227, 232
Downes, 247
Draine, 205
Drake, 259
Dravskikh, A., 4, 59, 107, 161
Dravskikh, Z., 4, 59, 107, 119, 161
Dreher, 167
Dressler, 259
Drexhage, 26
Duncan, 257
Dunning, 20
Dupree, 4, 38, 42, 43, 124, 278
Dyson, 39
de Geus, 233
de Vries, 219
Eberly, 20, 29, 30
Egorova, 2
Ekers, 250, 255
Elshin, 219
Encrenaz, 51, 52, 57, 101
Engelbrecht, 170
Epchtein, 241
Erickson, 10, 11, 109, 176, 181-183, 185, 186, 191, 192, 194, 195, 196, 197, 199, 200, 203, 204, 206, 208, 209, 219, 222, 281
Ershov, 10, 12-15, 57, 76, 106, 109, 112, 132, 137, 147, 150, 182, 191, 194, 195, 197, 198, 203, 204, 206, 219, 222
Escalante, 161, 162
Evans, 111
Fabre, 20
Faist, 20, 29, 30

Farley, 24
Federman, 199
Feld, 26
Felli, 161, 162
Field, 259, 265
Figger, 20, 22, 23, 29
Filipowicz, 20, 33
Finlay, 242
Finn, 115, 116
Flannery, 38
Flather, 5, 109
Foltz, 20
Forster, 135
Fortov, 46
Fowler, 220
Frisch, 220
Gabrielse, 26
Gaisinsky, 41, 44, 100, 103
Gallagher, 2, 6
Gallas, 29
Gammon, 106
Garay, 73-77, 143, 145, 147, 148, 152, 156, 159
García-Barreto, 141, 145, 149, 156, 159
Garden, 257
Gardner, 65, 76, 137, 162, 227
Garwood, 232, 233
Gathier, 143, 145, 147, 148, 152, 155, 156
Gatley, 257
Gautier, 219
Gayet, 43
Geballe, 67-69, 124, 257
Gee, 44, 177
Geneux, 20, 29, 30
Genzel, 84, 250, 253
Georgelin, Y. M., 239, 243
Georgelin, Y. P., 239, 243
Ghosh, 169
Glassgold, 105
Goad, 147
Goldberg, 4, 6, 7, 39, 43, 47, 93, 97, 107, 124, 161
Goldsmith, 259, 265
Goldwire, 95
Goodman, 78

Gordon, 7, 14, 44, 74, 94, 95, 99, 101, 161, 229, 242
Goss, 65, 67-69, 74, 76, 124, 135, 137, 198, 199, 206, 208, 225, 250, 252, 255
Gottesman, 14, 242
Goudis, 135
Goy, 20, 26, 34
Gómez, 75, 141, 145, 149, 159
Gómez-Gonzalez, 8, 137, 145, 153, 162
Graham, 34
Green, 52
Greisen, 206
Grenier, 233
Griem, 2, 5, 51, 52, 57, 93, 107
Gross, 20, 26
Gry, 219, 220
Gudnov, 4
Guilloteau, 135
Gull, 105
Gulyaev, 5, 44-47
Gundel, 45
Gurevich, 39
Güsten, 250
Haas, 109
Habing, 259, 265
Hall, 24, 257
Handa, 105
Harmanec, 170
Haroche, 20, 26, 29, 34
Harris, 135, 136, 250
Hart, 9, 10, 176, 226, 237, 239, 240, 260, 261, 263
Harten, 74, 76, 77
Hartmann, 167
Harwit, 106, 109, 111
Hasegawa, 102, 109, 111
Haslam, 220, 242
Hayakawa, 138
Haynes, 227, 229, 230, 232, 241, 243, 245, 247
Heiles, 199, 206, 208, 220, 225, 259, 265
Heinzen, 26
Helfand, 226
Henderson, 118, 246
Hieronymus, 20

Higgs, 124, 129, 130, 145
Hildebrandt, 20
Hilfer, 26
Hill, 62, 123
Hills, 192, 196, 197, 199
Hinds, 26
Hirabayashi, 105
Hjellming, 7, 42, 107, 108
Ho, 76, 77
Hoang-Binh, 8, 38, 39, 43, 44, 52, 53, 69, 100-102, 278
Hodge, 229
Hoffleit, 227
Hollberg, 24
Höglund, 4, 6, 107, 161, 225
Huchra, 167
Huchtmeier, 131, 133, 135
Huggins, 105
Hughes, 247
Hulet, 26
Hulse, 242
Hummer, 46
Humphreys, 1, 242
Ikeuchi, 138
Iljasov, 12, 15, 57, 187, 191, 194, 195, 197, 198, 203, 204, 219, 222
Innocenti, 26
Inoue, 105
Jacobovitz, 26
Jaffe, 111, 167, 250
Jahoda, 220
Javanainen, 20, 33
Jaynes, 20, 26, 29
Jäger, 106, 108, 109
Jenkins, 220, 259
Jhe, 26
Johnston, 8, 97, 98, 105, 106, 110
Joly, 43
Jones, 242
Juda, 220
Jugaku, 102
Kaberla, 198, 199
Kakar, 212
Kalberla, 208
Kallman, 171

Kardashev, 2, 4, 6, 39, 59, 107, 249
Kellert, 20
Kennicutt, 229, 231
Kerr, 246
Kesteven, 262, 263
Keto, 76, 77
Killeen, 256
Klein, 20, 26, 31
Kleinmann, 257, 275
Kleppner, 25, 26
Knapp, 118
Knapp, 212, 280
Knight, 20, 29-31
Knude, 220
Koch, 20
Kolbasov, 4, 107
Kolotovkina, 10, 212
Kondo, 220
Kondratenko, 229
Konovalenko, 10, 176, 180-186, 189, 191, 194, 195, 203, 204, 209, 214
Krause, 20, 33
Kraushaar, 220
Krisciunas, 257
Krügel, 62, 69, 123, 124, 129
Kuhn, 26
Kuiper, 111, 212
Kulkarni, 225, 259, 265
Kurucz, 102
Kutyrev, 219
Kwok, 156
Lacy, 251, 253, 254, 256
Lamb, Jr., 33
Lang, 115, 116, 118
Larson, 74
Lasenby, 256
Leahy, 251
Lee, 257
Leeman, 109
Lekht, 10, 12, 13, 15, 57, 182, 191, 192, 194, 195, 197, 198, 203, 204, 219, 222
Leuchs, 20, 22, 23, 29
Lichten, 133
Lilley, 4, 5, 95, 124
Lin, 233

Linke, 101
Liszt, 253
Lo, 250, 256
Lockman, 5, 14, 79, 94, 118, 131, 133, 137, 226-230, 232-234, 237, 239, 240, 245, 247
Lodge, 44
Low, 219
Lubenow, 135
Lugiato, 20, 33
Lutz, 159
Lütken, 20, 25
MacLow, 91
MacLeod, 124, 129, 130, 145, 247
Maddelena, 227
Magnani, 219, 220, 221
Malkan, 145
Marr, 76, 88, 91
Martin, 105, 135
Martín-Pintado, 8, 124, 129, 141, 143, 145, 146, 153, 162, 164, 167
Mataloni, 26
Mathis, 131, 135
Matsumoto, 138
Matsuo, 138
Mauersberger, 76, 162
Max, 78
May, 229, 231, 233
McCammon, 219, 220
McCarroll, 43
McCray, 171
McGee, 13, 14
McKee, 259, 260, 264, 265
Mebold, 192, 196, 197, 199, 267, 272
Megn, 178
Menzel, 4, 5, 40, 95
Meschede, 20, 26, 27
Meystre, 20, 29, 30, 33
Mezger, 4, 6, 73, 105, 118, 131, 133, 135, 145, 161, 225, 227, 240, 246, 250, 251, 256
Mihalas, 46, 102
Mikhalchi, 40
Mills, 112
Milne, 227

Milonni, 20, 29, 30
Minaeva, 5
Misezhnikov, 4
Moi, 26
Moore, 282
Moran, 74-76, 145, 156, 159, 161, 162
Moreno, 156, 159
Morris, 227, 250
Morton, 259
Moscowitz, 227
Mountain, 282
Mugglestone, 115, 116
Mundy, 219, 220, 221
Murakami, 138
Murphy, D., 233
Müller, 20, 26, 27
Münch, 105, 109
Myers, 74, 78
Nakai, 105, 282
Narayan, 259
Narozhny, 20, 29, 30
Nefedov, 45, 47
Newton, 13, 14
Novikov, 137, 138
Nussbaumer, 171
Nyman, 233
Olano, 220
Olnon, 161
Omont, 51, 52, 57
Ondrechen, 253
Oort, 232
Osterbrock, 5, 105, 107
Ostriker, 259, 260, 264, 265
Palmer, 4, 9, 95, 105, 106, 110
Palous, 226
Panagia, 161, 162
Pankonin, 6, 9, 62, 65, 66, 70, 106, 107, 109, 111, 118, 123, 124, 129-131, 133, 134, 137, 145, 210, 211
Paresce, 220
Parijskij, 119
Parrish, 9
Pauliny-Toth, 84
Pauls, 105, 107-110, 135, 227, 250, 256
Payne, 12, 141, 181-183, 191, 192, 194-

197, 200, 203, 204, 206, 208, 209, 219, 222, 281
Pedlar, 5, 9, 10, 175, 176, 192, 194, 237, 239, 255, 260-263
Peimbert, 102, 108, 109
Penfield, 4, 5, 9
Pengelly, 40, 44
Percival, 42, 44, 177
Peters, 169
Phillips, 141, 145, 149
Picart, 52
Pinnaduwage, 34
Pisma, 57
Pitaevkii, 39
Pitault, 131, 133
Plamsas, 8
Planesas, 137, 145, 153, 162
Plaskett, 40
Poppel, 220
Pottasch, 13, 14
Pound, 25
Purcell, 25
Puxley, 282
Qaiyum, 70
Quattropani, 20, 29, 30
Radhakrishnan, 10, 176, 185, 186, 191, 194, 195, 199
Radmore, 20, 29, 30, 31
Raimond, 20, 26, 29, 34
Raith, 20
Ravndal, 20, 25
Reid, 74, 75, 76, 77
Reif, 267
Reifenstein, 227, 240, 246
Rempe, 20, 22, 26, 31, 33, 34
Reynolds, 219, 220, 259
Richards, 44, 177
Ricker, 275
Roberts, 251
Rocchia, 219
Rodgers, 241
Rodríguez, 74-77, 133, 141, 143, 145, 147-149, 152, 156, 159, 161, 162
Roelfsema, 63-70, 74, 75, 124, 131, 133-137, 267, 273

Rogerson, 259
Rood, 279
Roth, 75
Rots, 135, 267
Rubin, 109
Rudnitskij, 10, 12, 13, 191, 194, 212
Ruf, 110
Rydbeck, 9, 44, 101
Ryle, 136
Ryzkov, 2
Safinya, 24
Salem, 8, 42, 47, 93, 99, 101, 118, 242, 274
Salter, 242
Sanchez-Mondragon, 20, 29, 30
Sancini, 220
Sanders, 220
Sandner, 24
Sargent, 33
Sato, 138
Schäfer, 26
Schmidt-kaler, 34
Schraml, 227, 240
Schucking, 105
Schwartz, 8, 97, 98
Schwarz, 74, 76, 77, 206, 249, 250-256
Scoville, 257
Scully, 20, 33, 34
Seaquist, 267, 272
Seaton, 5, 40, 42, 44, 76, 95, 96, 103, 107-109, 152, 164
Sejnowski, 7, 42
Serabyn, 251, 253, 254, 256
Serrano, 156, 159
Shaver, 7-10, 12-15, 42, 43, 60, 74, 76, 77, 94, 97, 98, 106, 108, 112, 161, 175-177, 183, 191, 192, 194, 195, 197, 200, 205, 214-216, 260, 261, 263, 267, 272
Shepelyanski, 34
Shinohara, 44
Shklovsky, 10
Sholin, 45
Simons, 242
Simpson, 8, 94, 98, 109
Sirko, 34

Slodkov, 57
Smilanski, 34
Smirnov, 6, 8-10, 12, 13, 15, 44, 100, 101, 103, 106-109, 112, 118, 132-135, 137, 182, 191, 192, 194, 195, 197, 198, 200, 203, 204, 219, 222
Smith, 20, 131, 135
Snowden, 220
Sobelman, 5, 41
Sodin, 10, 176, 178, 189, 194, 212
Sofue, 105
Solodkov, 12, 15, 182, 191, 194, 195, 197, 198, 203, 204, 219, 222
Songaila, 259
Sorochenko, 2-10, 12, 13, 15, 44, 57, 59, 97, 100, 101, 103, 106-109, 112, 118, 132, 137, 161, 182, 191, 192, 194, 195, 197, 198, 200, 203, 204, 219, 222
Spencer, 26
Spitzer, 109, 200, 213, 259
Stebbings, 20
Steeman, 169, 170
Steinschleiger, 4
Stenholm, 20, 29, 30
Stevens, 199, 225
Stoffel, 242
Storey, 125, 171
Straubinger, 20, 22, 23
Strittmatter, 161, 162
Stutzki, 250
Summers, 42, 43, 101
Tapia, 75
Tarafdar, 169
Taylor, 242
Tenario-Tagle, 62, 69, 78, 111, 123
Terzian, 141, 143, 145, 149, 151, 156, 159
Thaddeus, 226, 227, 229, 231
Thomas, 26
Thomasson, 62, 70, 123, 124, 130, 134, 137, 210, 211
Thum, 124, 129, 133, 141, 143, 145, 146, 153, 162, 164, 167
Tinbergen, 220
Tomasco, 213
Torres-Peimbert, 109, 141, 156, 159

Townes, 253
Tran-Minh, 52
Troland, 199, 206, 208
Tsivilev, 106, 109, 112, 132, 137
Ukita, 102
Ulrich, 275
Ungerechts, 44, 197, 233
Unsöld, 46
Vaidyanathan, 26
Vainshtein, 41
Vallée, 52, 132, 134, 214
Val'itts, 283
Van Buren, 91
Velusamy, 226
Viala, 199
Vidal-Madjar, 219
Vivekand, 259
Volk, 213
van Blerkom, 44
van de Hulst, 2
van der Hulst, 253
van Gorkom, 60, 65, 74-77, 79, 81, 159, 249, 250-253, 255
van Raan, 20
van Regemorter, 44
van Woerden, 220
von Forster, 20, 29, 30
von Hoerner, 105
Wade, 257
Waller, 275
Walmsley, 14, 44, 45, 62, 70, 87, 90, 94, 95, 99, 101, 106-109, 111, 123, 124, 130, 141, 143, 145, 146, 149, 151, 156, 161, 164, 165, 177, 181-183, 191, 194, 197-199, 201, 204, 206
Walther, H., 20, 22, 23, 25, 26, 27, 29, 31, 33, 34
Walther, T., 20, 33
Waltman, E., 8, 97, 98
Waltman, W., 8, 97, 98
Wamsteker, 220
Watson, 10, 43, 176, 177, 181-183, 189, 191, 194, 197, 198, 201, 204, 253
Waygaert, 169, 170
Weaver, 220

Webster, 105
Weisheit, 44
Weisskopf, 25
Welch, 76, 77, 88, 91, 167
Wendker, 161, 162, 227
Western, 43, 177, 204
White, 165
Whiteoak, 65, 137, 241
Wigner, 25
Wild, 2
Willner, 125
Wilson, O., 109
Wilson, T., 8, 62, 70, 76, 105-110, 123, 124, 130, 135, 162, 206, 227, 232, 240, 246, 247, 279
Wilson, W. E., 8, 97, 98, 105, 107
Wilson, W. L. 242
Wing, 24
Wink, 232, 247, 251
Witzel, 84
Wood, 74, 75, 76, 84, 85, 91
Wright, 33, 161, 250
Yakubov, 46
Yamashita, 282
Yoo, 20, 29, 30
York, 219, 220, 259
Yorke, 78
Yoshioka, 138
Yu, 182, 203, 204
Yukov, 41
Yusef-Zadeh, 250
Zel'dovich, 137, 138
Zhao, 255
Zhidkov, 219
Zuckerman, 4, 5, 9, 106, 123

SUBJECT INDEX

Absorption coefficients, 7
Abundances,
  primordial, 102, 137-138
  see also helium
Atoms,
  maximum size of, 8-13
  see also Bohr atom
Background radiation,
  quantum density, 13
Bohr atom, 3
  line frequencies, 1
  size of, 103
  see also Atoms
Carbon radio recombination lines, 51, 123-130, 189-201, 203-208, 209-217, 222-223, 280-281
  link with molecular clouds, 198-199, 206-208
  low frequency, 10-11, 175-188, 189-202, 203-208, 209-219
  near HII regions, 4
  planetary nebulae, 143, 153
  see also CII Regions
CII regions, 69, 123-130, 175-188, 189-201, 203-208, 209-217, 222-223, 280-281
  emission mechanisms, 69
  location with respect to HII gas, 62
  near Be stars, 169, 278
CIII regions, near Be stars, 169
Clumping, 136
  Galactic center, 250
  ISM, 259-265
  ultracompact HII regions, 89
Collisional broadening, see pressure broadening
Compact HII regions, 73-82, 279
  Alfvén waves, 78
  association with OH masers, 76
  definition of, 73
  density-size relation, 74
  internal kinematics, 76, 77
  line to continuum intensity ratios, 74, 79
  line widths, 74, 81
  line widths, variation with density, 77, 79
  magnetic fields in, 78
  motions with respect to host cloud, 76
  observed characteristics of, 75
  pressure equilibrium, 81
  stimulated emission in, 81
  turbulence, 77
  velocity variation with opacity, 76

Compact planetary nebulae, 155-160
  definition of, 155
  distances to, 159
  electron densities in, 155
Compton Effect, inverse, 138
Cosmic background, 138, 204
Cosmic rays, 219-221, 223
Decameter wave radio recombination lines, see Low frequency radio recombination lines
Degeneracy, in angular momentum, 38, 40, 95-96, 278
Departure coefficients, 7, 37, 40, 42, 44, 47, 87, 94-95, 108, 177, 197-198, 277
  effect of Lyman lines on, 100-101
  helium and heavy elements, 47
  stellar masers, 165
Diffuse component of the ISM, 259-265, 280-281
Dust,
  compact planetary nebulae, 155
  planetary nebulae, 141
Electron densities, 189-191, 197, 203-205, 214-217
  carbon radio recombination lines, 179-186
  CII regions, 175-188
  compact planetary nebulae, 155-157
  determining, 115-120
  diffuse ISM, 243-247
  planetary nebulae, 141, 143-147, 149, 152
  pulsars, 244, 259
Electron temperature, 64, 107, 189-191, 197, 203-205, 214-217
  calculation of, 95, 278
  compact planetary nebulae, 156, 158-159
  comparison with optical, 93-94, 97, 109
  diffuse ISM, 243-247
  Galactic center, 255
  mmwave radio recombination lines, 99, 161
  planetary nebulae, 150-151
  Ori A, 111
  variation with L/C ratios, 157-158
  variation within compact planetary nebulae, 158-159
  variation within HII region, 64, 97, 129
  variation with emission measure, 96
  variation with principal quantum number, 96
Emission measure, 155-156, 189-191, 197, 203-205

295

Emission measure, *(continued)*
  collisional determination, 38,42
  diffuse ISM, 243-247
  Lyman lines, 100-102
Excitation,
  Lyman lines, 44
  non-thermal radiation field, 38
  planetary nebulae, 143
  radiation, 38
Excitation rates, 38
Extragalactic radio recombination lines, 70, 119, 267-275, 282
Flares,
  Hα in Be stars, 169-173
  radio recombination lines in MWC349, 161-167
Galactic center, 249-257, 259-265, 282
  molecular gas in, 250-251
  separation of thermal from nonthermal continuum, 255
Galactic distribution,
  3-kpc arm, 231-234
  diffuse RRL emission, 14-15
  HII region temperatures, 13
  ionized gas, 14-15, 225-235, 237-247, 249-257, 259-265, 281-282
  ionized gas in galactic center, 249-257
  northern versus southern hemispheres, 229-231, 233-234, 239-240, 282
  surface density of ionized gas, 229-231
  spiral structure, 234-235
Galaxy, the, chemical evolution of, 131
Heavy elements, 132
HeII regions, 64, 66
  location with respect to HII gas, 62
  see Helium
  see Helium radio recombination lines
Helium,
  abundance relative to hydrogen, 4, 66, 102, 131, 137, 279
  galactic center, 249-257
  planetary nebulae, 143
  variation within HII region, 133
  velocity with respect to hydrogen, 112
HII regions, 63, 65-67, 105, 129-130, 237-247
  age of, 137
  blister model, 131, 134
  champagne flow models, 105, 111
  clumping, 108
  compact, 73, 83, 279
  densities of, 115-120

distances to, 229
expansion of, 109
extragalactic sources, 267-275
galactic center, 259-265
Galactic distribution of, 225-235, 237-247
internal kinematics, 66
mm wavelengths, 93-102
model of DR21, 131-137
model of Ori A, 106
models of, 60-62, 106, 131-137
morphology of, 241-242
opacities of, 65
relation to molecular gas, 229-231
relationship to continuum emission, 241-242
surveys, 225-235, 237-247
surveys, selection effects, 226-227
surveys, sensitivity of, 226-227
turbulence in, 109
ultracompact, 83-92
see Hydrogen radio recombination lines
Hydrogen, 21 cm line, 203,206
  towards Cas A, 192, 194, 197, 199
Hydrogen ionization rate, 200-201, 205
  local ISM, 219-223
Hydrogen radio recombination lines
  diffuse ISM, 14, 243-245
  early calculations of, 2
  early searches for, 2-4
  extragalactic sources of, 267-275
  first detection of, 2-4
  frequencies of, 2
  galactic center, 249-257, 259-265
  H^0 region, see H^0 regions
  low frequencies, 10
  theory, 2, 277
  Zeeman splitting of, 249, 256
H^0 regions, 66-67, 123-130, 280
  distribution of, 126-128
  internal kinematics, 68
  ionization of, 69
  location with respect to HII gas, 62
  relation to CII regions, 129
  stimulated emission in, 68
Infrared emission,
  cirrus, 219
  planetary nebulae, 141-149
Intensity of radio recombination lines, 94-95
  see Line intensities
Interstellar dust, 83
  see Dust
Lamb shift, 2,7

Line blanketing,
    effect on departure coefficients, 102
Line intensities,
    conflict with optical observations, 8
Line profiles,
    normalization of, 115-120
    see Pressure broadening
    see Profiles
Line widths,
    compact HII regions, 74,76
    ultracompact HII regions, 90
    variation with density, 77,79
Local ISM, 219-223
    definition of, 219-220
Low frequency radio recombination lines,
        175-188, 189-201, 203-208, 209-217, 222-
        -333, 280-281
    α/β intensity ratios, 182
    frequency separation, 183
    Galactic center, 259-265
    history, 9-10, 176
    models of, 176-177, 181-182, 194-201, 203-
        208, 214-217
    observing techniques, 178-180
Lyman discontinuity, 100
Lyman line excitation, 277
Maser effect, 94, 161-167, 197, 278
    size of stellar maser, 167
    time variability, 163-167
Maser radio recombination lines, 153
    see Maser effect
Matter ejection, 162
    Be stars, 169, 278
North Polar Spur, 219-220
Nucleosynthesis, 102, 137-138
    see also Helium abundance
Oscillator strengths, 95-96, 103
Oxygen,
    abundance in HII regions, 108-109
    temperature in HII regions, 93-94, 97, 109
Photon statistics,
    inside a cavity, 20
Planetary nebulae, 70, 141-153, 278
    compact, 155-160
    distances from radio recombination lines,
        149, 152
    electron temperatures of, 150-151
    evolution of, 141, 156
    expansion of, 143
    morphology of, 156
    radio recombination lines detected in, 143

velocity variation with quantum number,
    150
Pressure broadening, 4-6, 46-57, 65, 93-94, 115,
        177, 195, 203, 214, 277, 281
    (in) cavities, 28-29
    compact planetary nebulae, 157
    comparison of theory and observations, 5-6
    electrons, 4-6, 57
    formula for, 6
    inadiabatic correction, 5
    ions, 46, 51-57, 277
    planetary nebulae, 143-147, 152
    probe of ISM densities, 15
    Stark effect, 2
    ultracompact HII regions, 87,90
    see Stark shift
    see Widths of radio recombination lines
Pressure equilibrium,
    compact HII regions, 81
Primordial mass fractions, 102, 137-138
    see Helium
Profiles,
    Voigt, 8
    see Line profiles
Quasars, 283
Radial matrix elements, 52-53
Radial velocities,
    ultracompact HII regions, 88,91
    variation with principal quantum number,
        88-89, 137, 150
Radiation broadening, 177, 195
    see Line profiles
Radiation transfer, 107
Recombination, 38
    CI in Be stars, 171
    dielectronic, 169, 181-183, 204-205, 215, 277,
        280, 43
    local ISM, 220
Resonance,
    cavity, 25
Radio recombination lines,
    properties of, 59
    see Carbon radio recombination lines
    see Helium radio recombination lines
    see Hydrogen radio recombination lines
Rydberg atoms, 277
    definition of, 19
    effect of blackbody radiation, 21-24
    excitation in cavities, 29-32
    excitation of, 19-20
    levels, populations of, 21-22

Rydberg atoms, *(continued)*
 lifetimes in cavity resonators, 25-26
 one-atom masers, 27-29
 photon statistics in cavities, 33-34
 see Bohr atoms
Saha-Boltzmann equation, 37, 45
 see also Statistical equilibrium
Solar radio recombination lines, 278
Spontaneous emission, 37-50
 enhancement in cavities, 26
 from Rydberg atoms, 19-20
 inhibition in cavities, 25-26
 rate of, 21
Stark shift, 28
 induced by blackbody radiation, 24
Stars,
 Be, 169-173, 278
 maser emission from, 278
 (in) planetary nebulae, 141-143
 radial pulsations, 170
 with RRL maser emission, 161-167
 Xray emission, 170
 see Solar radio recombination lines
Statistical equilibrium, 37-39
 diffusion equation, 41
 effect of merged levels, 46-57
 solution to equation by cascade matrix technique, 40
 solution to equation by matrix condensation, 42
 solution to equation of, 39-57
Stellar maser, 161-167
 see Maser effect
Stellar winds,
 free-free emission from, 161
Stimulated emission, 20, 94, 129, 161, 277
 compact planetary nebulae, 159
 low frequency radio recombination lines, 177
 planetary nebulae, 143-144
 predictions of, 6-7
 Rydberg atoms, 21-24
Strömgren radii, 62, 279
 see HII regions
Turbulence,
 HII regions, 109
 see Line profiles
Ultracompact HII regions, 83-91, 162
 characteristics of, 85
 clumping in, 89
 definition of, 83
 line widths of radio recombination lines from, 90
 models of, 87
 motion with respect to host cloud, 88-89
 relation to ISM, 83
 variations of line/continuum ratios, 88
Voigt profiles, 65, 115-122, 195-197
 definition of, 115
 see Line profiles, Pressure broadening
Warm ISM, 259-265
Widths of radio recombination lines, 62, 115-120, 177
 carbon, 51-57, 194-195
 compact planetary nebulae, 157
 extragalactic sources, 271, 273-274
 Galactic center, 250-251
 planetary nebulae, 151
 see Line profiles
 see Pressure broadening
 see Voigt profiles
 see Zeeman splitting
Xray emission, 219, 223
 Be stars, 170
Young planetary nebulae,
 see Compact planetary nebulae
Zeeman splitting, 282
 Galactic center, 249, 256